Appropriate Technology
for Development

Other Titles in This Series

Credit for Small Farmers in Developing Countries, Gordon Donald

Boom Town Growth Management: A Case Study of Rock Springs—Green River, Wyoming, John S. Gilmore and Mary K. Duff

The Emergence of Classes in Algeria: Colonialism and Socio-Political Change, Marnia Lazreg

Strategies for Small Farmer Development: An Empirical Study of Rural Development in The Gambia, Ghana, Kenya, Lesotho, Nigeria, Bolivia, Colombia, Mexico, Paraguay and Peru (2 vols.), Elliott R. Morss, John K. Hatch, Donald R. Mickelwait, and Charles F. Sweet

Administering Agricultural Development in Asia: A Comparative Analysis of Four National Programs, Richard W. Gable and J. Fred Springer

Economic Development, Poverty, and Income Distribution, edited by William Loehr and John P. Powelson

The New Economics of the Less Developed Countries: Changing Perceptions in the North-South Dialogue, edited by Nake M. Kamrany

The Military and Security in the Third World: Domestic and International Impacts, edited by Sheldon W. Simon

Protein, Calories, and Development: Nutritional Variables in the Economics of Developing Nations, Bernard A. Schmitt

Technology and Economic Development: A Realistic Perspective, edited by Samuel M. Rosenblatt

Food, Politics, and Agricultural Development: Case Studies in the Public Policy of Rural Modernization, edited by Raymond F. Hopkins, Donald J. Puchala, and Ross B. Talbot

A Select Bibliography on Economic Development: With Annotations, John P. Powelson

Governments and Mining Companies in Developing Countries, James H. Cobbe

Westview Special Studies in Social, Political, and Economic Development

Appropriate Technology for Development:
A Discussion and Case Histories
edited by Donald D. Evans and Laurie Nogg Adler

This analysis of appropriate technology first explores the concept of development in terms of needs, characteristics, and theories and then examines the pivotal role of technology in the developmental process. The twenty contemporary case histories illustrate specific instances of applied technology, not necessarily as examples of successful application, but as subjects for critical review. They are followed by an analysis of the cases and an extensive annotated bibliography.

Donald D. Evans is deputy director of the Office of International Programs at the Denver Research Institute (DRI), University of Denver. Laurie Nogg Adler is research associate in the Office of International Programs at DRI.

Appropriate Technology for Development: A Discussion and Case Histories

edited by Donald D. Evans
and Laurie Nogg Adler

A Study Prepared by Denver Research Institute, University of Denver

Westview Press / Boulder, Colorado

*Westview Special Studies in
Social, Political, and Economic Development*

This project has been sponsored through contract No. AID/DSAN-C-0062 with the Office of Science and Technology, Agency for International Development, U.S. Department of State. The views and opinions expressed in this report, however, are those of the authors, and do not necessarily reflect those of the sponsor.

Published in 1979 in the United States of America by
    Westview Press, Inc.
    5500 Central Avenue
    Boulder, Colorado 80301
    Frederick A. Praeger, Publisher

Library of Congress Catalog Card Number: 79-5154
ISBN: 0-89158-567-2 (hardcover)
ISBN: 0-89158-750-0 (paperback)

Composition for this book was provided by the Editors.
Printed and bound in the United States of America.

Contents

LIST OF TABLES. xi

LIST OF FIGURES xiii

LIST OF PHOTOGRAPHS xv

FOREWORD. xvii

ACKNOWLEDGMENTS xix

CONTRIBUTORS. xxi

ABSTRACTS OF CASE HISTORIES xxv

PART 1 - APPROPRIATE TECHNOLOGY AND ITS ROLE IN
       DEVELOPMENT. 1
  Donald D. Evans
    The Need for Development 3
    The Meaning of Development 8
    The Relationship of Technology to Development. . . 24
    Concept of Appropriate Technology. 35
    Appropriate Technology--Limitations and Barriers . 52
    Appropriate Technology in U.S. Policy. 67
    Appendix 0.1: Excerpts from World Development
    Report-1978. 73

PART 2 - CASE HISTORIES OF APPROPRIATE TECHNOLOGY . . . 81

  1. Tunisia: Bottled Gas for Automobiles. 83
     Richard S. Roberts, Jr.

  2. Thailand: Cassava Pelletizing Technology. . . . 95
     Ronald P. Black, Wanawan Peyayopanakul,
     and Sachee Piyapongse

  3. Colombia: The Composite Flour Program 117
     James M. Miller

4. Haiti: The Coffee Roads 129
   Laurie Nogg Adler

5. Pakistan: Ball-Point Pen Manufacturing--
   A Case of Technology Transfer. 151
   James W. D. Frasché

6. Java, Indonesia: The Introduction of Rice
   Processing Technology. 167
   Melinda L. Cain

7. Central America: The Lorena Cookstove 181
   Suellen Sebald Edwards

8. Honduras: An Experimental Lime Kiln 195
   Judith Evans Blum

9. Malaysia: Small-Scale Brick Manufacturing . . . 209
   Ronald P. Black, A. Rahim Bidin, Woo
   Seng Khee, and Nik Ahmad Kamil

10. Colombia: An Innovative Approach to Rural
    Development. 223
    James M. Miller

11. Brazil: The Role of CRUTAC in Community
    Development. 235
    James P. Blackledge

12. Western Samoa: The Samoa Methodist Land
    Development Program. 249
    Ruth Lechte

13. Tanzania: Biogas Generator. 263
    Richard S. Roberts, Jr.

14. Brazil: Explosive Metalworking Program. 275
    James M. Miller, Jim D. Mote, and Henry E. Otto

15. Ghana: Small-Scale Sugar Processing 291
    Gary D. Kilmer and David L. Sussman

16. Thailand: The Introduction of Mint
    Agriculture and Processing 307
    Ronald P. Black and Sachee Piyapongse

17. Central America: Fungal Fermentation of
    Coffee Waste 329
    Suellen Sebald Edwards

18. The Philippines: Fish Preservation Techniques . 343
    Melinda L. Cain

19. Singapore: The Development of a Design
    Consulting Engineering Firm. 359
    Ronald P. Black and Chan Beng See

20. Sri Lanka: The Ceylon Institute of Scientific
    and Industrial Research. 369
    Donald D. Evans

21. Papua New Guinea: Micro-Hydroelectric
    Projects for Rural Development 397
    Ed Arata

22. The Republic of Korea: An Example of
    Integrated Regional Development. 411
    Donald D. Evans

Analyses of Case Histories 427

    Case Histories Matrix. 431

PART 3 - BIBLIOGRAPHY 433

    Literature Search. 435
    Entries. 436

Tables

2.1	Operating Costs of Thai Cassava Pelletizing Technology	100
2.2	Investment Requirement for Native Pelletizing Factory	101
2.3	Sample List of CCMW Products and Concomitant Prices	103
3.1	Composite Flour Formules	119
3.2	Nutritional Characteristics of One I.I.T. Pasta Formulation and a Straight Semola Pasta	121
3.3	Nutritional Characteristics of Raw Materials Used in Colombia for Composite Flours	121
4.1	Coffee Centers and Roads	133
6.1	Number of Sample Farmers Processing Rice with Hullers, and Numbers of Hullers in Sample Villages, 1970 and 1973	172
9.1	Uses for Old Rubber Trees	210
9.2	Capital Costs for Brick Factory Similar to Eng Huat	217
9.3	Monthly Operating Expenses for Eng Huat	217
11.1	Participation of UFRN Students in CRUTAC (1975)	242
15.1	Capital Investment and Labor Requirements	301
15.2	Income Derived	302
16.1	Projected Crude Oil Production and ASRCT Revenue at Different Price Levels	313
16.2	Location and Area of Farms Serving Distillation Plants During the 1976-1977 Season	316

16.3	Cost and Revenue of Mint Growing in Nan Province 1976	317
16.4	Production and Importation of Mint Oil in Thailand	318
16.5	Production Cost of Mint Oil	318
16.6	Fluctuation of Prices Offered by the Thai Chemical Company for Locally Produced Oil from 1974 through 1977 (Baht/kg)	319
16.7	Investment Requirement for Factory with Eight Stills	321
16.8	Production and Consumption of Mint Oil and Menthol in Thailand 1972-1977	322
16.9	Production, Import and Export of Mint and Menthol in Thailand 1973-1977	323
18.1	Annual Per Capita Use (Kilos) of Seafood by Region, May-June 1974-March 1976, Philippines	345
18.2	Variations in the Preservation Process	348
20.1	Ceylon Institute of Scientific and Industrial Research	383
21.1	Breakdown of Costs for Baindoang Micro-Hydroelectric Project as of 21 November 1978	404

Figures

2.1	Flow Diagram of Cassava Pelletizing Process	98
2.2	Marketing Routes for Cassava Tubers, Chips, and Pellets	98
2.3	Cassava Pressing Mechanism	102
2.4	Schematic Flow Diagram of Chonburi Casting and Machine Work Company's New Hard Pellet Mill	105
2.5	Flow Chart for the Krohn Brand Pellet Mill	107
3.1	Percentage of Recommended Allowances of Calories and Proteins Met by Food Purchased by the Two Lowest Income Groups of Colombian Population	118
3.2	Percentage of Recommended Food Consumption Met by Food Purchased by the Two Lowest Income Groups of Colombian Population	118
3.3	Flowsheet of Milling Operations in Making Composite Flours	123
4.1	Map of Coffee Centers, Coffee Roads, and Existing Road Network in Haiti	134
4.2	Typical Road Sections	144
5.1	Ball-Point Pen Production/Storage Relationship	157
7.1	Interior View of Corner-Built Lorena Cookstove	186
8.1	Cutaway View of Inclined Chimney Kiln Showing Limestone Stacks and Fire Chambers - Note Lower Placement of Square Fire Holes	201
8.2	Original Design of Inclined Chimney Kiln before Modifications	201
9.1	Eng Huat Kiln	213
9.2	Brickmaking Process at Eng Huat	213
11.1	Organization of CRUTAC	238
13.1	Solid Drum Generator	266

13.2	Seven-Drum Biogas Generator	271
14.1	Essential Elements of an Explosion Forming Operation	278
14.2	Schematic for Parallel Plate Cladding	278
14.3	Schematic of Colliding Plates Showing Jetting Action During Explosion Welding Process	279
14.4	Projected Industrial Demand for Clad Plate in Brazil	284
15.1	Raw Crystal Sugar Process	294
15.2	Schematic Diagram of Low-Polarizing Sugar Syrup Process	297
15.3	Furnace Ground Plan	299
17.1	Processing of Coffee Berries	332
17.2	Pilot Plant for Waste Water Treatment	333
17.3	Coffee Pulp Juice Processing	339
20.1	Organization Chart of CISIR	374

Photographs

Refillable butagaz tank in Mr. Keskes' car.	86
Women cultivating seedlings in a coffee nursery	136
Road supervisor taking measurements; instruments used are tape measures and Abney level	137
Laborers discussing need for new tools with Pettis.	141
Road construction showing Telford base and macadam.	141
Hand compacter designed for coffee roads.	143
Drying of rice before milling	173
Rice huller in Kepala Desa village.	174
Hand pounding of rice by women in Java.	175
Lorena cookstoves at the Choqui Experiment Station in Quetzaltenango.	189
Cooking surface of stove; note the way the pots "fit in" the stove and the position of the dampers.	189
Shed covering brick kiln.	214
Stacking bricks after firing.	215
Woman making buttermilk donuts for school snack program.	252
Women slashing weeds with machetes to make paddocks	253
Foreground, copra being sun dried; background, copra burners and shed where the workers cut copra.	255
Solar fruit dryer adapted from Brace Research Institute plans	256
As clad tubesheet composite	285
Tubesheet composite after machining	285
Tubesheet during fabrication into tube bundles.	286
Stainless clad plate after explosion bonding.	286
Harvesting of sugarcane by hand	295

Cane being crushed at plant site. 295

Primitive still composed of an oil drum and pan,
Prae Province 315

NTCL still--most common type in Thailand. 315

Coffee cherry depulpers and fermentation tanks at a
coffee beneficio in Guatemala 336

Coffee waste pile at a large beneficio in Guatemala,
to become compost or animal feed supplement 336

Large open-air drying patio for coffee beans. 337

Fish drying in the sun. 349

Smoking fish in a rural smokehouse in Cavite. 349

Woman processor packing fish for market 351

Baindoang villagers helping to lay power cables
from the power house to school area 402

Foreword

> The soul of wit may become the very body of untruth. However elegant and memorable, brevity can never, in the nature of things, do justice to all the facts of a complex situation. On such a theme one can be brief only by omission and simplification. Omission and simplification help us to understand--but help us, in many cases, to understand the wrong things; for our comprehension may be only the abbreviator's neatly formulated notions, not of the vast, ramifying reality from which these notions have been so arbitrarily abstracted. But life is short and information endless: nobody has time for everything. In practice we are generally forced to choose between an unduly brief exposition and no exposition at all.
>
> Aldous Huxley
Brave New World Revisited

The purpose of this publication is twofold: first, it seeks to present an overview--a synthesis--of contemporary thought on the relationship between technology and the economic development of those nations which, for various reasons, have not fared as well as others.

Second, it provides cases of the application of technology in locations throughout the world. This is done to present more detailed information than the case literature typically yields. And it is for the purpose of presenting subjects for group discussion or individual contemplation concerning the still misunderstood relationships of technology to development.

In undertaking this task some difficulties were encountered. For example, a large and far-ranging literature is available on the subject of "appropriate technology" and related matters. And, as Huxley comments, there is always danger in abstracting and reducing the sayings and thoughts of others. Second, the subject is complex, it involves many different considerations, and presents a plethora of perspectives. There is a great mixture of motivations and circumstances associated with the subject of development, and it involves the very roots of the human relationship and condition--to address it requires boldness.

Finally, in the effort to somewhat enrich the literature of empirical experience (and it is sparse in this subject area), the problem of the selection of cases to present proved more difficult than originally contemplated. One effort at matrix analysis disclosed that the possible varieties of basic factors characterizing development situations would permutate to over 2,800.

The result? Readers will inevitably identify serious gaps in the litany of the twenty-two cases presented and will lament the failure to provide, say, an example of the contribution to development wrought by the efforts of a multinational corporation in pursuit of its goals. The cases, however, as all cases must, do represent reality; they were recorded on site and were drawn from recent and palpable experience--as related by those who were themselves involved.

The cases, it is hoped, will serve as bases for discussion and consideration in the context that they are used in the case study method--for analysis, extrapolation, and the testing of personal theses.

The authors' aspirations in mounting this effort were to make what it is hoped will be a useful contribution in the area of development studies, through the combination of a synthesis of what others have already contributed, with the presentation of new material and examples of technology application in the developing countries.

This effort was supported by the Office of Science and Technology of the Agency for International Development, on behalf of the Office of the U.S. Coordinator for the United Nations Conference on Science and Technology for Development, of the United States Department of State.

Acknowledgments

As with any publication, there are many persons whose involvement makes the final product possible. Although it would be impossible to thank all of those who were part of this project, in addition to our supportive colleagues at the Denver Research Institute who traveled around the world to collect the case histories found herein, special thanks go to: Dianne Kedro, whose editing and unfailing good nature made all the pieces fit together; Sue Coldren, whose retentive mind kept the vast amount of literature in order; Judy Levin, Joseph Gitari, and Judith Evans Blum, whose research "filled the gaps;" Gail Young and Suellen S. Edwards, who provided constant assistance and encouragement; the Denver Research Institute Word Processing Center, especially Tina Johnson with the help of Carolyn Bauer, who patiently typed, envisaged and produced this manuscript; Geralyn Virtue at the Interlibrary Loan Department of the University of Denver's Penrose Library, who made obtaining many of the documents possible; Beverly Blum, who kept track of all the finances; Carolyn Rhoades of Congressman Clarence Long's office, Lois Hobson of Ambassador Wilkowski's office, and Wendy Schacht of the Congressional Research Service, who provided supplementary documents; Dave Clint and William McAninch, who helped with the art work; our publishers, Frederick Praeger and Lynne Rienner of Westview Press, who offered continual advice and expertise; our sponsors, USAID's Office of Science and Technology, specifically, William Eilers and Susan Owens, whose decisions made publication of this book possible; and the many persons in the countries visited who assisted in the compilation of the case histories. Many thanks to you all.

Donald D. Evans
Laurie Nogg Adler

Contributors

Laurie Nogg Adler is a research associate in the Office of International Programs, Denver Research Institute. Her background is in marketing, information science, and publishing.

Ed Arata is an industrial engineer for the Appropriate Technology Development Unit of the University of Lae, in Papua New Guinea. He works with small-scale turbine development and hydroelectric installations and previously worked as a pump company representative and as a teacher in the areas of metals, woods, and mechanical drawing.

A. Rahim Bidin is an engineer and acting director of research, Standards and Industrial Research Institute of Malaysia.

Ronald P. Black, whose formal education was in chemistry, is now assistant director (Programs for Management Development) and senior research scientist within the Office of International Programs, Denver Research Institute. He also coordinates DRI's programs in Southeast Asia.

James P. Blackledge is associate director of the Denver Research Institute and director of its Office of International Programs. He has a working knowledge of eighty research centers in approximately thirty countries in Latin America, the Far East, the Middle East, Africa, and Asia and is directly involved with the transfer of technology to Industrial Research Institutes (IRI's) in over ten developing countries.

Judith Evans Blum is an information specialist in the Office of International Programs, Denver Research Institute. Her experience is in the design and implementation of technical information systems and services for developing countries and in the training of LDC personnel.

Melinda L. Cain is a research social scientist at the Office of International Programs, Denver Research Institute. Her background is in political science and international relations.

Chan Beng See, a chemist, was a former deputy director of the Singapore Institute of Standards and Industrial Research. At present, she serves as a private consultant to international organizations.

Suellen Sebald Edwards is an information specialist in the Office of International Programs, Denver Research Institute. Her background is teaching and information science.

Donald D. Evans is deputy director of the Office of International Programs, Denver Research Institute. He is involved in experimental programs concerning technology transfer to developing countries, especially the organization and operation of institutions for this purpose.

James W. D. Frasché is a management specialist in the Office of International Programs, Denver Research Institute. His background is in international business management.

Nik Ahmad Kamil is a research officer at the Standards and Industrial Research Institute of Malaysia. He is conducting research in the area of ceramic technology.

Gary D. Kilmer is a senior program officer of Technoserve, Inc. He was the director of a Technoserve Program in Ghana from 1973 to 1977 and worked for two years in cooperative development in Kenya and the Peace Corps.

Ruth Lechte is South Pacific Area Secretary of the World YWCA. Her activities include women's and community projects in the South Pacific and involve appropriate technology research and dissemination with special emphasis on improving the status and responsibilities of women.

James M. Miller is a management specialist in the Office of International Programs, Denver Research Institute. He has thirty-five years of experience in university and Research and Development management.

Jim D. Mote is senior research metallurgist in the Metallurgy and Materials Science Division, Denver Research Institute. Over the past fifteen years, he has been involved in research, development, and commercial production using explosive metalworking methods.

Henry E. Otto is manager of the Metals and Metallurgical Engineering Division of Southwestern Laboratories in Houston, Texas. He was formerly senior research metallurgist, Denver Research Institute.

Wanawan Peyayopanakul, a chemical engineer, is a research officer in the Industrial Research Department of the Applied Scientific Research Corporation of Thailand.

Sachee Piyapongse is an economist and acting director of the Economic Study Division of the Applied Scientific Research Corporation of Thailand.

Richard S. Roberts, Jr. is responsible for the International Management Development program in the Office of International Programs, Denver Research Institute. He has been involved in foreign assistance activities during the past fifteen years, principally in the Middle East, Central America, and Africa.

David L. Sussman is a program officer of Technoserve, Inc. His background is in mechanical engineering, and he served as a Peace Corps volunteer in Colombia. From 1976 to 1977 he acted as operations manager for a sugar processing project in Ghana.

Woo Seng Khee is a physicist and research officer at the Standards and Industrial Research Institute of Malaysia.

Abstracts of Case Histories

The majority of the following cases were written by members of the Denver Research Institute, Office of International Programs professional staff. The authors were assisted materially by associates in developing countries who had firsthand knowledge of the technology being examined. Three of the cases were written by consultants commissioned by DRI because of their particular interest and expertise in the area of appropriate technology.

Tunisia: Bottled Gas for Automobiles
    Richard S. Roberts, Jr.

A Tunisian technician-entrepreneur reacts to rapidly rising gasoline prices by converting cars to run on bottled LPG (butagaz), normally used for household cooking. Advantages are numerous, and the potential appears to be considerable, but many challenges, particularly for the entrepreneur, are created with this imported technology.

Thailand: Cassava Pelletizing Technology
    Ronald P. Black, Wanawan Peyayopanakul, and Sachee Piyapongse

Indigenous technology for producing cassava pellets in Thailand is the subject of this case, which also examines related steps from the growing of the cassava through marketing the cassava pellets abroad. It addresses issues central to the vitality of the cassava pelletizing industry, such as related environmental problems and protectionist tendencies in foreign markets.

Colombia: The Composite Flour Program
  James M. Miller

Efforts were made by IIT, Bogotá, to reduce imports of wheat by substituting locally available flours, e.g., defatted soya and rice, partially in breads and totally in pastas. The pasta program has succeeded in its goals but the bread program still needs initiative from the Colombian government.

Haiti: The Coffee Roads
  Laurie Nogg Adler

The design and construction of 110 kilometers of new penetration roads is facilitating the production of coffee in previously barren mountain areas in Haiti. The roads are also allowing for increased maneuverability and are opening up communications channels to some Haitian villages.

Pakistan: Ball-Point Pen Manufacturing--A Case of Technology Transfer
  James W. D. Frasché

Ball-point pen manufacturing was initiated in Pakistan. This case demonstrates how the combination of entrepreneurial drive and foreign technology has been important in creating a goal of precision engineering expertise that may be beneficial to more than one sector of the economy.

Java, Indonesia: The Introduction of Rice Processing Technology
  Melinda L. Cain

New innovations have been made in Indonesia's agricultural development policy for rice production on Java. In particular, the use of mechanized rice hullers, the tebasan system of harvesting, and the use of sickles and scales have had a significant impact on increasing rice productivity. However, widespread labor displacement, has resulted, particularly among women, who have traditionally been involved with the rice harvest.

Central America: The Lorena Cookstove
   Suellen Sebald Edwards

The Lorena cookstove has been disseminated throughout Guatemala, Mexico, and Honduras. The clay and sand cookstove uses 50 percent less wood than cookstoves currently in use, will burn sawdust and cornstalks, holds heat well, and is modern and attractive-looking. A training program has resulted in an estimated 800 stoves being built in Guatemala.

Honduras: An Experimental Lime Kiln
   Judith Evans Blum

Through cooperation of several development assistance organizations, an experimental lime kiln was constructed in Honduras in an attempt to improve fuel efficiency and produce a better quality of lime.

Malaysia: Small-Scale Brick Manufacturing
   Ronald P. Black, A. Rahim Bidin, Woo Seng Khee, and Nik Ahmad Kamil

A typical, small-scale brick factory is described, and future challenges that the industry may have to face are related by the factory owner. For a different perspective, an industry forecast from an expert at the Standards and Industrial Research Institute of Malaysia is presented.

Colombia: An Innovative Approach to Rural Development
   James M. Miller

Young people with the greatest learning potential, regardless of educational level, were selected for six years of customized education in the broad fields of sanitation, engineering, medicine, public health, and agriculture. The objective was to retain these individuals as professionals in their rural environments.

Brazil: The Role of CRUTAC in Community Development
   James P. Blackledge

CRUTAC reaches out to the countryside by combining higher education curricula with practical experience at the village level. It is a reciprocal learning process in which students learn to apply academic knowledge to specific situations and villages benefit directly from programs developed and implemented by university students and faculty.

Western Samoa: The Samoa Methodist Land Development Program
    Ruth Lechte

A church-owned farm has been turned into an active resource for community development by the utilization of many village-level technologies for farm training and food production.

Tanzania: Biogas Generator
    Richard S. Roberts, Jr.

This response to one village's request for a fuel to replace firewood for cooking involved a technology imported from India. The technology was embodied in an apparatus that could be made in towns and villages at a price villagers could afford.

Brazil: Explosive Metalworking Program
    James M. Miller, Jim D. Mote, and Henry E. Otto

The introduction of a metallurgical technology--explosive forming, cladding, and hardening--has opened up a small but profitable industry in the state of São Paulo.

Ghana: Small-Scale Sugar Processing
    Gary D. Kilmer and David L. Sussman

A nonprofit assistance agency, Technoserve, aids in the development, modification, and replication of a labor-intensive, small-scale sugar-processing technology in Ghana.

Thailand: The Introduction of Mint Agriculture and Processing
    Ronald P. Black and Sachee Piyapongse

Mint agriculture and processing were introduced in Thailand; farming, mint oil extraction, menthol production, and related research activities were developed. The organizations involved in the process are also examined.

Central America: Fungal Fermentation of Coffee Waste
Suellen Sebald Edwards

The fungal fermentation of coffee waste is a program underway in Guatemala and El Salvador that will result in useful by-products (protein). The cultivation of fungi on the coffee pulp juice represents a useful technological advance for LDCs--it will help to control some of the pollution involved with coffee production.

The Philippines: Fish Preservation Techniques
Melinda L. Cain

Traditional fish preservation techniques of smoking, drying, and fermentation have been used by lower income people in rural and urban areas of the Philippines. Most of the methods are not scientifically based, but rather are passed down through generations. Women play an important role as fish processors, supplying a much-needed source of income.

Singapore: The Development of a Design Consulting Engineering Firm
Ronald P. Black and Chan Beng See

An indigenous design consulting engineering capability was developed in Singapore. This process is described in terms of the types of projects undertaken by the Singapore firm, its ventures into foreign markets, the development of its relationships with foreign firms, and the nature of the evolving design consulting engineering market in Singapore.

Sri Lanka: The Ceylon Institute of Scientific and Industrial Research
Donald D. Evans

The Ceylon Institute of Scientific and Industrial Research in its twenty-third year had many technological advancements to its credit, but over its history has had internal and external problems that are characteristic of research institutes in many countries. These involve staffing, training, equipment, government and industrial relationships, and finance.

Papua New Guinea: Micro-Hydroelectric Projects for Rural Development
    Ed Arata

A pilot hydroelectric plant in a mountainous region of Papua New Guinea is changing villagers' lives by supplying power for lighting and heating for the first time.

The Republic of Korea: An Example of Integrated Regional Development
    Donald D. Evans

Regional development on a 703-square-mile island includes Research and Development progams for producing a high-nutrition cattle feed, improved orange processing and shipping methods, design and partial testing of a small wind-powered electric power generating system, and other technical and economic studies.

Part 1
Appropriate Technology
and Its Role in Development

Appropriate Technology and Its Role in Development

Donald D. Evans

THE NEED FOR DEVELOPMENT

Development, as classically defined, included activity primarily concerned with increasing the Gross National Product (GNP) of nations or regional entities that were significantly less productive than the industrialized nations. In more recent years, especially during the 1970s, a more humanistic view of development has been taken, largely as a consequence of the growing number of formal studies in the field, disappointment with development progress, the evolution of thought and opinion resulting from extensive considerations in world fora such as the United Nations, and movements resulting from environmental and population issues.

In general, the need for development derives from the physical, economic, and social conditions of man, from the fact that great disparities and inequalities exist in this condition, and from the realization that the means for redressing such discrepancies are at hand and can be brought to bear on the problem effectively. Rationales for economic development can be predicated reasonably on either relativistic or humanitarian grounds; that is, the need for optimization of material growth in order to provide adequately and humanely for the human community affords ample justification for the development effort and, in fact, requires it.

One change in the perception of need for development comes from the realization that man will not necessarily benefit adequately and equitably simply as a consequence of the economic enlargement of political states. The economic gains of industrialization, for example, will not automatically "trickle down" to the benefit of all. The devolvement of the advantages of material advancement to all economic strata is not a natural occurrence--at least within an acceptable time frame--and the development process must therefore include functions that extend well beyond the enhancement of industrial productivity per se.

Particularly since the early to mid-1960s in the industrialized nations, there has rapidly developed a pronounced perceptual outlook emphasizing the physical limitations of the environment, the exponentially increasing depletion of natural resources, and the finite nature of most of the earth's endowments. Although throughout modern and ancient history speculations about these natural (or supernatural, for that matter) limitations to man's habitat have been made, a large-scale, systematic examination of the subject is a recent phenomenon, brought on in part by greater measurement capability and by availability of data and information on which to base analyses. The changing relative and absolute availabilities of natural physical resources already have had a pronounced effect on world relationships (e.g., impact of oil-production cartelization), and that they will have a more striking effect in the immediate future is irrefutable.

These concerns, of course, center around two considerations: (1) the efficient recovery and utilization of the earth's resources (including incident solar energy) constitutes a primary challenge; and (2) the impact of man's activity on his environment is progressively deleterious and must be brought immediately into balance. Most observers believe that environmental concerns and the utilization of the earth's resources are properly elements of development and should be included under that rubric.

The problems of earth resource utilization are compounded by the increasingly rapid growth of world population. Although there are pockets of population, especially in the industrialized nations, where the conditions for achieving essentially "zero population growth" exist, the fact remains that the overall increase in man's numbers on the earth poses the principal threat to political stability, to man's welfare, and to man's continuation as a species. It is believed that had the population growth of the developing world not been greater than that of the industrialized nations since World War II, then the rate of growth in GNP that the developing nations have already experienced would, today, have resulted in their essential economic self-sufficiency--at least in their ability to feed, clothe, and shelter their populations above minimum standards. The need for development as here discussed would then be largely nonexistent, because the present major material deficiencies would not exist. (Perhaps these needs would have been replaced with a complete set of equally portentous requirements, however, but at least the elemental prerequisites would have been met and the situational progress of mankind greatly accelerated.)

That the question of limiting population growth is a highly sensitive and consequential issue with such a large portion of the world's population naturally has great effect on development strategy and on the results of its implementation. Measures of long-term trend in attitudes towards population issues seem to indicate that greater numbers of persons accept that there is an absolute requirement for limiting the world's human population.

That the nature of the need for development is the subject of divergent views is well demonstrated throughout the literature. For example, various authorities analyze such differences in terms of the historic, religion-based perceptions of the role of man on earth. These contrast the linear progression and perfectability-of-man concepts which have characterized the West with Eastern cultures' cyclic concept that holds that the individual's presence on earth is but one phase of the being's existence, one to be accepted and endured in anticipation of attaining higher realms of perfection in subsequent states. To the extent that such philosophic/religious beliefs influence the individual's perception of the need for change in his material circumstances, views will differ as to the desirability and necessity for various types of development effort.

Recognition of these differences of objective, combined with disappointments in the efficacy of previous development strategies, has resulted in revised definitions of need. These definitions are more fully developed and are more "human needs oriented" than the monolithic, somewhat simplistic, criterion of increase in GNP which has served in the past.

The immense variety of circumstances that exists in the developing world must necessarily have considerable impact on perceptions of just what comprises development need. Not only are the economic circumstances of widely varying nature, but there are the readily observed differences in climate, topography, culture and social custom, religious traditions, political systems, and economic and colonial history. All of these enter into the determination of development need and character. Consequently, a rather pronounced danger exists in the tendency to generalize on the subject of development, to be too sweeping in pronouncements on the character of it, and a failing to take into account those variations that make "special cases" out of virtually every nation's circumstances. It is this differentiation that causes much of the difficulty in agreeing as to what constitutes "appropriate technology." It also helps to explain why the demonstration of the need for development and the definition of development itself prove difficult.

Certainly the most immediate and persuasive argument for development is the evidence of hunger and malnourishment in the developing countries (and in portions of the developed world as well). Not only are the statistics immediately disheartening, but the projections based on future population requirements vis-à-vis rates of increase in agricultural production paint a picture of greater deprivation for the future in the absence of immediate and effective action. It is estimated[1] that in the thirty-five LICs (low income countries--those with GNP per capita values below $250 per annum), 80 percent of the population is below the Food and Agriculture Organization's minimum food and nutrition standards, and that 25 percent has a serious food deficiency of over 250 calories per day. This situation is somewhat better in the fifty-seven middle income countries (MICs: $250 to about $3,500 GNP per capita), where 66 percent are above the minimum food and nutrition requirement level, and 8 percent are below the minimum daily caloric intake level.

In few areas of human endeavor have developmental results been as dramatic as in the fields of medicine and health care. Nevertheless, the benefits of these advances are today largely limited to those few affluent and fortuitously located persons who have ready access to medical practitioners on an affordable basis. Statistics[2] show that in the LICs in 1974 there were over 21,000 persons to every physician, whereas in the industrialized countries there was one physician for every 630 persons; the MICs had 2,430 persons per physician. Lack of medical attention, combined with low food and nutritional standards, no doubt contributes to world mortality figures. Although these ratios have improved dramatically in recent years, the life expectancy at birth in the LICs was only forty-four years in 1975, compared to seventy-two and fifty-eight, respectively, in the industrialized and the middle-income countries. These data argue effectively for increases in the numbers of medical personnel and services, and this becomes an important element of the development need of these areas of the world.

Education figures display equivalent discrepancies, with adult literacy rates in the LICs at 23 percent in 1974, versus 99 percent in the industrialized countries and 63 percent in the MICs. Clearly, then, education is another area of critical importance in the assessment of development needs.

One of the effects made possible by technologically enhanced communications is a widespread knowledge of other societies. This knowledge, of course, has many benefits, but perhaps one of its more consequential effects is that it has made people in remote areas of the globe aware of the

---

[1]The Development Coordination Committee. <u>Development Issues: U.S. Actions Affecting the Development of the Low-Income Countries</u> (Washington, D.C., April 1978).

[2]The World Bank. <u>World Development Report - 1978</u> (Washington, D.C., August 1978).

material progress that is experienced by other societies. Thus, an era of "rising expectations" of unprecedented dimensions has occurred in which the desire for increased material well-being has come to dominate the actions and concepts of large segments of the developing country populations. This may be beneficial to the extent that there is a reasonable prospect for the realization of desirable elements of such aspirations. On the other hand, many social observers have condemned this effect because of the frustrations it has caused, with consequent political impacts.

Perhaps at no previous time has there been as much concern with materialism and the loss and subjugation of spiritual values and traditional social structure, brought on by the advance and availability of modern technology (e.g., Iran's recent upheaval). This seems true in all the world's societies and political systems. It was more or less violently demonstrated in the United States during the 1960s and has had a pronounced impact on U.S. laws and regulations relating to the implementation of foreign assistance programs. Therefore, consideration of the need for equitable distribution of the beneficial application of technology, and a specific focus on the improvement of the circumstances of the poorer economic strata of the world's societies, are now manifest in the development assistance programs of the United States.

Consequently, determination of the need for concerted efforts at development in the world (by whatever definition "development" may be assigned) is based on:

(1) The gross deficiencies in the material circumstances of mankind.

(2) Threats of ever greater inequities and deficiencies as a result of burgeoning populations.

(3) Realization of the finite limits of the earth's resources and the requirement for rational allocation of these.

(4) Threats to the habitability of the earth as a consequence of man-initiated activities that degrade the environment.

(5) Inequitable opportunity for the social and intellectual development of the majority of the world's population through education, resulting in great waste of human potential and a denial of the fundamental right of each

individual to realize his or her maximum personal development.

(6) The widespread presence in the developing countries of debilitating and fatal diseases, contributing to differences of life expectancy at birth of nearly two to one between the low income and the industrialized nations.

THE MEANING OF DEVELOPMENT

Measures of Economic Development

Some of the principal issues facing the world in terms of the economic and social evolution of the developing countries are summarized in the following paragraphs. These reflect the views of the World Bank,[3] and are presented here as a summary of principal aspects considered important by that institution (see Appendix 0.1 for a series of tables excerpted from the Bank's report).

The developing countries have made progress in the last twenty-five years in terms of increasing the per capita incomes of their rapidly growing populations. Progress also has been significant in nonincome fields, such as the extension of educational opportunities, provisions of improved health services with a consequent reduction in mortality and morbidity rates, growth in infrastructure to support industrial development, and an increased ability to accept and adapt technology for development purposes.

However, despite this growth record, world poverty, especially in the LICs, has not been appreciably lessened. There are estimated to be 800 million persons presently living below the minimum subsistence level, that is, in "absolute poverty." These represent some 40 percent of the entire population of the developing world; most of them are found in the rural sectors of their countries.

This failure to relieve poverty satisfactorily may be attributed in large part to a world population that has continued to grow at an alarming rate, despite an inability to accept most new members into the society with the assurance of providing them a minimum standard of living. Population growth in the developing countries has expanded from 2.1 billion in 1975 and is expected to increase to 3.5 billion by the year 2000.

---

[3]The information in this section is drawn from World Development Report - 1978 (Washington, D.C.: World Bank, August 1978).

The economic impact of this population expansion continues to severely dissipate any advances otherwise made in the absolute productivity of developing countries. One conclusion is that it is imperative to continue efforts at increasing savings and investment in these countries. This can work as a basis for expanding per capita income and raising large sectors of humanity out of poverty. Nonetheless, extreme conditions require immediate relief, but it is necessary to concurrently maintain the effort toward growth in productivity and income as the basis on which permanent solutions to the poverty problem can be built.

The LICs particularly must increase agricultural productivity as the source of food and as a basis for generating capital to support industrial and infrastructure development. It is recognized, however, that it is difficult to attain this goal given the often severe lack of supporting agricultural infrastructure, plus the fact that conditions vary so widely among the nations that each requires its own specific, planned program.

An external restraint on achieving development is the decrease in Official Development Assistance[4] rates (as a percent of GNP) flowing to these countries. Thus, the relative amount of support accorded has decreased in recent years, and ways are being sought to raise it to higher levels. The condition of the MICs is much more favorable, however, as they continue to show encouraging improvement in their development patterns. The principal problem facing the MICs is how to respond, with internal trade and industrial development policies, to rapid changes in the international environment--a condition precipitated by increased costs of energy, worldwide inflation, and needs for increased sales of manufactured products to the industrialized nations.

The growing role of the MICs as sources of manufactured products for consumption in the industrialized countries is viewed as encouraging and as a precursor of the future. Development of such external markets with larger value-added increments is considered by many to be even more important in the economic development context than the internal consumption-goods production increases and improvements ("import substitution") that many of these countries are experiencing.

Also considered auspicious from the long-term development perspective is the growing role of the industrialized

---

[4]Official Development Assistance consists of disbursement of grants or loans made at concessional financial terms by official agencies of the members of the Development Assistance Committee of the Organization for Economic Cooperation and Development (OECD), with the objective of promoting economic development and welfare. It includes the value of technical cooperation.

countries as sources of economic "pull" for the developing countries. The growing interdependence of the industrialized and the developing countries is recognized and is considered favorable for increasing world stability, both economically and politically.

The past twenty-five years have demonstrated how slowly social institutions change and the persistent effort that is required to effect such change. It is more fully realized that many fundamental development relationships are imperfectly understood, and research into these will be necessary to progress toward more effective development systems.

In the view of the Bank, the major question is: will the international environment be as supportive of development efforts in the next twenty-five years as in the past, given the changing requirements and the (in many ways) more difficult circumstances?

Methods of Economic Development

This section is based primarily on the economic development approaches of Hla Myint of the London School of Economics,[5] Romesh Diwan and his associate, Dennis Livingston, of the Renssalaer Polytechnic Institute,[6] D. P. Ghai of the International Labour Office,[7] and Dr. Wassily Leontief and associates.[8] These publications represent a range of current thought on economic development, although numerous other authorities might have been selected as well, and all variations are not presented.

---

[5]Hla Myint. Economic Theory and the Underdeveloped Countries (London: Oxford University Press, 1971).

[6]R. Diwan, and D. Livingston. Development Strategies and Technological Choices In Developing Countries (Troy, N.Y.: Rensselaer Polytechnic Institute, 1978).

[7]D. P. Ghai, et al. The Basic Needs Approach to Development (Geneva: The International Labor Office, 1977).

[8]W. Leontief, A. P. Carter, and P. A. Petri. The Future of the World Economy (New York: Oxford University Press, 1977); several works in this section of the book are indicated as footnotes, but do not appear in the annotated bibliography. This omission is purposeful in that the bibliography contains works dealing specifically with appropriate technology while the works herein contain much more general development information.

Elements of Development Theory. Myint presents a structure of development theory based on historic elements. The "Missing Component" approach is predicated on the idea that development can be effected by identifying the critical weak or missing element in a developing country's assets and by subsequently providing this component. Such missing elements typically are identified as domestic savings, foreign exchange, and education.

The propensity for saving in the developing economy, including both public and private entities, is viewed as essential to development goals. It is noted that, despite views to the contrary, the developing countries have shown a significant savings record, even the LICs, where this is considered least likely. This tends to refute the conventional wisdom that minimum income economies present no opportunity for domestic savings, particularly in the private sector, and that extraordinary measures must be taken to create savings for capital investment. The fact that rapid population increases have tended to offset savings gains does not fully detract from the significance of this accomplishment.

However, domestic savings, even when abetted by infusions of foreign assistance, fall short of meeting requirements for industrial, agricultural, and infrastructural investment. Some of the principal difficulties in the LIC situations are the poor selections that have often been made for capital investment. For example, the concentration on developing industry to produce goods for import substitution has created demands for other scarce elements of production without providing additional employment or utilizing other indigenous resources.

Viewing foreign exchange as the most significant component in short supply has sometimes resulted in the borrowing of it to domestically produce suboptimal, subsidized, and noncompetitive products. The view of capital as comprising only currency has ignored the concept of "capital" as also including the natural resources of the country, including agricultural production capacity and infrastructure such as transport and communication facilities. Borrowing foreign exchange may sometimes be required to provide "sustaining capital" that will support the start-up and initial operating costs of projects utilizing such noncurrency capital.

The problem with a development strategy that emphasizes the acquisition of foreign exchange is that it leads to foreign monetary overcapitalization at the expense of a more balanced use of domestic resources. Resulting negative effects may include excess productive capacity, overborrowing with heavy debt servicing requirements, and underutilization of labor as a substitute for capital. (This latter condition is the basis for much of the argument supporting the "appropriate technology" concept.)

Education has been popular in the past as an area of concentration for development, based on the view that one of the critical shortages in the developing economy is trained manpower to meet needs in industry and infrastructure. While this may have been the case in some recently independent former colonies where an organizational structure needed "fleshing out" after the departure of the foreign incumbents, it did not apply for long in most development situations. The result in a number of instances was that many persons from new and enlarged higher education institutions became available in a society where few, if any, positions were available to them. This is not to say that a shortage of qualified persons in the developing economies does not exist, but that journeymen artisans, foremen, and managers with on-the-job experience are greatly needed.

The "surplus resources" strategy involves the presumed availability of under- or unemployed labor and the existence of unexploited natural resources. In the latter case, too often it is discovered after considerable investment that the development of the resource does not result in an exportable commodity--one able to compete in the world market with acceptable quality and/or at a profit--and that the domestic market is not sufficient to support it.

The assumption of the availability of nonutilized or underutilized labor is often confounded by the realization that the mobility of the force is not so great as thought due to its need to be near sustaining sources of food and shelter, as on the home farm, or that the availability is only seasonal, depending on the harvesting and planting seasons. Even when the labor is actually available, the capital needed to support the labor force until it can become essentially self-sustaining may not be available.

The "critical minimum effort" approach to development is based on the theory that in a given situation, the infusion of a massive assistance effort is required to initiate self-sustaining development. This view has its roots in the "take-off stage" theory of W. W. Rostow. The theory inferred from observations of apparently successful development effort that if certain preconditions of amount and rate of effort were met in the economy, development would be initiated and then proceed on its own momentum. A sufficiently large and diverse infusion of assistance from the outside was therefore required.

This approach has proven unreliable in practice because the offsetting effect of rapid population increases has made it difficult to attain the initial condition of "mass." It also assumed the desirability for an overall integrated approach, resulting in a complete economic development effort. In fact, however, many developing country capabilities are now viewed as inadequate to coordinate or sustain such a broad economic effort.

One problem is the shortage of experienced managers and entrepreneurs to meet the needs of new sectorial developments. This problem is further compounded by the difficulty of coordinating the simultaneous development of new public and private enterprises--a task that would be challenging to organizers even in a developed environment. Also, a "scale effect" indicates that only an autarkic society of sufficient size and diversity can be raised to the take-off point; thus, it may be possible in such large countries as India and Brazil but is not likely elsewhere.

In general, development strategies can be divided on the basis of their emphasis on the evolution of either the domestic industrial base to meet internal demand, or on the encouragement of a trading economy that depends upon the cultivation of specialized export markets for which a domestic industrial base can meet the test of international competitiveness.

The traditional arguments against the external trade emphasis turn on the conventional wisdom that international commodity markets are so cyclic and variable that they provide a very insecure base on which to predicate the principal economic effort of the country and that there is a trend toward decline in the importance and growth rates of primary commodity markets as international markets evolve toward more trade in manufactured products.

Although appreciable empirical evidence supports these arguments, a number of successful country development efforts also have been based on the idea of export market exploitation at the expense of extensive domestic industrial development. A case in point is Malaysia, which has been successful in promoting natural rubber even with the rise of synthetic products (now declining due to the increase in the price of crude petroleum).

Conditions for the success of the export development thesis include: (1) concerted effort must be made to control inflation so that world market competitiveness can be maintained; and (2) for the same reason, a realistic exchange rate must be maintained to assure that the export products are not artificially priced out of competition. This strategy for development also requires utilization of the best current technology to meet quality standards and to result in least cost of production (this has obvious implications for the "appropriate technology" approach, which may call for substitution of less-efficient labor for capital).

The argument for development of comprehensive domestic industrial capability acknowledges the need for a reliable source of cheap labor--and the record in this regard is not particularly encouraging. Experience shows that labor is likely to organize and drive up unit costs of production to some world level or that shortages will develop as other industry creates demand for skilled labor.

Other detracting effects of development of indigenous industry result from the protection that government often accords new industry to get it started and to maintain it. This subsidization of domestic consumption-goods industry is often accomplished at the expense of the agricultural sector or a potentially viable export industry, through the artificial discounting of capital cost by the government, protective tariffs, tax moritoria, and so on. Development of domestic consumption goods industry therefore may be achieved through the literal impoverishment of potentially productive sectors that could provide food or earn foreign exchange through fully competitive export sales.

A common argument against creating specialized export industries cites the effect of the plantation economies introduced during the colonial epoch. This resulted in the creation of "dual economies" in which those sectors directly identified with the export industry prospered and created labor markets that provided an elite few with higher incomes while the mass of the society was left out of the benefit cycle directly. However, the creation of subsidized, protected domestic consumer goods production facilities has had much the same effect, and this, in fact, accounts largely for the negative effects of the urbanization trend. The disguised unemployment of the rural areas, where the extended family system cares for the unemployed, becomes the manifest unemployment in the urban environment.

Myint advocates the development of labor-intensive, export-producing industry that nevertheless utilizes the best contemporary technology in whatever ratio is necessary to maintain production of a commodity that is fully competitive in the world market. Reinvestment of the profits will provide capital for the development of competitive domestic industry and infrastructure.

Other requirements are: increasing the absorptive capacity of the national economy to make efficient use of both domestic and foreign capital; encouraging technological innovation; developing the infrastructure for industrial support; effectively utilizing land through distribution and tenure policies and technology application; developing the domestic capital markets; maintaining a realistic cost of capital; and planning effective long-range goals.

In another analysis, Owens and Shaw[9] identify four "fads" that have characterized Western efforts at encouraging the development of the Third World. First was a period of postcolonial effort at instituting democratic concepts and procedures in the belief that these processes were

---

[9]E. Owens, and R. Shaw. <u>Development Reconsidered</u> (Lexington, Mass.: Lexington Books, 1972).

fundamental to a self-generating development movement. It failed for a variety of reasons, but generally due to the fact that these newly independent nations did not have the social traditions or concepts to utilize such a dramatically different political system. Second was a period of believing that education held the key, and this encouraged a spate of formal education-institution development programs that produced large numbers of graduates who were ill-equipped to deal with the real problems of their societies, for whom no adequate infrastructure existed to utilize their talents, and in whom had been generated a desire for opportunities and positions which could not be realized. Third was the industrialization phase in which it was concluded that the application of Western technology embodied in massive industry development programs would provide the motivating energy for development, but this has resulted rather in the various economic and social ills that the "appropriate technology" concept attempts to redress. Finally, and more recently, the idea has been adopted officially by the United Nations and other entities that the failure of development has been not so much a question of technique as a shortfall in quantity. Thus, if assistance levels could be raised to approximately 1 percent of the gross national product of the industrialized nations, this would be sufficient to bring about the metamorphosis. The validity of this thesis, of course, has not been tested because that level of financial assistance has not been approached, nor does it appear likely that it will be in the foreseeable future.

Alternative Development Strategy (ADS). Diwan and Livingston describe five situations in which the ADS strategy is believed to be effective: (1) in dealing with rural populations; (2) where (as is the case in many developing countries) the ratio of poor (defined as underfed and malnourished) ranges between 60 and 70 percent of the population; (3) where unemployment and underemployment exist in either or both the urban and rural environments; and (4) where the situation makes it virtually impossible for all the developing countries to achieve the economic standards of the industrialized countries, but the strategy effectively eliminates poverty and allows all persons to reach a minimum standard of living.

The ADS concept reflects the humanistic/conservator ethic that has become prominent in development strategy. The basic elements of the ADS comprise the following:

    (1) The production of goods and services that meet the fundamental needs of the poorer

strata of the population, both urban and rural;

(2) The generation of employment, both as a means of production and as a primary method of redistributing the wealth;

(3) The conscious, structured redistribution of wealth to the poor from the rich, through political intervention;

(4) The encouragement of production in rural areas to induce urban migrants to return to these regions;

(5) The encouragement of cottage and small-scale industry; and

(6) The encouragement of "local values," initiatives, and self-reliance.

This approach to development emphasizes certain social values and, while seeking a form of national self-sufficiency, gives primary consideration to the role of the individual rather than to the maximization of capital utilization.

The "Basic Human Needs" (BHN) Approach to Development. D. P. Ghai provides information that establishes the BHN as essentially similar to the ADS.

The BHN concept was expressed in the deliberations of the International Labor Organization and has since been discussed in the context of United Nations Conference on Trade and Development (UNCTAD) meetings and in other fora. Its basic similarity to the ADS is evident from a listing of its essential elements:

(1) The basic physical needs that the strategy is intended to meet are food/nutrition, education, health, housing, and sanitation;

(2) To take immediate exceptional steps to meet these needs and not wait for the present protracted development process, which, in the view of some, will never relieve poverty;

(3) To emphasize the basic human rights to equity, participation, and self-reliance;

(4) To provide for a dynamic, changing situation in the planning and execution of the strategy,

recognizing that conditions are fluid and that momentary needs are likely to change in the perceivable future;

(5) In the same vein, to recognize that national and even local conditions vary considerably, requiring the structuring of programs that are specifically suited to the situation at hand;

(6) That the successful conduct of this strategy will necessarily require new social structures and political forms for redistributing assets and incomes; and

(7) That to effect the strategy will require a distribution of political power to the affected social sectors.

Ghai also presents evidence as to why this approach does not, in his view, constitute a strategy *per se*.

The Attainability of Development

The potential for development is, of course, a basic question: is it feasible to achieve the far-reaching goals of the strategies?

Myint cites three reasons why the attainment of development, in the sense of essential equivalence with the industrialized nations, proves difficult if not impossible:

(1) In contrast to the conditions of the Western nations when they were beginning their industrial revolutions, the present-day developing countries for the most part lack a per-capita productivity base;

(2) The developing countries already are grossly overpopulated, given domestic resources. Unlike earlier Western development situations, an immense problem exists in simply generating sufficient resources to meet basic needs; and

(3) For the most part in the LICs, social and political bases on which to predicate development programs are weaker. Many LICs have emerged from colonialism and other forms of foreign domination just recently and so are unpracticed in the art of government.

On the other hand, some elements do assist development in pervasive ways. For example, the store of technology is generally considered adequate to achieve the material goals of development, be they in food production, health care, extraction and utilization of the earth's resources, social organization, communications, or trade.

Another factor assisting development is the "pull" exerted by the industrialized world in its own economic evolution. Not only does it create an enlarging market for the primary materials that have been the traditional products of the developing world, but to an ever greater extent it becomes a market for manufactured products from the developing world. It is this symbiotic relationship that many, possibly most, of those concerned with development wish to encourage.

Notwithstanding the controversial character of the thesis, many believe that, as a realistic matter, it is not possible in the foreseeable future for parity in economic development to be an attainable goal for the developing countries, overall. (When Herman Kahn of the Hudson Institute reached a similar conclusion several years ago concerning the prospects for such a strong industrializing nation as Brazil, it was greeted with considerable indignation and rejection by various Brazilians and others; consequently, it is dangerous to make assumptions or statements concerning the inevitability of anything concerning economic and social development.)

"The Future of the World Economy" - A Model for Development. Based on mandates from the United Nations and supported by funding from various institutions and foundations, Nobel Laureate Wassily Leontief and associates Anne P. Carter and Peter A. Petri in 1977 completed a study of alternative scenarios for world development to the year 2000. The study was based on an interactive "input-output" mathematical model of the world economy.

Their results suggest the feasibility of development, measured in quantitative terms, on a worldwide basis. This is not the first such world model, and the authors are quick to note its deficiencies and postulate improvements that will come with refinements.

One of the assumptions on which the scenarios of the UN study were based derived from the objectives of the International Development Strategy for the Second United Nations Development Decade as first promulgated in 1970 and as reaffirmed by the General Assembly in its Sixth Special Session in 1975. Under this plan, the growth targets for the developing countries as a whole were set at 6 percent per annum for gross product and 3.5 percent for gross product per capita, based on an assumed population growth rate of 2.5 percent. No targets were established for the developed countries, but they were based on an extrapolation that showed about 4.5 percent in gross product and 3.5 percent for gross product per capita. However, at these assumed rates virtually no decrease in the economic gap between the developing and the developed nations would occur by the year 2000; this gap would remain approximately at the twelve-to-one ratio that obtained at the start of the decade (1970), and this is inconsistent with the intent of the UN membership that it be narrowed appreciably by 2000, leading toward a goal of parity sometime in the mid-twenty-first century. Consequently, another scenario was adopted which had as its objective the reduction by approximately one-half of the ratio between rich and poor by 2,000. The table illustrates the effect of the different assumptions.

TABLE 1
HYPOTHETICAL SCENARIOS OF WORLD ECONOMIC DEVELOPMENT

	Scenario	Developed Countries	Developing Countries
Growth rates (percentage):			
Gross product	I^a	4.5	6.0
	C^b	3.6	6.9
Population	I	1.0	2.5
	C	0.6	2.0
Gross product per capita	I	3.5	3.5
	C	3.0	4.9
Income gap in the year 2000^c		12 to 1	
	C	7 to 1	

---

[a] I indicates scenario based on extrapolation to the year 2000 of International Development Strategy targets for gross product increase in developing countries and extrapolated long-term historical rates in developed countries.

[b] C indicates scenario based on substantial reduction of gap in gross product per capita between developing and developed countries.

[c] Average per capita gross domestic product of the developed regions, as related to the average per capita gross domestic product of the developing regions.

Another assumption was that as per capita gross product increases toward the levels of the industrialized countries, the average rate of increase will drop. This is sustained by observation and reflects the reasoning which says that exponential increases, including population growth, cannot continue indefinitely.

The study assumed that the higher rates of gross product growth could not be attained immediately in any event and so projected a gradually increasing rate over a thirty-year period. "Even under these relatively optimistic assumptions, by the year 2000 average gross product per capita would only reach about $400 (in 1970 prices) in non-oil Asia and Africa, while other developing regions would be able to enter the $1-2,000 range," the report stated.[10]

The model discloses the absolute importance of the achievement of sufficient food production to sustain whatever population levels are realized. Toward this goal, an increase in agricultural production through implementation of modern technology, plus bringing into production latent land resources, makes such food sufficiency possible. The possibility exists of bringing 30 percent more land into production throughout the world, although an investment of $60 billion would be required in Asia alone to raise the productivity of agricultural land to one-half of its theoretical potential. Thus, the problem of such development is immense.

Mineral resources are viewed as being increasingly scarce, requiring progressively more expenditure to extract unit values. This economic situation will work to the advantage of those developing countries that contain such resources but will obviously be detrimental to those that are resource poor. Within the time frame of the study, minerals, including fossil fuels but excepting lead and zinc, are considered to be sufficient to meet demand. Petroleum, for example, is purported to be present in the ratio of 1.3 times projected consumption. The developing countries will have to make strong efforts to develop their mineral resources, employing modern technology to do so (this also has relevance to the use of appropriate technology).

Although the problems associated with pollution will continue to concern the industrialized countries, until the year 2000 this will not be a major consideration or cost to the developing countries. Possible exceptions will be in the generation of urban solid waste. In any event, the cost of dealing with pollution in developing countries is not expected to exceed 0.7 to 0.9 percent of GNP.

The driving mechanism of development is considered to be the progressive industrialization of the developing countries, but this will require investment in productive capacity as high as 35 to 40 percent above present relative levels. Such investment ratios will necessitate dramatic changes in fiscal and monetary policies, taxation structure to generate internal savings, credit systems, and, most important, means

---

[10]Leontief, p. 3.

to achieve more equitable income distribution. Investment from foreign sources will have to grow but will remain a secondary source of development capital. If model parameters are to be achieved, a relatively slower growth of the industrial sector of industrialized nations and a greater importation of manufactured goods from the developing countries will be required. This suggests preferential tariff treatment and other inducements.

Heavy industry will assume a proportionately greater role in the industrial mix of the developing countries--at least among those that can assume this role. Sharing of roles in this evolution is suggested among regional groups. Light industry will be more important in the LICs.

Since the industrial infrastructure must be in place before it can produce export commodities, and since this requires a large amount of investment capital, trade deficits of developing countries will persist, probably declining relatively in 1980 and 1990 but assuming larger proportions in 2000.

The balance of payments discrepancy that the developing countries will incur as a consequence of the industrialization drive will need to be at least partially offset by increased sales of manufactured products to the industrialized countries, by stabilization of the world commodity markets, and by increased financial transfers such as direct investment. Earnings can be increased through higher agricultural and minerals prices, both of which appear justified given their increasing scarcity and the fact that greater technological inputs will be required to increase or maintain production levels.

Developing countries that are not well endowed with exploitable resources will have to develop their industrial sectors to stimulate export trade and concurrently work to reduce reliance on imported products for domestic consumption. It will be necessary to provide special tariff and other trade incentives for them in the world markets. More aid flows and the stimulation of direct investment will be important. But most important will be changes in the commodity markets and increased trade in manufactured products.

The development model, in conclusion, stipulates two general conditions: "first, far-reaching internal changes of a social, political, and institutional character in the developing countries, and second, significant changes in the world economic order.... Clearly, each of [these measures] taken separately is insufficient, but when developed hand in hand, they will be able to produce the desired outcome."[11]

---

[11]Ibid., p. 11.

Technological Issues In "The New International Economic Order"

In 1974 the Sixth Special Session of the United Nations General Assembly gave form and substance to the concept of The New International Economic Order (NIEO). In discussing this development, Karl P. Sauvant states: "...the NIEO program distinguishes itself from earlier international economic programs by virtue of its objective. Its objective is not merely to improve the functioning of the existing international economic system but rather to expand its purposes and to change its mechanisms and structures to suit the new purposes. The purpose to be added to the existing ones is development. Given the special situation of the developing countries, the acceptance of this additional purpose involves the acceptance of a number of principles. They include greater control by the developing countries over their own economies, especially as expressed in the principle of permanent sovereignty of every state over its natural resources and all economic activities; greater participation of developing countries in decision-making processes that affect their situation; international cooperation for development and active assistance to developing countries; and preferential and nonreciprocal treatment for developing countries."[12]

The NIEO, among a number of far-ranging proposals for changes in the international economic relationships between the industrialized countries and the developing world, specifically addresses aspects of the transfer and utilization of technology as one of the key elements of the development process and the effort toward world parity. The concepts put forward are highly congruent with the idea of "appropriate technology" and are compatible with the principles of the ADS.

The theme of the NIEO[13] is the creation of "a basic needs strategy of development," which calls for "an unprecedented expansion in the production of foodstuffs and of simple manufactured goods, using technology appropriate to the resource endowments of the developing countries themselves." The implementation of this strategy "would mean

---

[12]Karl P. Sauvant, and Hajo Hasenpflug. The New International Economic Order: Confrontation or Cooperation Between North and South? (Boulder, Colorado; Westview Press, 1977).

[13]UNCTAD, "Trade and Development Issues in the Context of a New International Economic Order" (UNCTAD/OSG/104/Rev. 1), Feb. 1976.

the emergence of a new pattern of industrialization in developing countries." It calls for the more equitable return to the host country (developing countries) of the benefits and proceeds resulting from commodities and products produced by transnational corporations. It is recommended that the developed countries assist the developing countries in attaining economic independence and exercising full control over their natural resources and that they introduce "some measure of global management of these resources."

A revised world patent system is recommended, as is the creation of a "code of conduct" to govern the relationships of transnational companies with the developing countries (the code presently is under consideration in the meetings and activities of UNCTAD.) It has been suggested that some subsidy of commercial-rate loans to the developing countries be provided and that reductions in tariff and nontariff barriers to trade with the developed countries on a preferential basis be implemented.

Specific references to technological subjects include: "Developed and developing countries should cooperate in the establishment, strengthening, and development of the scientific and technological infrastructure of developing countries. Developed countries should also take appropriate measures, such as contribution to the establishment of an industrial/technological information bank and consideration of the possibility of regional and sectoral banks, in order to make available a greater flow to developing countries of information permitting the selection of technologies, in particular, advanced technologies."

It is recommended that "the developing countries identify supplies of foreign technology and know-how, strengthen scientific and technological cooperation, and establish their own production, trade, and marketing networks. Inasmuch as in market economies advanced industrial technologies are most frequently developed by private institutions, developed countries should facilitate and encourage these institutions to provide effective technologies in support of the priorities of the developing countries." Regional transfer-of-technology centers are suggested, as is cooperation among the developing countries in establishing manufacturing capabilities to avoid unnecessary duplication of facilities and personnel. This is especially recommended for smaller, less-affluent nations where the development of full-scale technological capabilities is likely to prove too expensive, given other development priorities.

In the developed countries, certain aspects of the NIEO are criticized, particularly concerning the transfer and availability of proprietary technology on what are viewed to be concessionary terms that fail to accord the owners a fair return on investment. Also, the concept of a mandatory code of corporate conduct, such as is widely advocated by the developing countries, is criticized. Some critics have observed that the NIEO seems to benefit only the developing countries per se and that it is directed more to the strengthening of political entities and states rather than to the individual citizens of those countries.

However, the desirability of providing and applying technology in the most effective way is generally accepted. The efficacy of technology in the drive for development, it is agreed, is of paramount importance. Ultimately, it will prove to be the means by which the developing countries lift themselves from underdevelopment and achieve the goals that are exemplified in the basic human needs strategy.

THE RELATIONSHIP OF TECHNOLOGY TO DEVELOPMENT

"Technology" in the broader sense may be defined as the means by which man undertakes to change or influence his environment. In this sense, it is close to the meaning of the French word technique, which, as Jacques Ellul explains it, "does not mean machines... It is the totality of methods rationally arrived at and having efficiency in every field of human activity."[14] Thus, the concept transcends the engineering arts to include forms of organization of human endeavor, methods of rational analysis, and the structure of systems for achieving determined objectives.

In discussing the relationship of technology to development, the means must not become confused with the end. The great emphasis on technology in recent years has tended to enshrine it as the objective of development rather than as a principal means for attaining development. Also, it must be kept in mind that not much is known about the interactions of technology in the development process. Despite extensive study of "technology" and its effect on the social and economic evolution of nations and societies, no quantifiable relationship of the interaction is available as the basis for formulating a theoretical model of the process. Eckhaus notes: "Recognition of technology's essential role in development does not imply a technological determinism. Not only can alternative products and methods be chosen, but the wider effects of these choices depend strongly on the political and economic environments in which they are implemented.

"The use of any particular technology is not an end in itself. The criterion for an 'appropriate' technological choice

---

[14]Jacques Ellul. The Technological Society (New York: Alfred A. Knopf, 1965).

must be found in the essential goals and processes of development."[15]

Jéquier, in describing the situations in which technology is applied, defines the private, community, and public loci.[16] "Private technology," being that which is utilized in a smaller, domestically owned enterprise, is related to the production of consumer goods, and its introduction is largely a decision by the individual. "Community technology" concerns local infrastructure, such as water systems, sewerage, and health delivery services, which affects the citizen profoundly and which requires individual cooperation in production and operation. "Public technologies" are represented by the large industrial firms that produce consumer goods or capital equipment and by the national institutions that supply basic services, such as railway transportation, flood control and irrigation systems, power grids, higher education, and banking and credit systems. He states that the community technology situation is the most difficult area in which to effect change because the forms of organization are less well defined and are less likely to provide incentive to the individual participant. Jéquier believes, however, that the community technology area provides the greatest return on investment in "appropriate technology," as narrowly defined.

Another system useful for examining technological inputs consists of the following:

Infrastructure. (1) civil works such as sewerage, highways, water systems, and public buildings; (2) power supply from central generating stations; (3) transport, such as railways, airlines, etc; (4) communications, including telephones, microwave circuits, television, postal service, and, in some instances, newspapers and other periodicals; (5) systems of higher education; (6) public services such as health care delivery systems; and (7) the national banking system, for savings, to provide a domestic capital base, and so on.

Agriculture. The areas of technological concentration include virtually every significant subject: plant genetics, fertilizers, equipment, irrigation systems, pesticides and herbicides, cooperatives and credit systems, and extension services.

---

[15]R. S. Eckaus. Appropriate Technologies for Developing Countries (Washington, D.C.: National Academy of Sciences. 1977), p. 6.

[16]N. Jéquier Appropriate Technology Problems and Promises (Paris: Organization for Economic Cooperation and Development, 1976).

Manufacturing. The product areas that are the recipients of technology inputs are divided into two categories: (1) capital and intermediate goods production (including the production of materials, heavy industrials, manufacturing equipment and machines, and automotive products); and (2) consumer goods (including durables such as appliances, shelter and transportation items, and consumables such as food, clothing, and medicines).

Infrastructure (public technology) and agriculture (which comprises both the private and community-type categories) traditionally have received the most concentrated foreign assistance inputs. Foreign private investment is concentrated in the public technologies sector.

The methods of technology transfer to developing countries can be similarly categorized. Multilateral assistance media include the development banks, the specialized agencies of the United Nations (including the principal assistance funding agency, the U.N. Development Program), and the various implementing agencies such as the U.N. Industrial Development Organization, the International Labor Organization, the Food and Agriculture Organization, and the Environmental Program. Although not directly involved in giving assistance itself, the Organization for Economic Cooperation and Development (OECD) is an important source of research and policy studies on development assistance.

Official development assistance (ODA), provided by seventeen members of the OECD, amounted to U.S. $13.7 billion in 1976 (the latest fully accountable year). Total commitment by centrally planned economies was $2.9 billion. The Organization of Petroleum Exporting Countries (OPEC) provided $5.2 billion in that same year.[17]

The multi- and bilateral agencies provide ODA or concessional aid; that is, funding on a grant basis (although some are also sources of development loans, usually on a low-interest rate basis). Much of this aid results in technology transfer in one form or another.

A principal source of technology transfer to the developing countries is through private-sector direct investment by the industrialized nations. The capital flow of the private sector to the development countries in 1976 was $20.0 billion.[18] Although this comprises all categories of capital use, much foreign equity is also gained through contribution of specific technical knowledge, including proprietary technology.

---

[17]World Development Report 1978, table 12, p. 98.

[18]United Nations, 1977 World Statistical Yearbook (New York, 1978), p. 872.

In addition to direct investment on a sole ownership basis or, as is more the trend, in a joint venture, significant transfer is effected through the licensing of technology on a royalty or direct payment basis. Also, technology is "embodied" in the technical products that are sold to the developing countries so that the sale of production equipment constitutes a technology transfer in the physical sense. Finally, engineering and technological consulting firms are a significant source of transfer because they accept commissions for the design and installation of manufacturing facilities, infrastructure, and the specification of technical equipment and processes in the developing countries.

Preference for Foreign Technology In Developing Countries

Much of the current controversy concerning the means of providing technology to the developing countries centers on the nature and conditions of such provision and on the effect of these conditions on the recipient nations. Although it is not possible to rank the following bases of preference for foreign technology, they all are significant and vary in degree depending upon the particular situation.

For the most part, imported, as opposed to domestic technology, is viewed as being more efficient and reliable. This stems from the fact that generally the imported technology must meet world standards to be competitive and often is offered by firms with a market position and reputation to protect. It is proven technology, based on designs and processes that for the most part have withstood application successfully in a variety of situations.

With the foreign technology often comes a "package" of management services and information that is essential to the developing country user. This may range from quality control systems to production management methods, training, accounting systems, and marketing aids (such as advertising materials).

Maintenance is of great importance in economies short of services. The ability of the supplier to provide prompt maintenance and repair service, including the provision of spares and replacements for consumable supplies, can make the difference in technological choice. (The unavailability of such services is often the cause for some of the strongest complaints against foreign technology.)

With the consideration of the time value of money, coupled with local contractors and suppliers who are inexperienced in providing a given technology, it is often advantageous to contract with the foreign supplier for a "turnkey" project in which the complete facility, including trained staff, is turned over to the buyer in a fully operative condition. This consolidation of responsibility with one source is usually viewed as preferable in remote locations where coordination of suppliers is difficult and where the foreign investor or joint-venturer feels uncertain about local business methods.

Depending on the conditions of the technology transfer, the purchaser in the developing country may be assured of continuing access to technological advances in the product or process that he has acquired. This will help him to remain competitive in the local or export markets.

Again depending on the conditions of technology acquisition, the user in the developing country may gain wider access to foreign markets as a result of acquiring foreign technology. (However, a condition of the use of the foreign technology may be that the technology not be used to produce goods for export.)

A major consideration in acquiring foreign technology is the "built-in" market acceptance that such technology has with domestic customers. A well-demonstrated preference for foreign goods exists in many countries. The right to display a worldwide trademark, for instance, may often be worth more than the additional cost of utilizing the foreign technology.

It is maintained that in many developing countries the local cost of foreign exchange is artificially pegged by the government to make domestic investors want to acquire foreign technology. At the same time, the domestic cost of labor may be artificially high due to actions by labor unions or the government, which for political or other reasons finds it expedient to induce an inflated labor cost. In these situations, and given the sometimes inefficient characteristics of local, untrained labor, the domestic investor will opt for the more capital-intensive imported technology, with a lower net operating cost.

Another version of the subsidized capital situation occurs when a foreign government provides favorable credits against the sale of its own producers' technology, providing, in effect, low-cost capital to the developing country purchaser.

Disadvantages of Imported Technology

Against the array of reasons why foreign technology may be selected by a developing country are a variety of counterproductive effects that provide the basis for much contemporary controversy. Many arguments favoring a significant change in the modes of technology provision-utilization are cited by those advocating the "appropriate technology" approach to the development and utilization of technology in the developing countries.

Two types of problems are associated with the utilization of conventional foreign technology in the developing countries. One results from the introduction of contemporary technology per se, and the other is associated with the choice of technology.

In the first case, Owens and Shaw[19] point out the creation of dual (sometimes called "two-tier") versus modernizing societies. The dual society is characterized by a pronounced division between economic classes. The majority of principally rural citizens remain in an income stratum that may be several times lower than the economic class of those employed in the foreign-technology-based industry of the urban enclaves. The allocation of political power and wealth maintains this division, and it is characterized by the control and use of technology.

In "modernizing economies," such as those of the Republic of Korea, Singapore, and Taiwan, governmental policy has deliberately led to a more equitable distribution of income. Such effects will not usually occur naturally in an industrializing economy and must be introduced by governmental action.

More equitable sharing in national income/wealth is widely represented as the objective of development and is frequently mentioned in treatises on the subject and in the resolutions of UN bodies and other entities. Still, despite thirty years of development effort, this remains the most difficult problem of world underdevelopment.

Another, related effect of the dual economy resulting from an uneven application of industrialization and other policies is the worldwide trend toward urbanization. The growth rates of cities are approximately twice as great as the population growth rates of the developing countries overall. The consequent strain on public services in most instances exceeds the government's ability to compensate, leading to increasing squalor severe impacts on public health, and distortions in the social structure.

Lured, perhaps, by the prospect of higher incomes and the possibility of steadier employment, individuals from rural areas are entering the city, creating growing areas of substandard housing, fostering crime and the other detrimental effects of urbanization. These migrants make manifest the hidden unemployment and underemployment of the agricultural sector. In the city, however, the extended family system is not available to compensate for the problem.

This forms the background to the oft-heard recommendation that research and development be emphasized in the agricultural sector. For example, extending growing seasons by introducing irrigation and double-cropping would not only

---

[19]Owens and Shaw, p. 3.

increase food production but would provide greater employment. Such emphasis on agricultural sector development is one of the features of the appropriate technology concept.

The nonjudicious introduction of technology has often negatively affected a country's natural resource base and environment, such as when deforestation to meet timber export requirements or to provide energy for manufacturing occurs, or when mineral extraction disfigures the landscape and causes erosion and stream pollution. Unwise concentration in single cash crops to meet export opportunities can deplete the soil, causing progressive deterioration in the yield and quality of output. Introduction of foreign flora and fauna can upset ecological balances. Excessive drawing of the subsurface water reserve depletes the supply, while irrigation systems cause salinity and waterlogging problems in areas such as the Indus river valley in Pakistan. High concentration of chemical fertilizers can result in pollution of streams and lakes. Agricultural wastes create problems of sanitation and habitability. All of these environmental effects can emanate from the introduction or inappropriate application of technology.

Governmental policy in establishing artificially low capital costs (i.e., interest charges), especially for foreign exchange, is often cited as contributing to the introduction of technology that makes unwise use of this scarce resource at the expense of creating more jobs in an environment where the labor resource is in surplus. This distorts the economic balance in favor of those few who have access to capital sources. In effect, the rich get richer at the expense of the other, far larger portion of the population. This exacerbates the social disparities and leads to unrest and political turmoil.

Other governmental efforts to attract foreign technology include such devices as tax moritoria, protective tariffs, land grants, and labor training schemes. To the extent that such activity distorts the economy and in effect is accomplished at the expense of, for instance, the agricultural sector, can it then obviously result in imbalanced development, working to the detriment of the economy?

An area of particularly widespread effect is in import substitution efforts, where the result of acquiring foreign technology to produce consumer goods results in little net added value to the domestic economy and may actually increase trade imbalances by creating demand for foreign materials and production equipment. Often, supplying products that can be afforded only by the high-income elites merely accelerates social bifurcation. Further, this may be accomplished at the expense of developing potentially viable sectors, such as agriculture or the export market for some indigenous product that could generate foreign exchange, create productive employment, and help to develop a viable and self-sustaining economy.

Some development economists argue strongly for putting maximum effort into the exploitation of export markets for indigenous products, doing so at the expense (initially, at least) of the introduction of any other technologies. This generally means the procurement and use of the most advanced technology in order to effect production efficiencies and to produce a product for the world market that meets the quality standards of that market. This policy has great implications for labor utilization, for example, and may conflict with a development rationale based on the appropriate or intermediate technology concepts.

Further effects of the utilization of foreign technology may be the development of dependencies on such sources, with a consequent reduction, in the view of some, in the sought-after quality of national self-sufficiency. Some[20] advocate that a developing country should create a technological moratorium on foreign technology, instituting policies leading to development of an indigenous technological, self-sustaining capability. Or, this strategy might be undertaken on a regional basis, as is being experimented with in the Association of Southeast Asian Nations (ASEAN).

This objective of developing a diversified domestic economy based on indigenous capabilities contrasts with the view of the need for a high order of production specialization. These opposing approaches provide the bases for some of the most spirited debate on current development issues.

Multinational corporations (MNCs) (or transnational corporations) have been at the focus of great controversy, starting in the late 1960s and rising to a crescendo at present.

---

[20]Research Policy Program, University of Lund. "Technological Transformation of Developing Countries" (Lund, Sweden, Discussion Paper 115), February 1978.

Although the presence of the MNCs in the underdeveloped world has many positive effects on national development, the negative effects are the basis for the arguments in favor of a dramatic change in the nature of technology as utilized by developing countries.

One oft-cited criticism of the acquisition of foreign technology is that it is frequently overscaled for the domestic market economy, resulting in undercapacity operations and inefficiency in the utilization of the invested capital that might have created more jobs. The question of overscaling and of the willingness or propensity of foreign manufacturers to redesign units to be better suited to the market demands of a developing country has been examined recently.[21] One general conclusion is that the producers frequently feel that the cost of such scale-adaptation would not be recovered in most instances and therefore are unwilling to attempt this change. This is not universally true, however. Another criticism is that unethical producers and sellers of "embodied" (i.e., contained in the form of the product or process sold) technology have deliberately provided plant and machine capacity that is beyond efficient size to increase the amount of their sale. Situations of this sort, to the extent that they do exist, give a basis for the efforts by UNIDO to provide consultation and education in equipment selection and to establish information banks.

Foreign firms often are accused of not effecting transfer of technologic capability to the developing countries where they have operations. It is pointed out that if a foreign firm departed, no residual indigenous capability would remain to perpetuate the utilization of the technology. No one would have been trained in-country to operate the production system, much less to design and fabricate/assemble another unit. Although machine operators and other semiskilled labor are trained, no capability is created in the design or engineering functions. On the other hand, it has been noted that foreign firms are increasingly entering joint ventures in the developing countries for a variety of reasons, including the insistence of the host government. Also, local citizens are entering all levels of management and

---

[21]H. Pack. "Technology and Employment: Constraints on Optimal Performance," in Technology and Economic Development: A Realistic Perspective, S.M. Rosenblatt, ed. (Washington, D.C.: International Economic Studies Institute (to be published in 1979); and L.J. White. "Appropriate Factor Proportions for Manufacturing In Less Developed Countries: A Survey of the Evidence," in Proposal for a Program in Appropriate Technology, Washington, D.C., Committee on International Relations, 95th Congress, February 7, 1977.

are gaining the necessary experience to conduct their businesses autonomously. Thus, the technology of management may be effectively transferred in these cases.

The fact that very little research and development (R&D) is carried on in-country by foreign owners has been studied to some extent to discover what factors determine where R&D will be carried out. Although some technology adaptation effort is conducted in-country (some studies indicating rather more of this by MNCs than had been supposed), there seems to be little prospect for significant R&D being transferred to developing countries for a variety of reasons too extensive to be explored here. This has provided an argument for the creation of indigenous R&D capabilities in the developing countries, although existing technological institutions are criticized on the basis that they are ineffective and out of touch with their intended users. The maintenance of an in-country technological capability of fairly broad dimensions is recommended in the development literature.

Some feel that the superior marketing capabilities of MNCs have introduced and generated demand for foreign-designed products that are not well suited to the domestic markets of the developing countries; and that tastes for nonutilitarian consumer goods have been stimulated, causing drain-off of capital that would be better employed in other ways. The stimulation of demand for products that cannot be afforded by the general population contributes further to dualism. Dishwashers, tape recorders, and frozen TV dinners may find a viable market in a country but may be of questionable economic or social utility. Efforts to produce such goods domestically in an import substitution rationale may lead to further distortions of industrialization patterns. However, the high tax revenues realized by the government from domestic sales of such luxury goods are a deterrent to policies for reducing consumption of these items.

Foreign technology is being criticized in many developing countries for what is felt to be its excessive cost, brought on by monopolistic/oligopolistic market conditions. UNCTAD meetings on the NIEO and other sources are strongly recommending lowering the cost of technology or effectively subsidizing its cost to the developing countries through bilateral and multilateral assistance programs. Developed country responses to this reflect a free-market approach that advocates maintaining trade in proprietary technology as the result of open negotiation between buyers and sellers.

However, the developed countries have a strong disposition toward facilitating the availability of technology and providing concessionary funding to make its acquisition possible on a competitive basis.

Other complaints are that foreign suppliers of technology often have provided obsolete units and systems. MNCs have been accused of transferring wornout plants to developing countries at overstated value to their local subsidiary, inflating their capital share, or receiving greater repatriated income. Efforts to stimulate trade in used production equipment have encountered many difficulties, including problems in providing maintenance and spares. Host governments may complain that these are attempts to keep developing country industry noncompetitive in world markets.

Another chronic complaint is that the developing countries are exploited for their raw materials and basic commodities without gaining the benefits of additional processing and manufacturing occurring within their borders. Additionally, some claim that, because of the control exercised by the MNCs over the "transfer pricing" of semimanufactures produced for futher processing or assembly within other countries, the full potential economic value of the products does not accrue to the country of origin. Developing countries therefore seek additional technologic inputs in their home areas, the involvement of local secondary industries, and other efforts to raise the gross domestic product.

In many instances, state-owned and operated industries are created to accomplish some of these objectives. Due to other constraints, however (such as hiring requirements, price controls, and so on), these organizations are often run inefficiently, have difficulty maintaining product quality, and generally are less effective than private enterprises. Entry of the government into the industrial sector is, of course, a matter of basic public policy; it is found most often in infrastructure industry, such as electric power, railroads, communications, and other public service activities.

Finally, the MNCs are sometimes felt to exercise excessive control over the markets they serve in the developing countries. They gain this by benefit of their large and more efficient manufacturing and marketing organizations, their ability to have greater control over the cost of manufacturing inputs, and their superior technology. In some instances these firms were granted protected entry into the domestic market and now have gained so much prominence in them that there is virtually no opportunity for competitive, domestically owned firms to gain a foothold.

Of course, in some instances the host country government has nationalized foreign holdings, especially in the natural resources area, although this practice has not been common in more recent years (a notable exception has been the oil industry). The ability of the new public companies to operate and maintain these technology-based production facilities is often severely tested, and in some situations being cut off from the source of technologic know-how has proven disastrous. The intent of all parties is to establish international norms and practices for dealing with such situations.

CONCEPT OF APPROPRIATE TECHNOLOGY

Rationale: The Humanistic/Conservator Ethic

The individual who has come to personify the humanistic/conservator ethic as it relates to development is the late E.F. Schumacher of Great Britain. In his now classic Small Is Beautiful he said: "In the excitement over the unfolding of his scientific and technical powers, modern man has built a system of production that ravishes nature and a type of society that mutilates man. If only there were more and more wealth, everything else, it is thought, would fall into place. Money is considered to be all-powerful; if it could not actually buy nonmaterial values, such as justice, harmony, beauty, or even health, it could circumvent the need for them or compensate for their loss. The development of production and the acquisition of wealth have thus become the highest goals of the modern world in relation to which all other goals, no matter how much lip-service may still be paid to them, have come to take second place. The highest goals require no justification; all secondary goals have finally to justify themselves in terms of the service their attainment renders to the attainment of the highest. This is the philosophy of materialism, and it is this philosophy--or metaphysic--which is now being challenged by events.

"We shrink back from the truth if we believe that the destructive forces of the modern world can be 'brought under control' simply by mobilizing more resources--of wealth, education, and research--to fight pollution, to preserve wildlife, to discover new sources of energy, and to arrive at more effective agreements on peaceful coexistence. Needless to say, wealth, education, research, and many other things are needed for any civilization, but what is most needed today is a revision of the ends, which these means are meant to serve. And this implies, above all else, the development of a life-style which accords to material things their proper, legitimate place, which is secondary and not primary."[22]

The current needs for development derive from the condition of man on earth--the effects of his enlarging numbers, his impact on the earth, which is his present and likely only home, and on the conditions that are viewed as

---

[22]E. F. Schumacher. Small is Beautiful (New York: Harper & Row, 1973), p. 272.

necessary for him to attain a satisfactory, stable reconciliation with his environment, both for his physical and psychological well-being. This patent enlargement of the need for and content of the development function represents the evolution of this concept from its earlier, largely quantitative and economic-based definitions.

These needs are categorized in terms of population growth and distribution food supply, natural resource utilization, environmental pollution, and the intangible qualities of self-determination and political and social stability.

Evidence--or at least tightly reasoned argument--in support of the Schumacherian thesis is forthcoming from a variety of sources, both recent and ancient. As early as 322 B.C., Aristotle said: "Most persons think that a state in order to be happy ought to be large; but even if they are right, they have no idea of what is a large and what is a small state... To the size of states there is a limit, as there is to other things, plants, animals, implements; for none of these retain their natural power when they are too large or too small, but they either wholly lose their nature, or are spoiled." In this statement, Aristotle appears to anticipate some contemporary concepts relating to the need of limiting economic growth and controlling the drain on the world's resources.

The potential harmony of man with nature, and especially the relatively modest absolute material needs he has to lead a productive life, is no better illustrated than by Thoreau's idyllic repose at Walden Pond. Thoreau's descriptions were graphic, simple, and nostalgically appealing--not presenting a highly structured metaphysical argument, but reaching out to the latent quality in all of us for seeking simplicity and satisfaction in a direct basic relationship to nature. Maintaining this primordial relationship is one of the objectives of the appropriate technology concept as it is envisioned by many.

The problem, however, concerns man's tendency never to be satisfied with his material status, to be caught in a self-perpetuating cycle of ever-increasing wants, and to be progressively more isolated from fulfilling innate associations with his natural surroundings.

Jacques Ellul of the faculty of law at Bordeaux University produced in 1964 a reasoned, complex, and brilliant discourse on the effect of technology on man. Ellul describes how <u>la technique</u> has so engulfed man--and so unconsciously--as to completely subvert his nature and his option for self-determination. All of "civilized" humanity is preempted from achieving a rapport with nature that will finally bring modern man into a stable, acceptable relationship. "The aims of technology," he states, "which were clear enough a century and a half ago, have gradually disappeared from view. Humanity seems to have forgotten the wherefore of all its travail, as though its goals had been translated into an abstraction or had become implicit; or as though its ends rested in an unforeseeable future of undetermined date, as in the case of Communist society. Everything today seems to happen as though ends disappear, as a result of the magnitude of the very means at our disposal.

"None of our wise men ever pose the question of the end of all their marvels. The 'wherefore' is resolutely passed by. The response which would occur to our contemporaries is: for the sake of happiness. Unfortunately, there is no longer any question of that. One of our best-known specialists in diseases of the nervous system writes: 'We will be able to modify man's emotions, desires, and thoughts, as we have already done in a rudimentary way with tranquilizers.' It will be possible, says our specialist, to produce a conviction or an impression of happiness without any real basis for it. Our man of the golden age, therefore, will be capable of 'happiness' amid the worst privations. Why, then, promise us extraordinary comforts, hygiene, knowledge, and nourishment if, by simply manipulating our nervous systems, we can be happy without them? The last meager motive we could possibly ascribe to the technical adventure thus vanishes into thin air through the very existence of technique itself.

"But what good is it to pose questions of motives? of Why? All that must be the work of some miserable intellectual who balks at technical progress. The attitude of the scientists, at any rate, is clear. Technique exists because it is technique. The golden age will be because it will be. Any other answer is superfluous."[23]

These arguments that state the perversity and pervasiveness of present-day technology are reminiscent of Huxley's Brave New World Revisited.[24] Huxley writes about a future in which man is "happy," pleasantly anesthetized and stress-free, but with complete loss of volition in matters where the political masters wish that there be none. Huxley's world is governed through behavior reinforcement abetted by chemical means--the Orwellian "1984" by coercive, terrorist methods--but they both have in common the extinction of any true "freedom" that mankind may feel it possesses.

---

[23] Ellul, chap. 6.

[24] Aldous Huxley. Brave New World Revisited (New York: Harper & Bros., 1958).

This reference to the possible illusion of man's self-determination reflects some of the thesis of B.F. Skinner,[25] the noted behavioral psychologist, who states: "What we need is a technology of behavior. We could solve our problems quickly enough if we could adjust the growth of the world's population as precisely as we adjust the course of a spaceship, or improve agriculture and industry with some of the confidence with which we accelerate high-energy particles, or move toward a peaceful world with something like the steady progress with which physics has approached absolute zero (even though both remain presumably out of reach). But a behavioral technology comparable in power and precision to physical and biological technology is lacking, and those who do not find the very possibility ridiculous are more likely to be frightened by it than reassured. That is how far we are from 'understanding human issues' in the sense in which physics and biology understand their fields, and how far we are from preventing the catastrophe toward which the world seems to be moving." He speaks of a "technology of operant behavior," which suggests (at the risk of simplistically analyzing Skinner's thesis) the imminent possibility of conditioning human behavior through a technology based on demonstrable scientific principles that can lead mankind toward some predetermined condition of interpersonal relationships.

What the foregoing references to behavioristic aspects and possibilities of human direction have to do with appropriate technology is that activity conducive to economic and social development may be possible to induce--and, in the process, may become, by definition, "appropriate technology." Of course, the Ellulian-Huxleyan-Orwellian aspects of this are obvious. What is suggested here, however, is that the imperfectly understood process of development would almost certainly benefit from a more profound examination of the motivational aspects.

Arguments regarding the concept of appropriate technology relate to the utility of technology and to the feasibility of its application. Ralph Nader in the contemporary U.S. scene has had a great effect on U.S. attitudes toward environmental impacts of private business. Nader views the activities of MNCs as being commercially exploitive in the international scene. Such enterprise often engages in activity that does not work to the best advantage of the developing country. The implication is one of immediate, capitalistic gain at the possible expense of the best interests of the involved country.

---

[25]B.F. Skinner. Beyond Freedom and Dignity (New York: Alfred A. Knopf, 1971), p. 5.

Barry Commoner of the United States strongly emphasizes the use of natural products to offset the negative ecological impacts of man's use of synthetics. Although not against "growth," he supports the use of technology that involves such natural products and processes. In terms of the precepts of appropriate technology, Commoner is considered an effective thought leader.

John Kenneth Galbraith has written several books relevant to the application of technology to the problems of economic development. In his 1958 work The Affluent Society, he developed an argument supporting the idea that, especially in the United States, man has become so materialistic and removed from rational allocation of resources that the country is immensely overbalanced in favor of high-consumption consumer goods production. Public programs that have great potential for improving the "quality of life" are consistently and progressively deprived of support. One of his formulae for offsetting this situation is emphasizing a humanistic education process (another element of appropriate technology for development). "One branch of conventional wisdom clings nostalgically to the conviction that brilliant, isolated, and intuitive inventions are still a principal instrument of technological progress and can occur anywhere and to anyone. Benjamin Franklin is the sacred archetype of the American genius and nothing may be done to disturb his position. But in the unromantic fact, innovation has become a highly organized enterprise. The extent of the result is predictably related to the quality and quantity of the resources being applied to it. These resources are men and women. Their quality and quantity depends on the extent of the investment in their education, training, and opportunity. They are the source of technological change. Without them investment in material capital will still bring growth, but it will be the inefficient growth that is combined with technological stagnation."[26]

Although he was writing with specific reference to the U.S. scene, the view is equally applicable to the developing countries. Galbraith further states: "Finally, with better social balance, investment in human resources will be kept more nearly abreast of that in material capital. This, we have seen, is the touchstone for technological advance. As such, it is a most important and possibly the most important factor in economic growth."[27]

---

[26]J.K. Galbraith. The Affluent Society (Boston: Houghton Mifflin Co., 1958), p. 272.

[27]Ibid., p. 318.

Referring to the effects of crowded urban environments on the quality of human life, urban sociologist Edward C. Banfield notes: "If some real disaster impends in the city, it is not because parking spaces are hard to find, because architecture is bad, because department store sales are declining, or even because taxes are rising. If there is a genuine crisis, it has to do with the essential welfare of individuals or with the good health of the society, not merely with comfort, convenience, amenity, and business advantage, important as these are. It is not necessary here to try to define 'essential welfare' rigorously: it is enough to say that whatever may cause people to die before their time, to suffer serious impairment of their health or of their powers, to waste their lives, to be deeply unhappy or happy in a way that is less than human, affects their essential welfare."[28] It is in the Schumacher/Galbraith views of the misapplication of natural wealth that the world's urban growth crisis is manifested; and it is toward a rational application of earth's resources that they and many other observers strive.

One of those observers and his colleagues in 1972 produced a book on the mathematical modeling approach to the examination of the macroeffects of the progressive depletion of earth's resources. Dennis Meadows et al., in <u>The Limits To Growth</u>, presented highly conjectural models based on assumptions that were subsequently criticized, but from which he concluded: "The basic behavior mode of the world system is exponential growth of population and capital, followed by collapse."[29] The publication of the book precipitated a large and active response and served perhaps as much as any other single effort excepting Rachel Carson's <u>Silent Spring</u> to awaken and stimulate worldwide focus on the problems facing mankind as a consequence of depletion and misallocation of earth's resources. The study concluded: "Applying technology to the natural pressures that the environment exerts against any growth process has been so successful in the past that a whole culture has evolved around the principle of fighting against limits rather than learning to live with them. This culture has been reinforced by the apparent immensity of the earth and its resources and by the relative smallness of man and his activities. But the relationship between the earth's limits and man's activities

---

[28] E.C. Banfield. <u>The Unheavenly City</u> (Boston: Little, Brown & Co., 1968), p. 10.

[29] D.H. Meadows, D.L. Meadows, J. Randers, W. W. Behrans, III. <u>The Limits To Growth</u> (New York: Universe Books, 1972), p. 150.

is changing. The exponential growth curves are adding millions of people and billions of tons of pollutants to the ecosystem each year. Even the ocean, which once appeared virtually inexhaustible, is losing species after species of its commercially useful animals."

To summarize, the principal issues concerning the application of technology in the developing countries are identified as follows:

(1) The exponential depletion rate of the earth's natural resources, the finite nature of these resources, and the consequent requirement to immediately balance use with supply;

(2) The threat to the environment as a consequence of careless, wasteful development of resource conversion processes, based on rapidly depleting fossil fuels;

(3) Burgeoning world populations, especially in the poorer countries, and the impact of this on food and resources supply, and the effect on the quality of life, including the problems of large-scale urbanization trends;

(4) The inequitable distribution of the world's wealth, resulting in persistent poverty on a major scale and untenable discrepancies between the affluent and the poor;

(5) Inappropriate allocation of capital, with disproportionate shares going for production of consumer products of marginal utility while investment that would benefit more people, increase industrial productivity, and provide necessary public services, goes unrealized;

(6) The alienation of man from his natural habitat and his establishment in surroundings alien to his nature, critically short of meaningful employment and demeaning to the spirit;

(7) The questions raised by behaviorists concerning the self-determinative capacities of man and the implications this has for development strategies; and

(8) The oftentimes negative role that technology has played in creating these conditions and

the present requirement for redirecting existing technology and for developing new technology that will be consonant with development needs as they are perceived in the humanistic/conservator context.

These aspects of the utilization of technology form the theme of the appropriate technology concept.

Characteristics of Appropriate Technology (AT)

So very much thoughtful, useful, definitional information has been written concerning appropriate technology that it becomes difficult to review, let alone to offer additional or new information in this regard. Consequently, this discussion will consider explanations and definitions that are attributed to a few expert sources, seeking points of commonality or difference.

Denis Frost, an executive in the London-based Intermediate Technology Development Group, established by E. F. Schumacher, offers these insights to the rationale of the organization's founder: "Drawing upon these philosophies (i.e., Gandhian) and his own experiences and observations of developing countries, Schumacher concluded that: poverty is the over-riding problem, and the source and center of poverty lies primarily in the rural areas of poor countries; conventional aid policies are ill-equipped to tackle the problem of poverty since they draw their experience from and have developed their expertise within an industrial, urban environment far removed from the rural/agrarian conditions prevailing in poor countries; in tackling the problems realistically, the correct choice of technology to be applied to local situations is critical--not only to increase productivity and wealth, but also to increase general well-being and to preserve the quality of life and the environment; the most appropriate technologies in the prevailing circumstances of developing countries are more often likely to be a range of intermediate technologies which are more productive than the often highly labor-intensive but inefficient traditional technologies on the one hand but, on the other, are less costly and more manageable than the large-scale, labor-saving, and capital-intensive technologies of highly industrialized societies; and to be fully effective, these technologies will respond to local needs and factor endowments. In general, they will be cheaper and smaller, giving a wider, more equitable distribution of capital investment; they will create employment, providing work opportunities in areas where people live; they will foster the use of local capital, skills, and raw materials and reduce reliance on the importation of these factors; they will produce goods primarily for local consumption and use."[30] Schumacher's "intermediate technology" may be considered a subset or particular type of appropriate technology.

At a meeting at Winrock International Conference Center in December 1978, participants agreed that: "Appropriate technology does not offer an absolute prescription but rather a process of choosing from among a broad range of options. The appropriate technology approach seeks to optimize solutions, wherever possible, through reliance on problem-solving capabilities of local people as well as a sensitivity to environmental and cultural impacts."[31]

Elsewhere, emphasis is placed on the inclusion of all degrees of sophistication of technology within the definition. Schacht reports: "A balanced approach to appropriate technology requires that all types of technologies--advanced, intermediate, and basic--be considered in the promotion and policies of development. The choice of technology should be based not only on technological considerations and sophistication, but also on economic, cultural, environmental, energy, and social standards."[32]

In defining an appropriate technology that he terms "light-capital," Congressman Clarence Long explains the concept as follows:

(1) It is not primitive or obsolete, but is economic, 'culturally congenial,' and ingenious in its design;

(2) It should represent the 'least-cost' solution, taking into account the factor costs of production;

---

[30]D.H. Frost. Appropriate Industrial Technology: An Integrated Approach (Vienna, UNIDO (ID/WG.279/6) 20 June 1978, p. 2.

[31]Preliminary Report from a Workshop on Appropriate Technology, Morrilton, Arkansas, Winrock International Conference Center, December 1978, p. 9.

[32]W.H. Schacht. Appropriate Technology: Technology for Growth in the Developing Nations (Washington, D.C.: Congressional Research Service, Library of Congress, Issue Brief No. IB77092, November 1978), p. 1.

(3) Labor intensiveness <u>is</u> a necessary condition;

(4) It is defined by low capital investment per worker, preferably in small enterprises that can be managed with local talents.

In Long's version, capital cost per workplace should approximate $100. But, he notes: "Light-capital technology does <u>not</u> necessarily mean the displacement of large-scale infrastructure projects. Light-capital technologies can be developed in the rural areas or inserted into the interstices of urban sectors of poor countries simultaneously with capital-intensive infrastructural development, especially if the latter are designed to complement the light-capital development."[33] Schumacher, too, chose a monetary quantity (in his case, U.K. 100 pounds sterling) as being the probable best average investment figure per workplace. Some reviewers feel these to be arbitrary figures.

Factor endowments of the developing country are considered to be of paramount importance in selecting appropriate technology. Writing about an appropriate technology conference organized by the United Nations Industrial Development Organization in 1978, Gamini Seneveratne reported that the delegates identified eleven factor endowments: "size of potential market, availability of natural resources, exercise of national sovereignty in the exploitation of these resources, the role of the public and private sectors, scale of production, desirability of geographic dispersal, capital and labor intensity of various techniques and processes, 'appropriate' sources of energy, technical efficiency, availability of trained manpower, and the impact on the environment."[34] Although some of these "factors" might be better termed <u>characteristics</u> of the technology being considered, or <u>measures</u> of the effect of its use, the list does illustrate the current emphasis in international deliberations on the humanistic aspects and national sovereignty issues of technology employment.

Schacht also summarizes characteristics of factor endowment as follows: "Basically, the developing nations have a large unemployed, underemployed, and unskilled labor force, few capital assets (except for the oil-rich countries), and at best a limited technological infrastructure which is

---

[33] Letter from the Office of Representative Clarence D. Long (Maryland), 15 February 1978.

[34] G. Seneveratne. "Appropriate Technology Legitimized!" article appearing in the <u>Development Forum Business Edition</u>, United Nations, 26 February 1979, p. 5.

necessary to sustain the development and innovation necessary for growth. The economics of the less-developed countries are often characterized by poor income distribution, small domestic markets, and a lack of marketing mechanisms and networks. The capital-intensive technologies of the industrialized world do not address these conditions. These advanced technologies tend to be labor-saving and designed to promote efficiencies of scale. Yet the economies of scale inherent in Western manufacturing are not well suited to the developing countries which lack the established communications, distribution, and transportation networks fundamental to this notion."[35]

Probably one of the better, more definitive listings of the elements of appropriate technology is the result of research conducted by Peter W. Askin. Askin sought to structure and summarize the subject of appropriate technology for use in the curricula of the Foreign Service Institute of the United States Department of State. In his brief but encompassing definition, he noted most of the aspects identified by those individuals previously mentioned but included additional considerations and comments on ramifications. "Terminology is definitely a stumbling block to the analysis and understanding of the concept of intermediate or appropriate technology. It has been alternately referred to as appropriate technology, low-cost technology, labor-intensive technology, capital-savings technology, alternate technology, self-help technology, village-level technology, progressive technology, indigenous technology, peoples' technology, light-engineering technology, adaptive technology, light-capital technology and soft technology. There are probably more, and more to come.

"For the most part these terms are used interchangeably in the literature, but at times they become almost slogans, deliberately chosen it would appear to connote a particular bias or ideology or to emphasize a particular set of economic or social variables. For example, the term labor-intensive or capital-savings technology (regardless of country origin) emphasize the employment factor in technological choice. Indigenous, peoples', or village-level technology focus on the local development, modification and adaptation of technology--normally technology that derives directly from traditional tools and methods. Alternate and soft technology tend (but not always) to embrace more ecological and environmental considerations such as alternate and renewable sources of energy and the environmental impact of technologies. Adaptive technology generally means the modification of an existing technology--normally the

---

[35]Schacht, p. 2.

scaling down of large, sophisticated machines and processes to smaller, cheaper and simpler products and methods."[36]

Requirements for Successful Employment of Appropriate Technology

Notwithstanding the humanistic/conservator ethic as a basis for choosing technology, difficulties of choice do exist among the complex and hard-to-assess technological variables. No better analysis of the criteria for making such choices is contained in the literature than that presented by Eckaus. Consequently, and with acknowledgement, the following discussion is based on his structure of analysis and is a paraphrase of his conclusions.[37]

Eckaus identifies nine alternative criteria on which to base technological decisions. Historically and conventionally, maximization of net national output and income has been a prime development objective. The question of maximization turns on the questions of when and what. The implication is that if such maximization is desired at the earliest feasible moment, then present expenditures will likely be required that will result in suboptimal conditions in the future, when the economy might otherwise be more productive. The "what" component concerns which particular elements of the country's productive structure should be maximized. Consumer goods generally are produced at the expense of alternative investment in productive capacity that would otherwise be effective at some future time.

Also implicit in this maximization strategy is the belief that distribution of the increased national product can be effected according to predetermined goals. This may not be so feasible, given the typical restrictions in such distributive mechanisms in developing countries.

Maximization of consumers' consumptions is sometimes simplistically assumed to be the economic goal, overlooking the fact that this policy may reduce the savings that otherwise may provide the basis for increase in productive capacity. Therefore, a problem of trading present pleasures for future progress exists; although, of course, when present pleasures comprise the essentials of survival, such as food and clothing, then the immediate choice is easier. Additionally, this policy has implications for foreign trade activity and constraints, both with regard to the acquisition of immediate consumption products and the requirement for

---

[36]Peter Askins. Intermediate Technology: An Informal Survey (Washington, D.C.: Department of State, 1975-1976).

[37]Eckaus, chap. 3.

future export trade development as the basis for long-term productivity gains.

The opposite side of the same coin comes up when the goal is to maximize the rate of economic growth, in which restraint of consumption is required to create savings for productivity investment. This may be accomplished while maintaining some minimum level of consumption, as is often done in developing and centralized economies, although the period of time of the imposed austerity is subject to great public pressure. Also, the selection of this strategy results in very different capital investment choices in technology.

Increased employment currently is an oft-stated goal of development strategy, reflecting the basic belief that not only will more employment stimulate the economy overall, but it also will accomplish distribution-of-income goals. Low rates of employment are often considered to reflect choices of "inappropriate technology," either made locally or imported as a consequence of foreign assistance pressures. This, in turn, is followed by calls for more labor-intensive technologies. Eckaus points out, though, that labor use is a function of production method and that production efficiency, in turn, is most favored by the optimum choice of combinations of labor with capital. In one view, then, the choice of labor intensiveness should hinge on the productivity maximization objective, given the overall resource mix of the economy along with other production objectives.

Wells, in one of the few empirical studies on the choice of technology in developing countries, examined several industries in Indonesia. Using the term "economic man" to indicate the type of choice that an individual who sought to minimize cost would make, versus "engineering man," who wanted to optimize employment of technology, evidence indicated that the latter type was most prominent in monopolistic/oligopolistic firms.[38] The implication was that where cost competition is not so severe, then choices that optimize product quality and other technology-related benefits are made. Capital intensity also may be employed to reduce reliance on a labor force that may be inflexible in times of need to reduce or increase production in response to market change, whereas machine-based production can be easily varied. Choice is also obviously closely linked to capital

---

[38]L.T. Wells. Economic Man and Engineering Man: Choice of Technology in a Low-Wage Country," appearing in Public Policy, Cambridge, Mass., Harvard University Press, Summer, 1973.

cost and availability. However, enough anomalies exist to prevent making any clear-cut distinctions in the technology choice situation either sectorially or across industry.

There is also the intangible "status" aspect of employment and the individual attitudes of those who may be essentially <u>required</u> to assume employment as a consequence of government intercessions. In some instances, the individual may prefer the opportunity for employment, whatever its nature, and to consequently bear his share of the redistribution tax on his income. Alternatively, perpetuation of extended family systems of rural employment, at very low levels of cash income, may be the clear personal preference.

A counterapproach is to opt for productivity maximization based on, perhaps, capital-intensive technology, and then distribute the increased benefits through taxation. This, however, assumes the ability of government to function with sufficient efficiency and this frequently is not feasible. In any event, the trade-off of productive technology for employment generation requires careful consideration.

The redistribution-of-wealth criterion affects technology choice also in terms of which strata of the population it is sought to benefit. Provision of more employment for the higher-skilled and urban dwellers, who may be thought to contribute the most to productivity generation and income increase, would probably result in different technology choice than would an effort to increase employment of the unskilled rural underemployed. The latter is often the objective of "appropriate technology" approaches.

Technology choices related to regional development obviously depend on the resource base of the region, including the skills and characteristics of the inhabitants. Climate and infrastructure have an impact, as in northeastern Brazil and the hill regions of Thailand. Political issues are frequently important in border regions, and these may virtually dictate technology choice. In rural regions, the choice will tend to favor either small, individual landholders or larger corporate or communal farms. Region-specific choice may necessitate suboptimal technology use because of factors of scale, market size, and transport facilities.

The objective of balance of payments relief results in two clearly different options. One view holds that the most efficient and rapid development of viable export industries is the best direction to take. For this choice, of course, the most productive, efficient combination of technology with labor resources is required. Arguments against this strategy usually cite the instability and vagaries of the international commodities markets and recommend an import-substitution strategy, leading to maximum self-sufficiency. This direction results in labor intensity decisions and other choices typical of the intermediate technology approach. In either instance, the choice will have many side effects, and the policy decision as to which to emphasize should be studied carefully.

Although little study has been made of the subject, it is apparent that technology choices may often be made, or greatly influenced by, political consideration. These choices will also have an impact on the efficiency and management of capital, especially in the face of shortages of qualified administrators, or in the case of dissident political factions.

As mentioned previously, there currently is great concern for the "quality of life" in employment of technology, and this has given rise to the "appropriate technology" component that is most frequently cited. While acknowledging the advantages of intermediate technology to smaller farmers and artisans, Eckaus feels that there is little scope for widespread development and application of the concept. He states: "A criterion proposed for village-level technologies is that the cost of the required capital equipment should be low enough for a single farmer or artisan to afford it. This criterion is justified by the customary shortage or unavailability of loans necessary for larger-scale technologies. In effect, a financial market imperfection...is used as the rationale for a particular type of technology. But that implies adjusting to, and preserving the effects of, the financial market imperfection. It would be better economic and social policy to attempt to eliminate the market imperfections themselves instead of trying to create techniques that preserve the effects of such imperfections."

He maintains that there has been little demonstration of the efficacy of the intermediate technology concept beyond agricultural and isolated settings and is of the opinion that, given a choice, individuals would opt for the more modern technologies. He states: "The economic outcome depends on the relative effectiveness with which such technologies would use resources compared to other techniques. Beyond that, their acceptance and use depend on other relevant criteria and the preferences for the life-style they make possible. But there is really not much point in debating at length the merits of intermediate technologies that do not now exist and whose contrivance remains to be demonstrated."

Research and Development of Appropriate Technology

In a paper presented in conjunction with the aforementioned conference in India in November 1978, Ross Hammond of the Georgia Institute of Technology set forth seven steps that he believes are necessary in carrying out appropriate technology research and development.[39]

> "(1) <u>Problem and Need Identification</u>. The selection of appropriate technology must be preceded by recognition of a problem or need."

Generally, in speaking of research, one of the greatest problems is that of problem definition. This is compounded in the developing country environment, where there is typically much less formal experience with the scientific method and with the utilization of technology. There also is confusion on occasion between the identification of <u>need</u> as differentiated from <u>demand</u>. Research institutions in developing countries sometimes confuse the two with the result that many research reports set forth solutions to problems for which there are no users. Given the very limited resources for R&D in developing countries, it is especially important that what is worked on is relevant, that the efforts are likely to be successful, and that the results will find application.

> "(2) <u>Available Alternative Technologies and Resources</u>. Some determination of the technologies which are known and, hence, available must be made in the light of the available materials and resources."

It is often difficult to systematically explore the technical literature seeking information on research that already has been performed, especially when one is isolated from the libraries and data bases of the industrialized countries. Given the relative scarcity of the literature on appropriate technology from the "how to do it" standpoint, this need is especially prominent. Consequently, it becomes the researcher's judgment as to how much effort should be devoted to seeking such data.

> "(3) <u>Analysis</u>. Analysis of the alternative technologies that may be available to solve a recognized problem is essential. The analysis for a developing country must consider educational, social, cultural, economic, infrastructure, and political aspects to the maximum extent possible."

---

[39]R.W. Hammond. <u>Systematic Approaches to Appropriate Technology</u>, a paper prepared for the International Forum on Appropriate Industrial Technology, New Delhi/Anand, India, 20-30 November 1978; issued by the United Nations Industrial Development Organization, Vienna.

This stipulation is an echo of the criteria of selection that Eckaus enumerates. It emphasizes the need for familiarity with the domestic situation in its economic and social aspects (as well as in its technology), as a prerequisite for successful technological innovation. This is the reason why appropriate technology R&D must be done primarily in situ and cannot be accomplished entirely remotely, even in regional institutions.

"(4) Design, Including Adaptation. Technology from the developed world usually requires modification, adaptation, or redesign when utilized in the developing countries. This is especially true in the small industry sector."

While some experienced observers might take exception to Hammond's use of the word "usually," it nevertheless remains true that much of the technology imported from the industrialized countries is not ideally suited to the developing country environment, and that, in fact, is one of the basic tenets of the appropriate technology concept. Emphasis is perhaps best placed on adaptation, because there is little need to reinvent something when existing technology can be modified. A pronounced tendency exists among less-experienced scientists and engineers to so reinvent, and many developing country laboratories are replete with such redundancies. It is an important function of R&D management to prevent this misutilization.

"(5) Prototype Development. When modification and adaptation take place, the question must then be asked, 'Will it work?' To answer this question, it usually is necessary to build a prototype and to analyze its operating characteristics and performance."

This function, because of a shortage of funds or facilities, is sometimes bypassed in the developing country environment, with the consequence that in the final stages the cost in time and money is greater than if the technology were modeled in the first place.

"(6) Testing, Evaluation, Modification. The prototype must be subjected to testing and evaluation. Modifications and adaptations, however small they may be, can significantly alter capabilities and performance of the technology."

This is the continuation of the process started in step five and is to be emphasized in the developing country environment, where the user of the technology is frequently without the means to adequately pretest technology applications; thus, the temptation is to go too readily into the production phase. The temptation is enhanced when, as is common in cases when capital is short and the investor requires a quick return, and when there is a receptive market, ready to take all the production at profitable prices.

"(7) Replication (Manufacture). When analysis indicates that the prototype has been debugged and appears commercially feasible, the final step may include the encouragement of manufacture of the prototype in sufficient volume to supply the market needs. This may involve the creation of a new venture for the specific purpose of manufacturing the appropriate technology but more usually, it would involve interesting existing manufacturers in adding the prototype as a new product."

Hammond's use of the word "replication" reflects the importance of a technological innovation being usable and reproduceable in similar environments. Some practitioners advocate that before much effort is devoted to steps four, five, and six, an understanding be reached with the ultimate user of the technology as to what his interest in the research results is likely to be. This user should be intimately involved in the succeeding steps so that the ultimate market test can be made efficiently under minimum risk conditions and with his full understanding of the research output.

APPROPRIATE TECHNOLOGY--LIMITATIONS AND BARRIERS

The following observations on the limitations and barriers facing use of appropriate technology, are arrayed in seven primary categories, although in a given instance a comment may be applicable in more than one such category.

1. Efficiency of Capital Utilization.

    A. The artificially low cost of capital in some countries encourages use of capital-intensive over what might otherwise be a choice of a more labor-intensive technology.

    B. Labor-intensive technology may not represent the most efficient use of capital when there are considerations of maximizing capital productivity and stimulating maximum growth of the economy.

C. Possible diminished returns on investment and the higher risk of appropriate technology utilization make attracting capital for it more difficult.

D. Appropriate technology ties up working capital longer per unit of output, thereby making it a less efficient user of this scarce resource.

E. Capital-intensive industry is more efficient and generates more "downstream" or secondary industrial requirements (which may be more amenable to appropriate technology approaches) and has a greater multiplier effect.

2. Aspects of Labor Utilization and Development.

   A. In some countries, government and/or labor unions artificially raise the cost of labor for economic or political purposes, thereby making a disincentive for its more intensive use versus capital.

   B. Labor in some situations is less dependable and causes production interruptions and quality maintenance problems, which encourages the use of more capital-intensive technology to reduce labor usage.

   C. Skilled labor and experienced management are also in short supply in developing countries, suggesting their combination in the most efficient relationship with capital rather than simply the maximization of labor use.

   D. Minimization of modern, capital-intensive technology use may retard the development of skilled labor and management that will be required subsequently as the country's industrial base grows and more modern technology is required.

3. Market Considerations.

   A. Use of labor-intensive technology may result in variations or subquality in products in comparison with that produced with contemporary capital-intensive technology, thereby limiting domestic market acceptance (vis-à-vis imports), and especially export market potential.

B. Employment of labor-intensive technology may result in overall limits on production such that domestic and/or export market demand cannot be met.

4. Logistic Considerations.

A. Dispersed manufacturing sites, combined with inadequate transportation systems, may severely interfere with production scheduling and materials supply, resulting in an unacceptable degree of inefficiency and cost.

B. These same transport and communications restrictions may make it impossible to meet customer delivery requirements, causing costly delays in their own further transportation and production schedules.

5. Technological Considerations.

A. Foreign technology often cannot be easily "unbundled" to facilitate the introduction of more labor-intensive/capital-saving elements in the production process.

B. Dispersed industrial production sites may result in subcritical conditions, not sufficient to foster inventiveness and innovation and to provide supporting services.

C. Technology from industrialized countries is more readily available, more dependable, more complete in its accoutrements, and often is more profitable overall.

D. Absence of an adequate technologic infrastructure in the developing country may make development of indigenous, labor-intensive technology difficult on any but a primitive scale.

E. Use of modern, capital-intensive technology may be required for the future industrialization of the country, to enable it to become or remain competitive and permit growth of the economic base of the nation. Extensive reliance on less modern technology may divert needed energies and capital from this industrialization effort, thereby retarding ultimate development.

6. Governmental Policy Effects.

   A. Often, governmental policies with regard to economic and social development prove inconsistent when used to evaluate the opportunity for the utilization of technology. Following governmental economic growth policies may militate against the use of appropriate technology.

   B. If governmental policy is conducive to adopting appropriate technology practices, it may result in economically suboptimal choices that put additional strains on other sectors of the economy. Thus, government must make conscious and consistent counterbalancing decisions if it intercedes in making technology choices; this may be beyond the capability of some governments.

7. Public Attitudes.

   A. In some instances there is a feeling that appropriate technology is a "second best" strategy of development and may actually reflect an effort to keep developing countries from adopting technologies that will afford them the best opportunity for development and the attainment of higher incomes.

   B. Prior experience and history among the nation's industrial and political leader may cause them to lean strongly toward maintaining traditional methods of technology acquisition from the industrialized world and to resist the admittedly difficult task of instituting appropriate technology approaches. (It is notable that apparently the strongest and most numerous boosters of the appropriate technology concept do come from the industrialized nations, not from the developing countries.)

C. Urbanization may, in fact, be a conscious policy on the part of the government, aided by public attitudes, that identifies city life with desirable personal goals and the drive to escape the numbing drudgery of subsistence farming. Efforts to make the rural life more humane and productive may not prevail over the lure of the city. The country's developers may view urbanization as laying the groundwork for entry into the modern industrialized era, where industry functions most efficiently in concentrations, gaining "critical mass" in urban environments for maximum efficiency.

Policies for Appropriate Technology

This discussion concerns four topics: the requirements for study and assessment of national resources; the elements to consider in development of national science and technology (AT) strategies and plans; the methods of implementation of those plans; and the special role of foreign assistance in this process.

At the Winrock Conference,[40] the participants identified four "strategy factors" in the determination of a plan for implementation of appropriate technology: people; finance; training, education, and skills; and organization. These elements must be understood in terms of their character and presence in the developing country for effective planning. Ranis[41] describes a typological approach to the problem of assessing the mix of national elements that are central to plan-development and cites such considerations as the size of the country, labor supply, dispersion of population, transportation, agricultural productivity, plus others, as essential determinants of policy formulation. He stresses particularly that when considering policy for appropriate technology it is important to be able to link it to a comprehensive rural development plan in order that appropriate technology activity not go forward independently of those policies guiding the overall development especially of the rural areas.

---

[40]Winrock Conference, pp. 21-23.

[41]G. Ranis. <u>Appropriate Technology in the Context of the Redirection of LDC Industrial Development Strategy: Concepts and Policies</u>, Vienna, UNIDO (ID/WG. 279/2) 13 June 1978.

Chebbi[42] writes of the importance of first determining strategy and formulating a plan to effectively apply appropriate technology in any system, and acknowledges the importance of marshalling data concerning the basic nature of the country.

These references all convey an impression of lack of adequate consideration of the role of appropriate technology in much of national planning carried out thusfar; of a tendency to consider technology as an end in the development process rather than a means. And this, of course, is the point that Ellul and others have made concerning the place technology has assumed in human life overall. There has been consideration of technology as a function separate from the society within which it has effect--a certain detachment of the scientist/technologist from those who are planning macro and micro-economic and social development. The effort now, clearly, is to integrate the employment of technology in the overall evolution of national economies. The fact that this has not been done very effectively in any instance thus far, either in the developed or developing world, does not relieve the desirability or necessity for doing so in the view of most who have studied the relationship of technology to development.

Some have suggested the preparation--in the developing country context--of a "technology development plan" (which is something more operative than simply "a science and technology policy" in the usual sense) and that this be a subset of the national development plan itself, reflecting the concomitant directions that technological development shall take.

Eckaus points out the very limited information available on the relationship of political interactions and technology choice and suggests that this basic relationship be researched. Further understanding of alternative means of encouraging technology utilization is needed; such factors as the utility of rural cooperatives, credit associations, trade associations, and the programs of government to aid small business would add to comprehension of how to use these means in designing plans for integrating technology in development. Eckaus makes the point that quite probably what is required at this time is more research on the policy aspects of the employment of technology and not so

---

[42]V.K. Chebbi. "Management of Appropriate Technology," New Delhi/Anand, India, UNIDO (ID/WG.282/3) 20 September 1978.

much on technology itself. Others believe that sufficient technology presently is available "on the shelf" as to pose no shortage; what is required is more information on the design and implementation of systems for applying it.

Eckaus recommends that an industrial technology inventory be taken in each country as a preliminary to planning; that this present a state-of-the-art appraisal of the status of industrial technology, its weaknesses, shortfalls, and strengths. (Such a procedure was followed in Korea in the mid-1960s as a preliminary to the design of the Korea Institute of Science and Technology project.)

Too little is known about the basic interactions of technology with development, and too little is known in most instances about the basic characteristics of the developing country to permit construction of an optimum plan for use of technology in development. In the absence of such information, planning for the integration of technology tends to become capricious, wandering, and inconsistent.

In planning for technology integration in development--the formulation of a "national technology plan"--the ministerial-level meeting in India[43] recommended the following: an evaluation of infrastructure, guidance in the acquisition of foreign technology; identification of technologic needs; information dissemination; provision of R&D capabilities, testing, pilot plant, and design facilities; technology assessment; and overall coordination of R&D in the country.

In Ranis' view, the technology plan must pay particular attention to the social impact of technology application and to be sure that solutions are not simplistic or emotional but reflect a rational approach. He notes that any plan is likely to encounter resistance from the vested interests in industry and government who perceive the continuation of the status quo to be to their advantage. He describes this in terms of the change from a "satisficing" to a "maximization" strategy of development, where certain sacrifices in traditional ways of doing things are required in the interest of future productivity. Others note that the technology strategy must derive from clear objectives, and that too frequently there is confusion and uncertainty in this regard, as motivations and actions on the political level are at cross-purposes.

---

[43]"Report of the Ministerial-Level Meeting," International Forum on Appropriate Industrial Technology, Anand, India, UNIDO, 30 September 1978.

Perhaps in recognition of this typical condition, Eckaus advocates the development of a "mixed portfolio" of policies, to permit flexibility and to accommodate changing conditions and thought. The question is raised about the compatibility of an appropriate technology implementation plan with an existing science policy structure and organization; whether it is better to try to incorporate such a program under the existing system or to create a separate one, given considerable differences in problems and technique with which the two areas are concerned. However, it appears inadvisable to dismantle an existing science policy structure to accommodate enthusiasm for some new appropriate technology approach--especially since there are questions about the nature of the concept and whether it represents anything really new or different from previous approaches.

A great deal is being written to the effect that developing countries are often at a disadvantage in negotiations for foreign technology because of unfamiliarity with alternative choices of technology and with what terms may be fairly negotiated. UNIDO provides some assistance in this area, but the suggestion is made that each country undertake to develop its own cell of expertise in negotiation.

In summarizing the elements of the policy formulation and planning function, the consensus is:

(1) Do adequate research to become knowledgeable of as much of the background of the country's external and internal environment as is required and feasible;

(2) Ascertain the national development goals and objectives, both tacit and explicit; and

(3) Develop a plan for technology utilization that is consistent with factor endowments, national objectives, and political constraints.

Notwithstanding the obviousness of this procedure, it seems that it is only rarely done; the tendency is to short-circuit these functions and jump to immediate conclusions and programs based on conventional wisdom and intuition.

Following a more-or-less well-developed technology utilization plan comes the implementation process. Winrock participants believed that three basic concepts must be kept in mind in the implementation phase: (1) it may be accomplished either institutionally or in a people-to-people mode so long as the problems are identified and agreed to; (2) it is vital to keep the implementors close to the users in all respects; and (3) every entity involved should be the object of coordination and direction. Ranis emphasizes the need for governmental participation in some significant way, given the large number of differing aspects and complexities of policy implementation, the large scale of the need (which cannot be met, for example, by private institutions alone), and the requirement for a long, sustained effort. He emphasizes that "there's no quick fix" when it comes to the implementation of appropriate technology.

Prime Minister Desai at the UNIDO ministrial-level meeting in his country made three observations concerning plan implementation: (1) the first objective is that the entire plan be guided by humane considerations (in the Gandhian tradition); (2) that technology does not stand by itself--there is an equal requirement for the management function in its application; and (3) that he believes large-scale, sophisticated technology applications should exist symbiotically alongside village-level technology.

In a trilogy of principles, Wells presents a formula for implementation that calls for removal of impediments (to application of appropriate technology), institution of incentives for it, and the careful balancing of technology choice with the instruments of policy and direction that are available to government (fiscal and monetary controls, tariffs, etc.). Reflecting the "grass roots" approach of the Winrock members, Jéquier[44] considers an important principle of implementation to be participation of those affected, plus the involvement of a wide array of concerned organizations and institutions. Finally, in terms of generalities, Eckaus calls for well-planned and consistent efforts at evaluations and the running adjustment of implementation programs as inevitable changes in circumstances and objectives occur.

Specifically, in the area of the identification, acquisition, and application of technology, a number of observations can be made. One is that a wide range of technologies are involved--there is no specific list of technologies that can be definitively called "appropriate" but rather that appropriateness is a function of the objectives and of the circumstances that obtain at technology selection time. Several make the point not

---

[44]Jéquier, p. 111-112.

to hold back development simply in the interest of using labor-intensive technology--that it is important to select technology on the basis of cost/benefit analysis unless under exceptional circumstances a deviation is made, and then government must be prepared to redress the imbalance caused by such economic suboptimization. For the same reason, arbitrary choice of "light-capital technology" will not be wise but should be utilized only when rational assessment of all pertinent factors suggests it. This is an argument against adopting a "cost per workplace" criterion, which seeks to maximize the effect of capital by limiting its use to narrowly defined labor-intensive situations.

The meeting of ministers concluded that a strong governmental role in the control of technology utilization was necessary and that this should consist of development of guidelines for selection of foreign and domestic technology and the provision of a screening process in each specific instance. This method was pioneered in India and has since been adopted in similar ways by Brazil and Korea, among others. It was suggested that means of "unbundling" technology to facilitate introduction of domestic technology into imported systems be explored systematically.

To facilitate the use of appropriate technology, Wells recommends that authorities do the following: (1) examine the legal constraints that affect choice; (2) analyze labor regulations to determine if the cost of labor is being artificially raised; (3) see if conditions discriminate against the small enterprises (who are the most likely users of intermediate technology); and (4) determine if there are impediments to free exchange of appropriate technology with other developing or developed countries. Generally, the government should remove barriers to operation of an essentially free environment for appropriate technology selection, including tariff restraints, tax discrimination, subsidized capital, etc. On the other hand, importation and use of undesired capital-intensive technology could be impeded and restrained through the use of tariff devices, licensing requirements, excise taxes, capital restraints, etc.

Jéquier notes that there is "a vast store" of technology in the world but that there is limited knowledge of it and how it may be acquired; he advocates the establishment or enlargement of in-country organizations expert in this function, to be complemented by international groups. There are beginnings of this sort of activity by various private voluntary organizations such as Volunteers in Technical Assistance (VITA) in the United States and the Intermediate Technology Development Group in Great Britain, but there is an absence of the detailed sort of economic analysis and case histories needed to evaluate appropriate technology applications. Also, the demand far exceeds the supply of such information.

In the area of funding and financial controls, the Winrock group recommended that a multiplicity of sources be developed within the country, using both public and private sources. Funding arrangements should be flexible to meet the variable conditions of appropriate technology application, and all parties to the activity should be expected to contribute to the project. It was felt desirable to be as direct as possible, with no unnecessary intermediary organizations.

Ranis recommends the judicious use of preferential funding and other encouragement such as tax moratoria, but points out that these devices tend to become enshrined. Knowing when to withdraw them to let a firm or industry stand on its own is very difficult. He further feels that financing agencies could do much to favor appropriate technology by how they administer their loans. The Asian Development Bank[45] as well as the World Bank[46] now have complete policies concerning how they will integrate appropriate technology considerations into their operations. They appear to have done this wisely while retaining the banker's traditional pragmatic view and quantitative/qualitative tests of feasibility. The World Bank has made an intensive study of the trade-offs on labor for capital in road-building projects, for example, while the ADB lists a dozen or so instances of how labor intensivity has been introduced into projects. There is also emphasis on projects that will more directly benefit the poorer economic sectors and favor agricultural development, with the objective of increasing world food supplies. The Interamerican Development Bank is undertaking a series of specific projects directed at appropriate technology-inspired projects.

Chebbi also advocates initial special protection for appropriate technology projects because of their more

---

[45]Asian Development Bank, Appropriate Technology and Its Application. In The Activities of the Asian Development Bank, Manila, April 1977.

[46]Central Projects Staff. Appropriate Technology and World Bank Assistance to Poor. Washington, D.C., World Bank, March 28, 1978.

complex nature and the fact that they often will involve persons in new experiences and seek to introduce unfamiliar practices--this increases risk.

All those authors reviewed emphasized the need to develop infrastructure within the country to facilitate appropriate technology adoption. The functions of this infrastructure Chebbi lists as: providing information; identifying and aiding local technological and development capabilities; evaluating and assessing technology; coordinating development efforts; identifying needs; cooperating with external agencies; and generally acting as a catalyst between concerned entities.

Specifically with reference to R&D institutions, Ranis points out that these are often rather static and that they require better contacts both within the country and externally. Many feel that they need assistance across the board including finance, technical information access, better contact with industry, assistance in provision of extension services, specific technical support and advice from outside sources, and improved access to specialized equipment. It would be well to develop relationships with "sister" institutions in the industrialized countries; MNC subsidiaries could perhaps make some of their parent organization's expertise available on a mutually beneficial basis to the local R&D institute. Some countries may find it advisable to have two separate systems for research, one being the traditional university-related research function and the other specifically devoted to the application of appropriate technology.

Throughout the literature appear references to the work of the International Rice Research Institute in the Philippines, not only on its plant genetics work and the "Green Revolution," but also its program on the application of appropriate technology principles to the design and introduction of agricultural equipment innovations.

Training and education are broadly viewed as areas of particular promise in gaining acceptance of appropriate technology principles. This is one of the principal "strategy factors" that the Winrock meeting stressed. The recommendation was for building on existing human capabilities but importing what was necessary in the way of missing skills. It also was recommended that training be pragmatic, avoiding an academic approach.

Questions are raised concerning what is known about the nature of innovation, and at least two authors point out that much, if not most, of what is useful in the intermediate technology area comes from the untutored (and thereby presumably less-inhibited) person who has intimate working knowledge of the problem. There is advocacy of establishing vocational training institutions where they do not exist and of making such training a prominent part of all primary and secondary school curricula, getting away from the more abstract and academic approaches that come from patterning schools on foreign models.

There is a basic lack of familiarity with technology in its practical aspects, and this needs to be overcome in many countries. More investigation of the nature of inventiveness and identification of entrepreneurial characteristics should be done. (One such effort is being conducted at the Institute for Technology in Bandung, Indonesia, for example, but this is a new field everywhere.) Interesting work also is being done with "innovation centers".

Most feel that there is a large deficiency in the methods of providing needed information, whether it be within the developing country or internationally. New channels are required because this is a time in the evolution of new concepts in the utilization of technology in which exchange of experience and cross-fertilization are particularly necessary. The basic functions of collection, classification, retrieval, and dissemination—all could be improved by known methods and to the considerable advantage of those working in the appropriate technology field. There is special need for case history information and data on hardware resources. (On the other hand, the extensive literature available on the theoretical aspects of appropriate technology suggests that perhaps this will suffice for the time being!)

Again, Eckaus suggests in-country research on the mechanisms of information flow and dissemination and asks what can be done to improve the way in which technical information gets to the potential user. He also asks how the user's experiences might be retrieved for the benefit of other potential appliers of appropriate technology.

On the other hand, there is considerable sentiment against the establishment of large, centralized appropriate technology data bases, although this question appears far from resolved. UNIDO has been conducting studies of this subject and has perhaps the most experience in the pragmatics of attempting to operate central systems (as also does the Organization of American States and Technonet Asia in Singapore). Specialized-subject data bases are advocated by some, such as on dryland agriculture. But the best information resource, according to most, is one that is integral with a leading technical institution, where experts can interact in the process of information provision, thus increasing the efficiency of the system.

The availability of financial services to faciliate appropriate technology is addressed by some. The general conclusion is that this is "grease for the wheels" but that the system is not well articulated and requires development. The regional and world development banks are becoming involved, but there appears to be little focus on the subject at the local level. Development finance institutions, which are found in most countries and are part of the worldwide networks connected with the regional and international development banking system, are not, for the most part, very oriented toward the appropriate technology concept and continue to direct their project appraisal efforts down traditional avenues. They usually end up financing conventional technology. This is a difficult subject, given a bank's primary requirement of making prudent investments. It is clearly a matter of special risk, but one which observers feel should be consciously grasped and dealt with.

There are several notable examples of the catalytic effect of credit extension services of rural cooperatives on the adoption of new technologies. The subsistence farmer is least likely of all to be able to risk trying a new technology, even if he were by some means able to borrow the required funds. Consequently, in too many instances the benefits of technological advance have gone only to those large and prosperous landowners who could afford the improved technology, thereby defeating one of the basic purposes of the appropriate technology approach and the Schumacherian vision.

Finally, in the subject of infrastructure development, is the question of dispersion, it being a tenet of appropriate technology that the use of technology should be spread through the developing country to mitigate the urbanization trends and to gain income spread and relieve regional inequities. While accepting this concept in principle, several observers point out that there are necessary preconditions. One is that the local resources should match the technology that is being introduced or, put another way, the technology should be designed to fit the local needs and capabilities. Dispersal simply for its own sake is not likely to be successful; there must be fertile local soil in which to plant the activity. Jéquier, for instance, suggests more experimentation with introducing new but related technologies in areas of traditional artisanship. This is being done in the carpet-producing regions of northern Pakistan with the introduction of an improved loom.

The traditional ceramic arts of Korea are being revived, and a historic foundry industry in Indonesia is receiving technical assistance.

It is also pointed out that a great abundance of labor does not exist universally in all the developing countries. In some cases, surplus labor may be highly seasonal. So in predicating new activities on the assumption of availability of labor, careful assessment should be made before dispersion of industry into the hinterlands is attempted.

Another consideration is the presence of sufficient local infrastructure--housing, water, sanitation, markets, schools, power. Chebbi's paper suggests the establishment of integrated industrial estates, a concept that has been attempted with varying degrees of success in different settings.

Finally, the role of bilateral and multilateral programs should be considered. The financial burden of promulgating appropriate technology--at least to the extent that it represents a departure from conventional usage or does not meet the literal tests of feasibility applied by conventional investment and lending institutions--will necessarily fall on the development banks. Although commercial banks have become increasingly involved in developing-country finance in recent years, they are unlikely to become a major factor in the appropriate technology movement.

The appropriate technology concept is obviously a "grass roots" effort in which the essential element is heavy involvement of those individuals and organizations that will be affected by its application. The central theme is one of including those large masses of people, essentially poverty-stricken, in the developing countries who have thus far been left out of the development process. Each appropriate technology advocate stresses that these "users" must be involved in the most fundamental sense. Implicit in the appropriate technology concept is its location-specificity, the fact that it is "custom tailored" to fit the unique circumstances of the place of its application. These characteristics mean that the more remote a development assistance agent is from the site of the identified problem, the less likelihood it will be able to directly aid in the appropriate technology development process. For this reason, foreign assistance efforts will usually have to be funneled through the domestic implementation infrastructure, although it has been suggested that certain specific technological subjects can be researched in places remote from the site. Also, the concept and the means of appropriate technology application can effectively be <u>taught</u> in distant places.

Consequently Eckaus, for example, sees a requirement that the developed country provider-of-assistance be in close cooperation with the developing country. He states that the delivery system for appropriate technology is more important, probably, than the technology itself. Wells believes, however, that the opportunity is limited for the aid donor country to directly participate. But he also sees an important donor role in financing the "costs of transition" from a conventional technology strategy to appropriate technology, recognizing that economic losses will certainly be incurred and that these must be met. Further, he advocates, as did the ministers meeting in India, that there could be considerable advantage from bringing other developing countries together in consortia to take common advantage of what is known and being developed. Chebbi would encourage international additional conferences and other means of exchange of experience; additionally, he felt that there is great scope for joint research projects between developing countries, although the author's recent experience with such attempts has not been particularly encouraging.

At the UNIDO appropriate technology meeting in India in 1978, in addition to these it was also advocated that a program of cooperation and exchange be developed which would involve the medium and small-scale industrial sectors in the industrialized nations in the appropriate technology effort, it being felt that a considerable latent value could be realized--that the smaller firms have certain knowledge and experience lacking in the larger MNCs which would be valuable in the developing country environment.

Some have recommended that foreign assistance be carefully evaluated to prevent its having a "distortion effect" on domestic efforts to integrate appropriate technology, that the donor nations must be sensitized to these needs and to act accordingly. Finally, it was pointed out that, potentially, the MNCs could do much to assist the appropriate technology effort, not only with regard to their own operations, but also through cooperation with developing country organizations in a noncompetitive relationship.

APPROPRIATE TECHNOLOGY IN U.S. POLICY

> "While the causes of underdevelopment are . . . complex and obscure--by no means as simple as the 'exploitation' theme some Third World rhetoric would suggest - people in the Third World seem to

have increasing faith that the condition is remediable, even as some of their problems become more acute. The advanced industrial powers--with the United States in the vanguard--have acted on the same principle; the concept of assisting the Third World with its development process has not been seriously challenged. While the various motivations for giving assistance have evolved, the main questions have been: how much and in what form?"[47]

Reflecting this concern for form, for example, current U.S. policy involving science and technology for developing countries (DCs) displays a major shift toward more appropriate technologies. While policies do not abandon large-scale, capital-intensive industries, policy makers now emphasize great interest in appropriate technology directions. Internally, these new policies have been shaped by the various perspectives of government, business, science and engineering organizations, the academic world, and private voluntary organizations, as well as externally by international momentum recognizing appropriate technology. Although so many diverse elements shape U.S. policy that it is impossible to paint a picture that is both all-encompassing and brief, a general policy emerges that attempts to meet the challenge of assisting DCs.

Implicitly, U.S. policy embraces many concepts inherent in appropriate technology. The basic thrust of current policy is: (1) to aid DCs in indigenous capacity building; (2) to promote research and development within DCs; (3) to assist with education and training; (4) to improve information systems; (5) to link public and private development activities within and among nations and regions; and (6) to promote more effective use of the United Nations.[48] Without even mentioning

---

[47] The United States and the Third World, Department of State Publication 8863, Office of Media Services, Bureau of Public Affairs, Washington, D.C., July 1976.

[48] These policies are gleaned from: Department of State, Office of the Coordinator, U.N. Conference on Science and Technology for Development, Science and Technology for Development: U.S. National Paper. Washington, D.C.: U.S. Government Printing Office, 1979.

appropriate technology, these central issues of U.S. policy show a careful analysis of needs and a desire to determine their development needs. Policy also indicates specific areas of concern: increasing food supplies, health and related needs; urbanization and industrialization; and management of resources.[49] The specific recommendations within these individual areas reflect a determination to find suitable alternatives and to learn from the past.

U.S. policy has turned away from the 'trickle-down effect,' which contended that income generated by large-scale, capital-intensive enterprises would quickly permeate all economic strata. Now, emphasis is on direct aid to the "poorest of the poor."

Today, U.S. policy seems to mirror the attitude that science and technology policies increasingly depend on the policies and priorities of the DCs themselves. The tone of U.S. policy reflects more efforts at suitability, responsiveness, and balance, and, therefore, also reflects the influence of appropriate concepts.

Explicitly, U.S. policy often mentions appropriate technology. Various governmental programs are specifically intended to promote appropriate technology. Section 306 of the International Development and Food Act amends Section 107 of the Foreign Assistance Act of 1961 and mandates a program of appropriate technology to promote "grants to support an expanded and centralized private effort"[50] for the development and dissemination of appropriate technology for developing countries. Funds are directed to the Agency for International Development (AID) to expand its own

---

[49]National Research Council. U.S. Science and Technology for Development: A contribution to the 1979 U.N. Conference. Washington, D.C.: U.S. Government Printing Office, 1979, p. 6.

[50]United States Agency for International Development. Proposal for a Program in Appropriate Technology, rev. ed. Washington, D.C.: U.S. Government Printing Office, 1977, p. 1.

program of involvement with appropriate technology, and funds are channeled into AT International, a new organization. AT International is a private, nonprofit corporation, created and funded by the U.S. government in response to perceptions of need to enhance appropriate technology activities. Its objectives include, among others, the development and application of technologies that result in increased employment and income among the poorer members of the developing nations. The purpose is to help meet needs for adequate food and nutrition, shelter, health care, and education--all conducive to establishing a quality-of-life standard that is the basic requirement for the world. In these goals, AT International is in consonance with the various objectives of the public and private bodies that are involved in the movement towards a more conscientious application of technology for development. Although relatively small in scale at the outset, AT International is a governmental embodiment of an institution for aiding appropriate technology.

The U.S. Department of Agriculture cooperates with state universities and on its own to provide technical information and expertise in various areas of importance. The Department of Commerce conducts programs through the National Bureau of Standards and the National Technical Information Service directly related to appropriate technology. The Peace Corps has always been active in appropriate technology. The National Science Foundation will be increasing its already significant involvement in appropriate technology, and the Department of Energy cooperates with AID in various LDC energy projects. President Carter has initiated an Institute of Scientific and Technical Cooperation (originally called the Foundation for International Technological Cooperation), which will consolidate and focus U.S. efforts in technical assistance throughout government.

Several acts of legislation as well as congressional hearings, reports, and documents give weight to the current governmental interest in appropriate technology.

Input from the business and academic communities was gained through a study undertaken by the National Research Council of the National Academy of Sciences which showed strong national interest in appropriate technology, especially in areas of increasing food supplies, health and nutrition, industrialization, and resource management. Among other things, the study indicated that rather than abandon capital intensive technologies, appropriate technologies' effect in this area should be to make massive industrialization efforts more balanced in domestic factor utilization than has often been the case in the past.

Private voluntary organizations within the United States have been assisted by AID and are very active in appropriate technology projects--and could be even more so if more effective sharing of information could be facilitated and more assistance given by government.

Although it seems that every internal governmental element involved in shaping U.S. science and technology policy for developing countries is eager to encourage appropriate technology, it must be reiterated that interest in appropriate technology is only part of the total picture. The United States supports, also, large-scale infrastructure development and, where appropriate, sophisticated and capital-intensive technology.

Thus, although the United States has increased its emphasis on labor-intensive, small-scale industries, it broadly defines the concept of appropriate technology "to include the whole range of technologies, whether capital-intensive, labor-intensive, large-scale, small-scale, sophisticated, or simple." Furthermore, the U.S. position paper for UNCSTD clearly respects the value of suitable large-scale applications of technology for DCs and advocates financial support of such projects.

A study of the various U.S. papers prepared for UNCSTD leads to the conclusion that U.S. policy for science and technology for DCs reflects appropriate technology in two ways. First, the movement toward appropriate technologies, in the broadest terms, has influenced overall policy to become more responsive to human need, and more environmentally suitable, including the large-scale, capital-intensive technology applications. Second, there is a significant effort to emphasize appropriate small-scale technologies alongside larger-scale, more capital-intensive technologies. Appropriate technology is now an important component of U.S. development assistance in DCs.

Appendix 0.1
Excerpts from World Development Report-1978

These selected excerpts from Chapter 2: The Development Experience of the Bank's report are intended to give an impression of the signal statistics and observations it contains; they are not intended to present a contiguous whole. The reader is referred to the complete report to gain an accurate view of its content.*

Developing Countries: Relative Size, 1960 and 1975

	Developing Countries (billions)		Industrialized Countries (billions)		Percentage Share of Developing Countries in Total[a]	
	1960	1975	1960	1975	1960	1975
Population	1.4	2.1	0.6	0.7	70	75
GNP[b]	460	1,048	2,071	3,841	18	21
Value Added in Industry[b]	120	350	745	1,483	14	19

[a] Share in the total of developing and industrialized countries.

[b] These data are in 1975 US dollars, using official exchange rates between national currencies which may not properly reflect differences between countries in purchasing power. For a further discussion of this problem, see the Notes to Table 1 in World Development Indicators.

---

*The information in this section is drawn from World Development Report - 1978 (Washington, D.C.: World Bank, August 1978).

- The developing countries have grown impressively over the past twenty-five years: income per person has increased by almost 3 percent a year, with the annual growth rate accelerating from about 2 percent in the 1950s to 3.4 percent in the 1960s.

- In countries accounting for half the population of the developing world, income per person has risen by less than 2 percent a year.

- In general, experience suggests that the distribution of income is likely to worsen in the course of economic growth. However, even if income disparities increase, the income of the poor can rise. Particularly where people are at the margin of survival, it is their income levels, rather than their relative position in the distribution of income, that require the most urgent attention.

Population--Urbanization

Developing Countries: Population, 1950-2000
(Billions)

	1950	1975	2000[a]
Low Income Countries	0.7	1.2	2.0
Middle Income Countries	0.5	0.9	1.5

[a]The assumptions on which these projections are based are described in the Notes to Table 16 in World Development Indicators.

- The progress made by developing countries is the more impressive considering that their populations have been growing at historically unprecedented rates.

- The pressure that rapid population growth exerts on resources, and the difficulties it imposes for raising income and employment levels, make the spread of effective family planning programs an urgent matter.

- Though the population in developing countries will continue to grow for decades ahead, effective action now can shorten the time required to achieve a stationary population and reduce its ultimate size.

Developing Countries: Urban Population, 1960-75

	Percentage of Total Population		Average Annual Growth Rate
	1960	1975	1960-75
Sub-Saharan Africa	14	19	5.0
North Africa and Middle East	32	44	5.0
Latin America	49	61	4.3
Asia	17	22	4.0
Southern Europe	40	51	3.2

Source: Selected World Demographic Indicators by Countries, 1950-2000, (New York: United Nations, 1975)

- Far more people have migrated to urban areas than could be absorbed, and despite large investments in urban infrastructure, the result has been a severe strain on urban services and labor markets.

- Little has been done either to deal with the appalling inadequacy of essential services, such as sanitation, in these settlements, or to assist the large part of the urban economy that consists of small-scale and informal production activities, which operate at low levels of productivity.

- While part of the growth of services in developing countries is in response to growth in demand, it also reflects the inability of the industrial sector to absorb fully the additions to the urban labor force.

Investment and Savings

Developing Countries: Investment and Savings Rates,
1960 and 1975
(Percentages of gross domestic product, at current prices)

	Low Income Countries		Middle Income Countries	
	1960	1975	1960	1975
Gross Domestic Investment	14.7	19.1	20.2	26.4
Financed by:				
Gross Domestic Savings	11.6	15.6	17.8	22.1
Net Foreign Resource Inflows	3.1	3.5	2.4	4.3

Note:
| Net Foreign Resource Inflows as a Percentage of Investment | 21 | 18 | 12 | 16 |

- In many countries an important reason for the difficulty in raising savings rates is a continued reliance on commodity taxation, which makes the revenues generated less sensitive to increases in incomes than if progressive income taxes and taxes on value added were used.

- The types of difficulties in raising investment levels differ among individual Low Income countries, but they essentially reflect the shortage of entrepreneurial and managerial talent and the difficulties of increasing savings at low levels of income.

- Private direct investment has also been an important channel for the transfer of technology and the introduction of more modern management techniques.

- Flows of capital to the developing countries on both concessional and market terms have played a crucial role in supplementing their import and investment capacity. The past twenty-five years have seen the establishment of bilateral aid programs in virtually all industrialized countries and a growing volume of increasingly concessional assistance. The number of international agencies concerned with various aspects of development has increased, as have the resources channeled through them to the developing countries.

Production

Developing Countries: Structure of Production, 1960 and 1975
(Median values, at current prices)

	Distribution of Gross Domestic Product (percent)					
	Agriculture		Industry		Services	
	1960	1975	1960	1975	1960	1975
Low Income Countries	52	43	12	23	35	45
Middle Income Countries	26	15	23	38	46	47

Note: Sectoral shares do not add to 100 percent because median values have been derived separately for each sector.

- In the poorest countries, with their slow average rate of growth, the incomes and consumption levels of the poorer half of the population have stagnated. Worse, in countries where agriculture has expanded more slowly than population (parts of South Asia and Sub-Saharan Africa), the incomes of some of the rural population have probably declined.

Developing Countries: Growth of Production, 1960-75
(Median values, at 1975 prices)

	Average Annual Growth Rates (percent)			
	Gross Domestic Product	Agriculture	Industry	Services
Low Income Countries	3.1	2.1	5.4	3.7
Middle Income Countries	6.0	3.5	7.9	6.7

- The scope of the economic changes of the past twenty-five years is perhaps suggested better by the fact that many developing countries have modernized their agriculture and sustained high rates of growth in agricultural production, while a number of them now manufacture technologically sophisticated equipment (electric power generators, for example). Many have sizable capacities in engineering industries; and some now compete effectively for turnkey projects internationally.

- One of the most important shortcomings is in agricultural research, specifically in the common failure to build up sufficient national capacity for the adaptive research suited to local agroclimatic conditions that is fundamental to achieving sustained gains in agricultural productivity.

Exports

Growth of Merchandise Exports, 1960-75
(Average annual percentage growth rages, at 1975 prices)

	World Trade	Industrialized Countries	Developing Countries
Food and Beverages	4.1	5.2	2.8
Non-food Agricultural Products	4.5	5.6	2.6
Non-fuel Minerals and Metals	3.9	3.1	4.8
Fuel and Energy	6.3	4.2	6.2
Manufactures	8.9	8.8	12.3
Total Merchandise	7.1	7.5	5.9

Sources: World Bank; United Nations Yearbook of International Trade Statistics, 1960, 1976; and Handbook of International Trade and Development Statistics, 1976 (op. cit.).

- The developing countries' exports have increased more slowly than those of the industrialized countries over the last twenty-five years, although there have been very important differences in growth rates among countries.

- Growth of population and incomes have raised domestic demand while the incentives for raising productivity have often been insufficient, and hence developing countries' exports have failed to keep pace with the growth of world demand for agricultural products.

- At the other extreme, where manufactures have accounted for a large share of exports and policies have not been biased against exports, growth has been much more rapid. Eight countries had exports increasing faster than 10 percent a year.

- Those following export-oriented policies have generally fared better than others.

Part 2
Case Histories of Appropriate Technology

1
Tunisia:
Bottled Gas for Automobiles

Richard S. Roberts, Jr.

INTRODUCTION

As a result of the efforts of a young Tunisian technician-entrepreneur, by late 1978 approximately one hundred taxis and government, private, and company automobiles were operating in Tunisia on butagaz (LPG)--a locally produced butane-propane mixture that also is used as the main cooking fuel in the country. Using butagaz as a primary fuel in place of gasoline lowers operating costs, prolongs the life of points and spark plugs, and has a less harmful impact on the environment.

THE TECHNOLOGY

Liquified petroleum gas (LPG) is taken in liquid form from a tank in which it is stored under pressure; from the tank it moves through high-pressure hoses and fittings, a fuel filter, and an electric solenoidal safety valve to a vaporizer-regulator-heat exchanger, where the LPG is reduced to a vapor in two stages of regulation. The vapor is made available to the engine in accordance with engine requirements through a metering valve. In Tunisia, this part of the system is mounted just below the butterfly valve of the gasoline carburetor and linked to the throttle. The Tunisian system makes it possible for an automobile to operate on either gasoline or LPG: a dashboard switch converts operations from one to the other. The system bears the Century trademark and is manufactured by the Marvel-Schebler/Tillotson Division of the Borg-Warner Corporation in the United States.[1]

LPG is a by-product of the manufacture of gasoline and also can be produced from natural gas. It is a mixture of

---

[1] 2195 South Elwin Road, Decatur, Illinois 62525.

gaseous petroleum compounds, principally butane and propane, and is in surplus in the world. In fact, it has long been common practice at refineries and in oil fields to burn off LPG in the atmosphere.

ADAPTATION AND INTRODUCTION IN TUNISIA

The introduction and adaptation of the technology in Tunisia began in 1973 when Ahmed Keskes acquired an Italian system and with it converted his Renault R-8 to operate on butagaz. His action was provoked by rapidly escalating post-1972 gasoline prices and by his recollection of an experience in Belgium a few years earlier. While a student at a technical institute in Belgium between 1967 and 1971, Keskes had one day noticed a slightly unusual odor in a taxi and had mentioned it to the driver. The driver explained that the taxi was running on butagaz, described the advantages of the fuel, and gave Keskes the address of the Italian company handling the system. Keskes visited the company and seriously considered its product as a possible subject of the thesis that he was obligated to prepare to complete his studies. He eventually chose another topic but did not forget the butagaz system.

With his technician training, which involved more "hands-on" work and somewhat less theory than an engineer would have, Keskes returned to Tunisia in 1971 to a job with a parastatal enterprise. Less than two years later, when his father was injured in an accident, Keskes left his job to return home and help run his father's woodworking business in Sfax--an industrial city on the coast in central Tunisia. He also became the local service agent for a number of washing machine manufacturers.

From this business base, Keskes travelled to Belgium in 1973 and visited the agents of the Italian butagaz-conversion system. He persuaded them to provide a sample of their equipment for a minimal cost. Keskes then installed the system on his Renault R-8 and tested it in Tunisia. After a number of months of adjusting and testing in his spare time, he concluded that the system was not suitable for Tunisia. The best that he had been able to accomplish was a savings of 20 to 30 percent, which he judged inadequate. He found that the system wasted fuel and was inconvenient in that the car had to be started on gasoline and warmed up before the switch to butagaz could be accomplished.

With the help of his brother, then a student in France, Keskes learned of another manufacturer--the Marvel-Schebler/Tillotson Division of the Borg-Warner Corporation--with a somewhat different system. Visiting their representative in Belgium, Keskes was well received but was told that no equipment was available at that time for him to test. Six months later, early in 1975, Keskes finally received a sample of their Century equipment that would modify cars to run on butagaz. According to Keskes, the Century technology that he first obtained differed in a major way from the Italian conversion system. The latter allowed the vehicle to run on either butagaz or gasoline but required a warm-up period on gasoline, as already noted. With the Century system, a car would run only on butagaz, which eliminated not only the warm-up period but also the possibility of operating the automobile on gasoline.

In Keskes's view, eliminating the warm-up period was an advantage, but losing the gasoline option was a decided disadvantage in Tunisia, where butagaz supplies were occasionally interrupted. Thus, he says, he decided to modify his sample system to use both fuels. Essentially, it was a matter of fabricating a plate that could be placed between the gasoline carburetor and the intake manifold and that could serve as a base for the butagaz metering system. The result was a base-mount, dual-fuel system used earlier by Century in the United States and still used by the company in Europe and in other countries. This Century system is what is now supplied to Tunisia.

Keskes installed the modified Century equipment on his car. Throughout 1975 he worked at improving system performance and operating economics. He obtained books and journal articles to learn more about the characteristics of the fuel, sent questions to the dealer in Belgium, acted on the answers, and added initiatives of his own. Sometimes only adjustments were necessary, while in other cases new parts had to be made.

A major problem was obtaining the proper mixture of air and fuel for both systems. Since the butagaz carburetor feeds gas just below the butterfly valve of the gasoline carburetor, both systems "breathe" air through the original, or gasoline, carburetor. Achieving an appropriate combination of fuel being fed through the butagaz metering valve and air coming from the original carburetor, with the latter set for normal gasoline operation, involved considerable trial and error. Keskes would drive for a time, keeping a record of performance and fuel consumption, make adjustments, then repeat the process. He eventually concluded that his problems were largely caused by the differences between Tunisian butagaz and the European fuel for which the system had been designed (a key difference being the ratio of propane to butane, and thus the Btu value of the fuel).

The butagaz-fueled car had promise from a business perspective. It was also a challenge from a technical point of view and thus a source of great personal satisfaction to Keskes as problems gradually were solved. However,

Refillable butagaz tank in Mr. Keskes' car

developing its potential was of necessity only a part-time activity for Keskes. Most of the time he was busy with the woodworking business and the washing machine repair service from which he and his family derived their income. Nonetheless, by the end of 1975 he had managed to tune the system to his satisfaction. His operating expenses were 58 percent of what they would have been with gasoline, and the performance of the car was adequate--a slight loss of power at around seventy-five to eighty miles per hour being the only noticeable imperfection. Keskes was ready to show people the fruits of his effort.

On January 1, 1976, Keskes displayed his butagaz car to the press, which gave the system encouraging support. That year it was viewed at an industrial fair, where it drew the attention of the prime minister. Additional press articles have appeared each year since that time. However, Keskes has done almost no advertising and has had no promotional assistance from his supplier or the manufacturer (for whom his volume of business is very small).

During his first year of selling the systems, Keskes had a number of problems. Three of the first ten installations had serious imperfections, caused partly by the varying conditions of the cars in question and partly by the fact that Keskes was learning and adjusting with each different vehicle. In one case considerable repair to the motor was needed, and the customer wanted the system removed; Keskes did both at his own expense. In fact, Keskes removed several other systems at the clients' request, regardless of the reason, and charged no fee for the work. Some of the complaints were less "real" than others; in one car the radio had stopped working not long after the conversion to butagaz, and the owner refused to believe that the failure was not due to the changeover.

Acceptance and spread of the technology have been slow. Keskes points out that Tunisians often still consider what is done in France before embarking on something new. Until early 1978, it was not legal to drive on the roads of France in a butagaz-powered vehicle. It was permitted in Tunisia and was legal and not unusual in Belgium, Holland, and Italy, but Keskes was convinced that the French law created market resistance in Tunisia. He is hopeful that a change in the law as of January 1978 and an order that year to a French firm for 300,000 Century-equipped vehicles for French police and other public authorities will do much to eliminate the resistance in Tunisia.

Growth also has been impeded by Keskes's lack of financial backing; he has had to finance the business himself. His attempts to obtain governmental or bank assistance have been to no avail, making him depend on his other businesses for income and for working capital to finance the butagaz-conversion business. Keskes's main need is for a parts inventory, the value of which he says has been as much as $20,000.[2] Keskes believes that he has difficulty in finding financial support because local sources do not have confidence in a Tunisian technical specialist.

Efforts to have the dual-fuel Century system introduced on a large scale through local auto importers and governmental agencies have not worked; Keskes has found them insufficiently interested. More successful has been his approach to private taxi owners. His first taxi installation, in his hometown of Sfax, was done at no charge. As the word spread, demand developed, and more than thirty of the approximately one hundred installations by late 1978 were done on taxis. The others were done on private cars, company cars, and a few government vehicles.

---

[2] His supplier points out that Keskes's inability to finance a sizable parts inventory makes it very difficult for him to carry the various sizes of pieces needed to fit the range of vehicles brought for conversion. As a result, Keskes spends a certain amount of time adapting pieces to fit. The fact that vehicles in Tunisia tend to be several years old also means that even the supplier does not always have the pieces needed, again resulting in custom-tailoring of the equipment.

Keskes has extended his reach by training mechanics in other towns to install and maintain the systems. However, he has avoided the capital city, Tunis, and the problems that could come with moving into a major metropolitan area. Mechanics with their own garages and two assistants in each case have been trained in two towns other than Sfax; service in a fourth town was foreseen before the end of 1978.

Keskes still maintains his other business activities, but the butagaz-conversion operation takes an increasing part of his time. He imports and distributes the systems (approximately 75 to 80 percent of the parts are imported, representing about 50 percent of the cost to the client), trains the mechanics in other towns, installs systems, gives technical assistance to the mechanics elsewhere, and provides service to his customers. He also acts as the technical contact between the supplier--still in Belgium--and the Tunisian market.

The nonimported parts made by Keskes are not complex and require no sophisticated equipment; nor does the installation of the systems. Only hand tools are needed, although Keskes has been advised to use--and to have his agents use--an exhaust analyzer to aid in tuning each installation. The analyzer is an imported electronic device that costs $600 to $700 plus import duty. Keskes has one and considers it a help but notes that obtaining them and their replacement parts is a major headache because they are imported.

THE IMPACT

Ahmed Keskes installed the conversion system for 255 dinars, or approximately $640, in October 1978. The mechanics handling the system in other parts of the country charged up to 10 percent more.

The precise benefits obviously vary from one vehicle to another. In general, however, after monitoring 30,000 kilometers on his own car and 33,000 kilometers on that of a relative in Tunis, Keskes has found that one liter of butagaz is consumed for each liter of gasoline previously required by the vehicle. The main saving to the owner is the difference between the price of gasoline and that of butagaz; in late 1978 gasoline cost approximately three times more than butagaz. Also, ignition points and spark plugs last longer, and oil changes are needed less frequently (one-third as often, according to Keskes, although one taxi driver reported cutting his oil changes only by one-half while another said that he cut them even more than Keskes recommended).

Keskes observes that where and how one drives influences consumption and thus savings. However, given late 1978 gasoline and butagaz prices, the butagaz system on the average reduces operating costs by approximately 50 to 60 percent. On a car that consumes eight liters per one hundred kilometers, (e.g., a Simca 1000), the savings over about 26,000 kilometers would pay for the installation of the system. The attraction of the system for those whose cars are used intensively is particularly great. For example, in one Tunisian town several taxi drivers who use the system report considerable economy. A taxi with two drivers (not unusual) travels an average of 400 kilometers daily, seven days a week, for up to six months of the year (the tourist season) and rather less the rest of the year; therefore, in approximately one-half of the tourist season, the system will pay for itself, and savings (additional gain) equivalent to approximately 50 to 60 percent of preconversion operating costs will be realized thereafter. Taxi drivers interviewed in Hammamet report that the savings from a change result in a 50 percent increase in their net daily income.

Thus, for the vehicle owner, costs are those of conversion, and benefits are principally savings on fuel. Experience with Keskes's Renault R-12 over 250,000 kilometers, his Fiat pickup truck over 500,000 kilometers, and client vehicles over lesser distances has not yet indicated any appreciable change in maintenance costs or durability other than those mentioned.

At the present rate of conversion (nearly one-half of the roughly 100 installations by late 1978 were done that year), the creation of new jobs has not been significant; system installation is a part-time activity for existing auto mechanics and technician-entrepreneur Keskes. The increased business for Keskes and the mechanics is at present only modest. The system brings them additional activity and income but is a small part of their business.

The impact on the Tunisian economy of the current level of conversions is insignificant. Moreover, calculation of the impact would be fairly complex. Approximately one-half of the client's cost is for imported equipment, but this import expense is offset partially by a reduction in the consumption of imported gasoline in favor of locally produced butagaz. Gasoline taxes provide considerable income to the state treasury; therefore, a small part of this income is largely or wholly lost each time that a driver switches to butagaz. On the other hand, the government could reduce its own costs by converting its sizable vehicle fleet to butagaz operation; the savings possibly could match or exceed the revenue lost through reduced private consumption of gasoline by others converting to butagaz.

Finally, it is noted that butagaz is a much cleaner-burning fuel than gasoline and is less of an atmospheric pollutant. Also, it is an underutilized fuel and is readily available next door in Algeria.

According to Keskes, his introduction of the Century system in Tunisia may have influenced the growing interest in the technology in neighboring Algeria. He has heard that Algerian officials read about his work in the Tunisian press and asked the supplier to convert a small fleet as a test of the system in Algeria. If the results are satisfactory, a project involving the conversion of several thousand vehicles per year over five years may be initiated. (Algeria, be it noted, is a member of the Organization of Petroleum Exporting Countries (OPEC) and is a major producer of natural gas, from which LPG can be made). The Belgian supplier confirms that the Algerians are very interested and will be testing the system in 1979.

PROSPECTS

Whether the technology will have a significant impact in Tunisia remains to be seen. The form of any such impact is also unknown. However, Keskes sees a number of possibilities and problems.

Keskes feels that the system would be more accepted if refueling was more convenient and if the economy of the system was improved. The technology for such changes is available. It is a matter of installing in the vehicle a fixed and properly equipped gas bottle and of establishing refill stations to replenish such built-in bottles. This would offer the added convenience of not having to physically lift (the twenty-seven-kilogram, or sixty-pound) gas bottles out of and into the car. Operating economies would result because the gas would be taken from these bottles through a tube running to the bottom of the bottle instead of coming directly from the neck. In current installations, which use household bottles, the gas is taken from the neck of the bottle, and a significant amount (20 to 40 percent) of the gas cannot be used. (The household bottles are also said to be unevenly filled.) The built-in bottles would eliminate this waste and thus reduce operating costs substantially. The bottles also could have a fuel gauge, which is not practical with the present arrangement.

The special bottles (tanks) are common on installations in Europe, where the filling stations are available. These bottles present no technological problems; Keskes has adapted one and tested it successfully. However, he has not been able to obtain governmental authorization to establish a refill station, and such stations are not available elsewhere in the country.[3] (Stations that refill household

---

[3]Later information indicates that Keskes has received authorization to have a refill station.

bottles are not equipped to refill bottles that are built into vehicles, although they could be modified to do so.) Keskes is hoping that some of the organizations now trying the system may decide to convert their fleets of vehicles and perhaps be successful in obtaining authorization to operate a refill station, at least for their own cars. In the long run, however, he recognizes that if the system is to enter widespread use, facilities must be established for refilling the fuel tank rather than replacing it, and the tank must be designed for vehicular application rather than for household use.

The importance of these considerations is reflected in the fact that at least two taxi drivers are known to have replaced their butagaz taxis with diesel taxis since October 1978. Although a new diesel taxi costs more to purchase than a new gasoline taxi converted to butagaz, operating costs are about the same. Moreover, the diesel does not have the refueling inconvenience of the butagaz system as currently available in Tunisia. The built-in butagaz tank would give butagaz an operating advantage, and service stations would eliminate the inconvenience problem.

Keskes also observes that having special filling stations for vehicles using bottled gas would make it easy for the authorities to tax the fuel. This is done to a considerable extent in the case of gasoline. At present, there is no way to distinguish between gas used for home cooking (by all strata of society within reach of a village or town) and that used in cars; as long as fuel is kept at a low price for home use, it stays low for vehicular use. Special filling stations would make it possible to distinguish between the two uses, and if the government were to tax butagaz used for cars, the difference between the cost of butagaz and gasoline obviously would be reduced. Thus, the appeal of butagaz to vehicle owners also would be lessened.

At present, Keskes estimates that 75 to 80 percent of the parts for the system are imported. He believes that he could reduce this, at most, to 50 percent (by value) by manufacturing virtually all of the parts except the pressure regulator in Tunisia. He would like to do this for export as well as for local use, and his supplier comments that it would make sense, if Keskes could do it. Such a project would involve employing additional people, reducing the import content of the system, and possibly producing a new export item, all of which would change the impact on Tunisia of the technology and its introduction.

Neither the manufacturer nor the supplier seems to be interested in financing such a venture, and as noted earlier, Keskes has tried to borrow through the Tunisian banking and investment promotion system with no success. Although this has left him pessimistic about such channels, he recently learned that investment promotion rules have changed; therefore, he may try again. He estimates that he would need approximately $20,000 for new equipment (a milling machine, a lathe, and miscellaneous tools) in addition to what he already has. He offered no estimate of other needs, but these would include working capital to complete the unfinished building that he has constructed and to bring in power. (His electricity now comes from a portable generator left in Tunisia during World War II; he has refurbished it and converted it from gasoline to butagaz operation.)

Conclusions

This is a case in which neither the government nor an organized technical facility was needed to bring about the introduction of a new technology that offers a number of advantages to individuals and, potentially, to the country. It is a case in which the businessman might have avoided much of his adaptive experimentation if he had known of and had access to organized, effective information services. It is also a case in which a measure--a limited measure--of government support could make a considerable difference in both the rate of adaptation of the technology and the nature and extent of the beneficial impacts on the country and its economy.

Thus far, Keskes has had limited assistance. He sees himself as an isolated individual, as a local technician who does not have the confidence of local financial sources because he is not foreign. He--and other Tunisians--comment bitterly on what they perceive as a widespread local refusal to believe that Tunisian engineers and technicians are capable of competent and serious technical work. Yet, the obvious technical competence, imagination, initiative, readiness to take risk, business sense, and general entrepreneurial spirit and ability of people like Ahmed Keskes are extraordinarily valuable local resources.

PERSONS INTERVIEWED

Mr. Ahmed Keskes

Mechanic in Hammamet who installs conversion kits

Three taxi drivers using biogas

Mr. Buon, Distributor in Belgium

QUESTIONS TO CONSIDER

1. What is the importance of inventiveness and entrepreneurship in the application of appropriate technology? Is this a definable quality and is it feasible to identify it in developing country environments? Is it in any way significantly different than the corresponding quality as it is found in an industrialized environment?

2. Given the financial problems that Keskes encountered, can any conclusions be drawn concerning the need for credit/finance systems especially designed to facilitate appropriate technology efforts on the small-business scale?

3. What is the reader's view of the reactions of the supplier of the conversion equipment? Does the possibility exist of considerable missed opportunity, or was the attitude toward the potential of the Tunisian market as represented by Keskes a reasonable and judicious one?

4. Does Keskes make the most of his opportunity as a result of the way the conversion business has developed? What other efforts might he have made?

2
Thailand:
Cassava Pelletizing Technology

*Ronald P. Black, Wanawan Peyayopanakul,
and Sachee Piyapongse*

INTRODUCTION

During 1978, cassava became Thailand's largest foreign exchange earner, and Thailand was the world's largest exporter of cassava products. Thai production of cassava had increased fivefold in the preceding decade. This rapid growth, however, had not been without associated problems. And some were forecasting further and perhaps more serious troubles for the future.

BACKGROUND

The cassava plant, also known as tapioca, originated in South America. It was reportedly introduced as a food plant in the Far East in the early part of the nineteenth century and began to be used as an animal feed in Europe in the 1940s. During the 1950s, Thailand exported small quantities of cassava chips to Europe, and in 1967, pelletizing technology was introduced in Thailand, allowing for less bulky and thus less costly shipping of the product. By 1971, Thailand was exporting most of its cassava to Europe in the form of pellets. In 1978, cassava surpassed rice as Thailand's largest export earner. During that year, 5.25 million rai[1] of land were planted in cassava, producing 11.2 million tons of cassava roots. Ninety-five percent of this was exported, earning Thailand over U.S. $500 million. It is reported that about eight million people, or one-fifth of the country's population, are directly or indirectly involved in cassava production and trade.

One of the characteristics of cassava is that it can grow in very poor soils and tolerate drought. It therefore has become a very popular crop for farmers in northeast

---
[1]One rai equals 1,600 square meters.

Thailand, the country's most impoverished region. During 1978, 2.8 million of Thailand's 5.25 million rai of cassava were located in this area.

Processing Cassava into Pellets

While the cassava plant can be harvested at any time after it matures, oversized roots and a high-fiber, low-starch content will result if the plant is allowed to continue to grow. This, in turn, may cause marketing problems. In Thailand, the root normally is harvested by hand when the plant is ten to twelve months old. The stalks of the plants are cut off forty to sixty centimeters above ground and are then pulled up by hand, by both women and men laborers. Next, the stalks and their tubers are piled at the side of the field where the tubers are cut from the stalk and collected by truck.

From the field, the tubers are taken to a chipping facility. Most pelletizing factories have their own chipping machines, but they also purchase chips from companies that specialize in this business.

After the tubers have been chipped, the chips are wheeled, usually in a wooden cart, to an outdoor cement drying floor that occupies two to twelve rai. They are then spread on the floor by four or five workers, although this is sometimes accomplished by a tractor-like vehicle known as a payloader. The chips are turned during the day every couple of hours by rakes; at the end of the day, they are heaped into piles under shelter. The drying process continues for two to four days, depending on the amount of sunlight and the size of the chips.[2] After drying, the chips are put into burlap sacks (if the chipping is not done at a pelletizing factory) that can hold seventy to eighty kilograms of chips each. Women and children usually are employed to turn and sack the chips.

From drying, the chips are moved via truck to a pelletizing mill, unless the chipping facility is part of the pelletizing company. Upon arrival, the chips are moved to the mill hopper by a payloader.

The next step involves the production of pellets (see Figure 2.1). In Thailand, two types of pellets are made: brand pellets and native pellets. Brand pellets are produced

---

[2]The length of drying also depends on the desired moisture content. Thai processors are often accused by European buyers of shipping pellets with a moisture content that is too high, which may cause the growth of bacteria or mold. There is, however, considerable disagreement on this question.

largely by subsidiaries of European Economic Community (EEC) companies, but two Thai companies also manufacture this kind of pellet. All native pellets are produced by Thai companies. At present, virtually all companies producing brand pellets use imported equipment.

Brand pellets reputedly have a higher quality, less moisture, less foreign matter, and greater hardness, all of which are achieved through better quality control and through changes in the processing procedure. One extra step, the preconditioning of the chips prior to their being pressed into pellets, and one modified step, the use of greater heat and pressure in the formation of the pellet, are employed in the manufacture of brand pellets. Therefore, the next step depends upon which pellets are to be produced. The native process produces 80 to 90 percent of all pellets exported from Thailand. In the production of native pellets, the chips are moved via conveyor belt from the hopper to a presser. Heated pellets are formed and are moved by a conveyor to a cooler, where the finished pellets emerge. Two well-trained workers operate the pellet mill, which is often equipped with two pressers and is powered by diesel oil or, more often, electricity.

Finished pellets are moved to a trans-shipment facility to be loaded on a ship for export.

The marketing channels that move cassava from the farm to the ship are varied and are often complex (see Figure 2.2). Agents, or traders, may be involved in moving the cassava between each of the steps in the process. While moving cassava products from the site of one step in the process to the site of the next step adds value to the products, this would not be done by agents unless they were able to make a profit in so doing. Theoretically, the farmer, chipper, or pelletizer should be able to make a similar profit in moving their products to the next step in the process. To the extent that they can do this, they increase the profitability of their activity. Some are able to accomplish this.

Economic Aspects

The operating cost for producing one ton of pellets at a Thai pellet mill was found in a recent study to be 1,653 baht (see Table 2.1). In another Applied Scientific Research Corporation of Thailand (ASRCT) study, the investment requirement for a native pelletizing factory was found to be 1,384,500 baht (see Table 2.2).

From a survey of owners of Thai pelletizing factories, ASRCT learned that to set up such a factory an investor would need, in addition to finances to cover capital costs, enough money to operate the plant for approximately 1.5

FIGURE 2.1
FLOW DIAGRAM OF CASSAVA PROCESSING STEPS

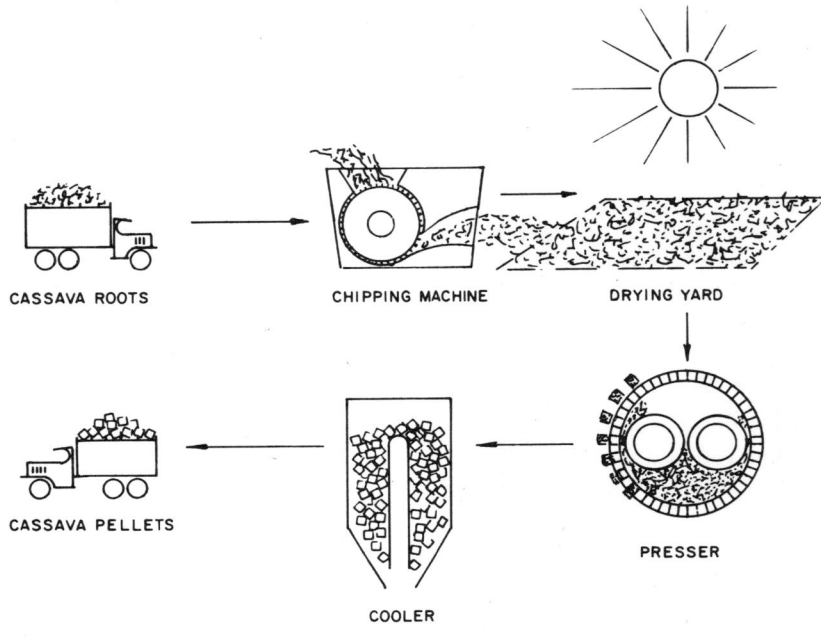

FIGURE 2.2
MARKETING ROUTES FOR CASSAVA TUBERS,
CHIPS, AND PELLETS

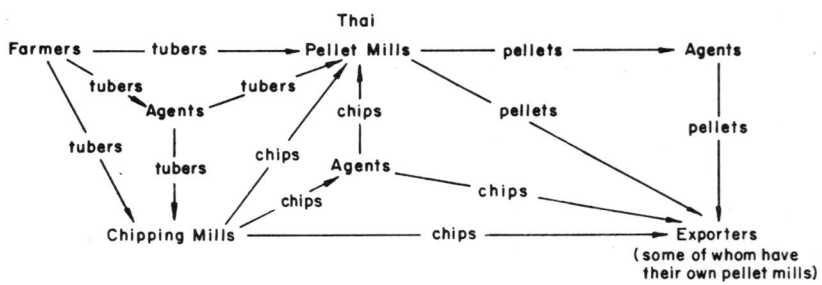

months. It also was found that a factory would produce three tons of cassava pellets per hour and would operate eight hours per day, two hundred days per year. Based on this information, operating expenses for 1.5 months are calculated to be approximately 1 million baht.[3] Assuming that one-half of the required capital could be borrowed and that interest would not be paid until income was generated (reportedly both reasonable assumptions for loans from Thai commercial banks) then an investor would need approximately 1.25 million baht to open a pelletizing factory in Thailand. About seven persons are required to operate a typical pelletizing mill. Therefore, an investor may create one work place for approximately 180,000 baht.

INTERVIEW WITH A THAI MANUFACTURER OF CASSAVA PROCESSING EQUIPMENT[4]

Pornsak Pusanguansit is the manager of the Chonburi Casting and Machine Work Company, Ltd., (CCMW), a family-owned and managed company on the edge of the city of Chonburi, approximately sixty kilometers southeast of Bangkok, bordering the Gulf of Thailand. The following account was given by Pornsak.

CCMW was founded in 1949 by Pornsak's father, Luan Pusanguansit, to manufacture rice-milling equipment. In 1968, this was still CCMW's main line of equipment; during that year, however, Luan was approached by an old friend, Yep Yong Nguan, who described a difficulty that he and other Thai cassava processors had encountered--they could no longer obtain cassava.

Two European companies, Krohn and Company, Bangkok Ltd. (KCBL), and Peter Cremer, Ltd., had opened subsidiaries in Thailand to process cassava in 1967. These two firms had introduced a radically new form of processing technology that produced a cassava pellet rather than the starch that was produced by the Thai industry. Pellets sold for 67 percent more than starch, which meant that the European firms could pay the farmers a higher price for their cassava roots. Consequently, these companies essentially had cornered the cassava market in Thailand.

---

[3]One U.S. dollar equals approximately twenty baht.

[4]As the titles of this and subsequent sections indicate, they are based on interviews in February 1979. The authors do not attest to the accuracy of the information provided by the persons interviewed. In fact, a close reading will reveal a number of discrepancies.

TABLE 2.1
OPERATING COSTS OF THAI CASSAVA PELLETIZING TECHNOLOGY*

Requirements	Cost/Ton of Pellets (baht)
Raw materials:	
Chips	1,450.00**
Material lost during processing	101.50
Sacks	20.00
Vegetable oil	2.75
Utilities:	
Electricity	29.25
Lubrication	3.50
Diesel oil for payloader	2.70
Labor:	15.83
1 Supervisor @ 100 baht/day	
1 Switchboard operator/quality controller @ 100 baht/day	
1 Pit person @ 60 baht/day	
1 Conveyor operator @ 60 baht/day	
1 Driver @ 60 baht/day	
Repair and maintenance	13.77
Administration expenses (two persons)	7.27
Depreciation	10.50
Transportation (ranges from 55 to 100 baht per ton)	77.50
Total operating costs per ton of pellets	1,734.57

*This technology produces soft, or native, pellets. The factory produces three tons of pellets per hour and is operated eight hours per day, 200 days per year. The annual output of the factory is 4,800 tons of pellets.

**The price of cassava chips is the average price for 1975 through 1978.

Source: Unpublished results from an ASRCT study--still in progress in February 1979.

TABLE 2.2
INVESTMENT REQUIREMENT FOR NATIVE PELLETIZING FACTORY*

Requirements	Cost (baht)
Land: one rai	100,000
Land improvement	20,000
Factory building: 1,300 square meters @ 500 baht/square meter	650,000
Machinery and equipment:	
Pressing machine and accessories	160,000
Motor, 160 h.p.	120,000
Motors for conveyor (5)	45,000
Transformer	200,000
Magnetic switch	70,000
Payloader	170,000
Scales (2)	6,500
Carts (2)	3,000
Total fixed capital	1,544,500

*Investment costs for factory with one pressing machine.

Source: Unpublished results from an ASRCT Economic Analysis Department study--still in progress in February 1979.

Yep wondered if his friend could make a similar machine. Luan worked on the problem and was able to determine how to make everything but the most essential mechanism--the one that produced a pellet from a chip. Not to be outdone, Luan contacted Thai workers employed at the Krohn Company. From these discussions, Luan deduced that the essential mechanism was a die containing two rollers (see Figure 2.3).

Luan returned to his metalworking shop, and within six months from the day that Yep had brought Luan the problem, CCMW had a prototype pelletizing mill. Yep and other Thai cassava processors were sufficiently impressed and advanced Luan payment for twenty mills. With this money, Luan purchased new metalworking equipment that he would need to manufacture cassava pelletizing mills. He also continued experiments with his prototype.

Approximately one year after the prototype had been constructed, CCMW went into commercial production. Three months later, twenty-eight mills had come off the assembly

FIGURE 2.3
CASSAVA-PRESSING MECHANISM

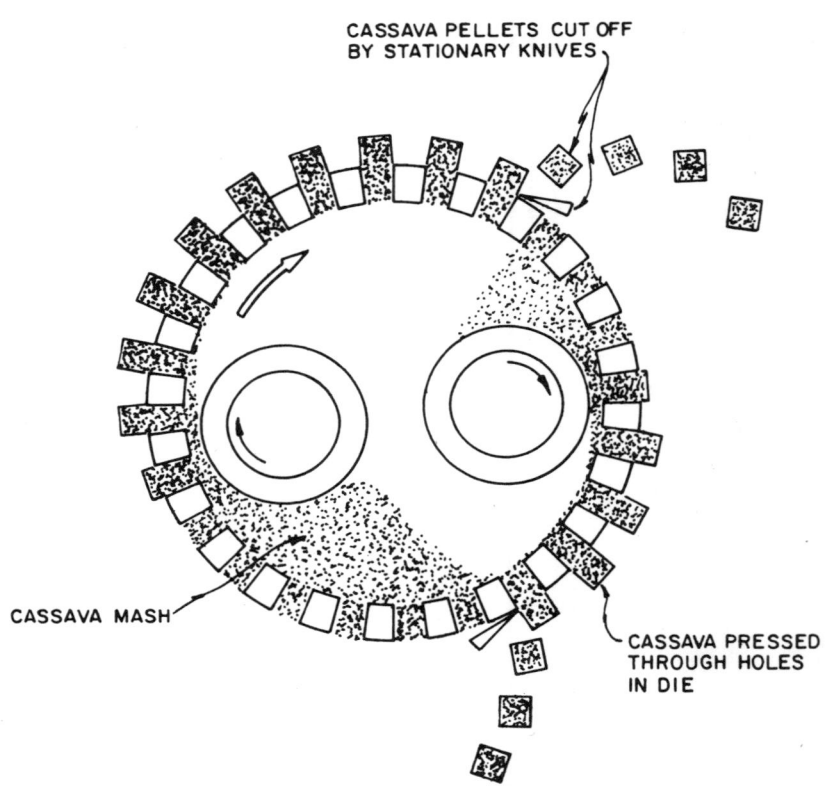

(1) Cassava mash

(2) Cassava pressed through holes in die

(3) Cassava pellets cut off by stationary knives

Source: N.C. Than, S. Muttamara, and B. N. Lahoni, "Technological Improvement of Thai Tapioca Pellets," Thai Journal of Agricultural Science II (April 1978), p. 21.

line. Pornsak noted that in 1979 they still could not do much better than was achieved in that first commercial run.

Over the years, CCMW expanded into other lines of cassava-processing equipment, including the production of a cassava chipper and a payloader designed especially for cassava processing (see Table 2.3). Pornsak noted that he designed the payloader and that 60 percent of its components are manufactured in Thailand.

TABLE 2.3
SAMPLE LIST OF CCMW PRODUCTS AND CONCOMITANT PRICES

Item	Price (baht)
Chipper (10 tons per hour)	13,000
Chipper (50 tons per hour)	35,000
Electric-powered pelletizing mill (4 tons per hour)*	350,000
Diesel-powered pelletizing mill (3 tons per hour)	250,000
Payloader	175,000

*This requires a 200,000-baht transformer.

Source: Chonburi Casting and Machine Work Company, Ltd., based on an interview with Pornsak Pusanguansit, 21 February 1979.

Two problems helped to launch CCMW into a new, innovative phase in its production of pelletizing mills. First, environmental laws in several EEC countries were causing problems for the exporters of the soft pellets produced with Thai equipment. These pellets crumble easily and lead to meal dust problems during unloading operations. Also, in mid-1978 the Thai government set regulations preventing the opening of new pellet mills or the expansion of existing ones. The regulations were a result of a perceived overproduction of cassava and cassava products in relation to what the world market was absorbing. The Thai regulation caused a serious reduction in CCMW's business, and it became necessary to cut the working hours of the 175-member labor force from twelve to eight hours per day. At that point, CCMW's activities associated with pellet mill production had to focus mainly on the replacement of old equipment.[5]

---

[5]According to the Factory Control Division of the Thai Ministry of Industry, there are 1,297 cassava pellet mills in Thailand.

Because of the increasing pressure in Europe for less pollution, the Pusanguansit family decided to design a mill for producing hard pellets and thereby reinvigorate their business. Pornsak also noted that the hard pellets would bring Thai processors 0.2 to 0.3 baht per kilogram more on the European market.

As of 1979, the several European subsidiaries are producing hard pellets with imported equipment. Two Thai firms also are using imported equipment and are producing brand pellets.

As of February 1979, CCMW has nine orders for its mill, and the first one will come off the assembly line in two months. The pellets from this mill will withstand fifteen kilograms of pressure as opposed to four kilograms of pressure that the soft pellets produced by Thai equipment can tolerate (see Figure 2.4 for a flow diagram of the new CCMW mill). Each new mill will cost 1 million baht. Pornsak estimates that, based on the present price for hard pellets, a processor will get his money back in one year's time using this mill. He noted that CCMW is prepared to help potential customers calculate the economic advantages of a hard-pellet mill as opposed to a mill that produces soft pellets.

Pornsak said that the CCMW hard-pellet mill would cost only one-third the price of a similar imported mill. He was, however, uncertain about relative operating costs. The pellets, he said, would be of equal quality to the brand pellets now being produced.

Pornsak noted that thirteen other companies were now producing pelletizing mills in Thailand. Of these, he believed that only one was capable of designing and manufacturing a machine similar to CCMW's new hard-pellet mill.

CCMW relies on its reputation for obtaining purchase orders; no special marketing effort is undertaken. Pornsak noted that the firm has produced 60 percent of the Thai Improvement pelletizing mills currently in operation; further, the reputation of the company has resulted in export sales; there are presently seventy mills in Indonesia and two in Laos.

INTERVIEW WITH A THAI OWNER OF A CASSAVA PELLETIZING MILL

In 1955 Niyom Chulaserekul began work in a Chonburi starch factory. As he became proficient in his job, he began to realize the potential of the cassava-processing industry, and five years later he started his own cassava-chipping business. In 1970 he opened a pelletizing factory. Today Niyom has a beautiful home located next to his mill.

Niyom purchased his first pelletizing mill from CCMW because he believed that the company had the ability to

FIGURE 2.4
SCHEMATIC FLOW DIAGRAM OF CHONBURI CASTING
MACHINE WORK COMPANY'S NEW HARD-PELLET MILL

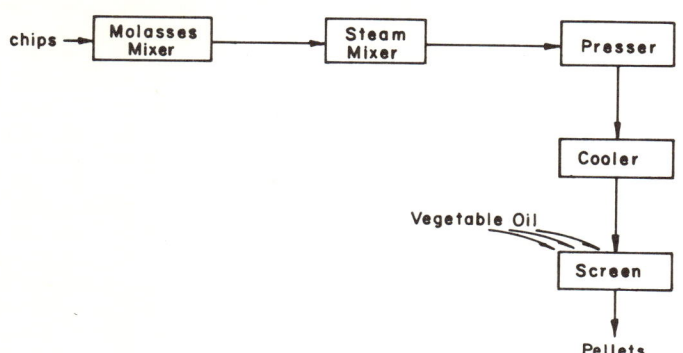

Source: Chonburi: Casting and Machine Work Company, Ltd.

service his machinery. At the time he entered the pelletizing business, three other Thai firms were manufacturing pelletizing equipment. Niyom also could have purchased a Taiwanese machine, but for some reason its capacity was too low. A U.S. mill was another choice, but the price was too high. Niyom paid 400,000 baht for the CCMW mill, which had three pressers and operated off a diesel engine. In comparing his first mill with the machine manufactured by the Krohn Company, Niyom said that the output was 30 percent less but that the pellet quality was comparable. His first mill from CCMW lasted for fifteen years.

Indeed, Niyom produced higher-quality pellets at that time than he does now. This is due, he claimed, to the quality of chips. When he first started his business, he could get top-quality chips. Now, as a result of the stiff competition, he must take what he can get. He is now able to operate only at one-third his 300-ton-per-day output capacity. The reason, he noted, is because a precipitous drop in the price of cassava occurred in early 1978. (The price of roots dropped from 520 baht per ton in December 1977 to 360 baht per ton in April 1978.) As a result, many farmers switched to growing pineapple.

Niyom now has five pressers and fifteen workers. He noted that if he were operating at full capacity this would still be a sufficient number of laborers.

Niyom is aware of the environmental problem caused in European ports by the unloading of soft pellets. He also is aware of the price differential between soft and hard pellets. However, he does not believe that he can produce hard pellets at a marketable price.

Two years ago, assisted by a professor from the Asian Institute of Technology (AIT), Niyom began to experiment in producing hard pellets. He consequently has been able to produce pellets that will withstand a pressure of twelve kilograms.

He has offered such pellets to exporters at a price that is 0.33 baht per kilogram higher than the going rate for soft pellets. His offer was last refused in January 1979. The cost of producing hard pellets, he stated, made the product noncompetitive with maize because of the higher nutritional value of the latter.

Niyom also noted that the 0.33-baht-per-kilogram increase would not yield him a greater profit. Therefore, even if the exporters would settle for this price, processors had no incentive to convert their equipment to produce hard pellets. He was confident that no one, including the engineers and technicians at CCMW, could reduce the differential cost of producing hard pellets rather than soft ones more than he had.

Niyom said that he did not know what price the exporters received for their pellets; however, he pointed out that the export companies were the same firms that were now producing the brand pellets. According to Niyom, these companies were interested in protecting their market and thus had little reason to encourage Thai producers to compete. Niyom went on to say that while a number of exporters would buy his product they all would offer the same price.

Five years ago, the small Thai pelletizing companies attempted to find a way to sell directly to the EEC buyers. Subsequently, the president of the Thai Tapioca Trade Association went to Europe to discuss this possibility with a feed company. Niyom noted, however, that the costs would have been prohibitive, since the association would have had to rent a ship to transport their product.

When queried about the Trakulkan Company, Ltd., a Thai firm that produces brand pellets, Niyom noted that the firm has been in business for a long time. It has a U.S.-manufactured mill with four presses capable of producing 500 tons of pellets per day. The firm made direct contact with a European buyer and worked out an arrangement with him. A small ship is hired by the company to transport the pellets. Niyom noted that the owner's son lives in Germany and makes all the arrangements for the Thai firm.

Niyom said that the future of the cassava-processing industry in Thailand depends on EEC policy. An EEC team will soon be coming to Thailand to negotiate a limit for the export of cassava products from Thailand to the EEC. Niyom himself does not expect this limitation to be too severe. He said that Thailand probably would be permitted to continue exporting products at about the present level.

INTERVIEW WITH THE MANAGER OF A EUROPEAN PELLETIZING FACTORY IN THAILAND

Peter Hudde is the general factory manager of Krohn and Company, Bangkok Ltd. (KCBL), a cassava-pelletizing factory. He has been with the company in Thailand for five years.

The head office of Krohn and Company is located in Hamburg, Germany. The firm is a trading organization that concerns itself primarily with grains and grain substitutes. The Thailand subsidiary, KCBL, is involved mainly in the exportation of cassava products to Europe. In addition, however, the Bangkok-based firm has a shipping line and a cassava-pelletizing factory that has been in existence since 1967.

The pelletizing factory has two electrically powered mills, both imported from Switzerland (see Figure 2.5 for a flow diagram of the Krohn mills). Each mill can produce 300 tons of high-quality hard pellets per day for twenty-five days per month. Hudde estimates that such a mill would cost 20 million baht if imported today.

For its own pelletizing factory, Krohn buys only chips. These are purchased through agents or from large-scale chip producers. Purchasing may be done through forward contracts or through agents who deliver on a daily basis in response to a price that is revised and quoted each day.

FIGURE 2.5
FLOW CHART FOR THE KROHN BRAND PELLET MILL

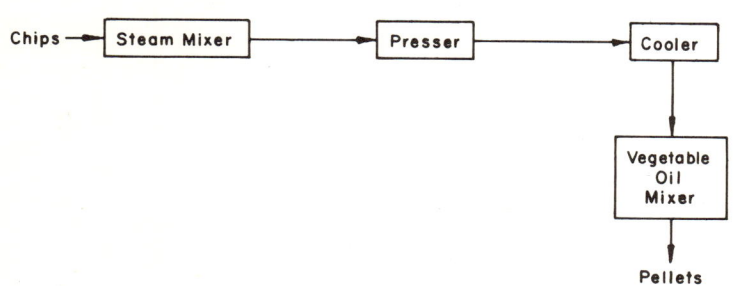

Source: Krohn and Company Ltd., Bangkok Ltd., based on information obtained in interview with Peter Hudde, 23 February 1979.

Hudde noted that trading is a very secretive business. Normally, daily quotes are lower than the buyer is actually willing to pay--with additional payment being made "under the table." This is done to keep the price low and to prevent other buyers from knowing exactly what is being paid. Hudde said that contracts between agents and chipping plants or between agents and buyers, such as the Krohn Company, are kept very confidential.

While Krohn usually buys through agents or, on occasion, through forward contracts with large-scale chip producers, different procedures are used by other buyers. For example, the European subsidiary Tradax has its daily price announced over the radio each morning. Any chip producer may then load his truck with chips and take them to Tradax for selling. According to Hudde, Tradax does not work with agents but will draw up forward contracts with chip producers, including very small operators.

Hudde said that his company has no problem obtaining chips because it pays a high price. In fact, the price it was having to pay at the time of the interview was higher than normal. Also, KCBL buys only high-quality chips. The quality of the chips is checked as the seller brings them to the factory. This is done during unloading because good chips may be on top, and poor ones underneath. A final laboratory analysis is made after the chips are unloaded. If the product does not meet Krohn's standard, it will be refused, or, if it is not too inferior, a deduction will be made in the quoted price for the day.

KCBL sells its pellets on the EEC market with guaranteed delivered quality. Quality specifications include starch, meal, raw fiber, moisture, and ash contents. For the delivery of high-quality hard pellets, KCBL receives a premium of about ten U.S. dollars per ton, although this can go as high as fifteen U.S. dollars per ton. Hudde noted that pollution caused by meal dust during unloading is beginning to be restricted at some EEC ports. Only hard pellets are now allowed to be unloaded at inland ports, and on several occasions the unloading of soft pellets has been halted at seaports. Hudde said that pollution was becoming a major problem at the Port of Rotterdam, where 40 percent of the Thai pellets for EEC markets are removed. Another problem with soft pellets and their concomitant high meal content is that the machinery at some compound feed factories cannot process them.

Hudde noted that over the last six months exporters have begun to purchase hard pellets from Thai pelletizing firms at a premium of 0.2 baht per kilogram for delivered quality. If the pellets do not meet specifications on delivery, the premium is reduced accordingly. While Hudde believes that a 0.2-baht-per-kilogram premium is fair, particularly for small-scale, family-run operations, the supply of hard pellets has been erratic at this price. He notes that significant maneuvering is still done to determine an acceptable price.

If port authorities were to exclude all but hard pellets, Hudde contends that the Thai mills could adapt their technology to produce the hard pellets in two months. He says that the operators of the Thai mills already know how to produce such pellets and that they would require no help in doing so. Hudde saw very high-quality hard pellets produced in Khon Kaen and Nakhon Ratchasima (Korat), both cities in northeast Thailand.

On the other hand, Hudde said that if the Rotterdam port were to be closed to soft pellets, the price of hard pellets would cause some compound feed mills to shift to the use of more corn. In practice, the closing of EEC ports to soft pellets will occur gradually so that all Thai pellet producers will not be required to shift to hard pellets "overnight." While it presently is not difficult to sell soft pellets, Hudde believes that within one year only hard pellets will be acceptable on the European market.

Hudde contends that cassava will continue to be a big foreign exchange earner for Thailand. He also believes that the pollution problem can be solved. A bigger uncertainty is what the EEC will do to protect the compound feed industry market for European maize farmers. As of February 1979, the EEC has a high duty on the import of maize but only a small duty on cassava. Hudde claims that no one knows what, if any, limits the EEC will place on cassava imports.

INTERVIEW WITH MEMBERS OF A EUROPEAN INDUSTRY MISSION TO THAILAND

As a result of the dust pollution that is created when soft cassava pellets from Thailand are unloaded, existing antipollution laws in several EEC countries that import large quantities of Thai cassava (particularly Holland, West Germany, and Belgium) may be used to prevent such unloading. Local special interest groups already are demanding that their governments take action. Because this directly endangers the cassava trade between Thailand and the EEC, the compound feed and stevedoring industries of two EEC countries dispatched a mission to Thailand in May 1978 to determine a solution to the problem. The mission concluded that the native Thai cassava-pelletizing mills could be converted to produce hard pellets at the same capacity as for soft with no extra cost. In February 1979 a second European industry mission visited Thailand to work with Thai pelletizing mills and the ASRCT to demonstrate the feasibility of the first mission's postulation.

The three members of the second industry mission were T. Tuinenburg of the Geraan Elevator Maatschappij, Rotterdam; Jos A. M. Trum of Mengvoeder W. T. Delfia, Maarssen, Holland; and Paul E. van de Venne of Wessanen, also in Holland. Tuinenburg works for a stevedoring company; Trum and van de Venne are employed by compound feed firms.

They noted that the first mission to Thailand was concerned with the soft pellet pollution problem in general, whereas the job of the second mission was to demonstrate, with Thai technology, that hard pellets could be produced for the same cost as soft pellets. Trum noted that before he left Holland he knew that this would be difficult.

The joint European mission/ASRCT team was able to modify the Thai pellet mills so that they could produce hard pellets, but only at 60 percent of their capacity to produce soft pellets. This could be achieved without substantial new investment. According to the European mission's information, there was sufficient unused processing capacity in Thailand such that no additional investment in new mills would be required to convert Thailand's entire pellet output to hard pellets.

In February 1979 the members seemed pleased with their work. While not completely achieving the goal that had originally been set for them, they believed that their results would allow the Thai pellet mills to switch to hard pellets without incurring an increase in pellet costs of sufficient magnitude to price the Thai cassava product out of the EEC market.

It was their opinion that this switch was inevitable and that it would have to occur in a matter of months rather than years. It was pointed out that the person who had the authority to close the Rotterdam port to soft cassava pellets was coming to Thailand for discussions with interested Thai parties. These discussions would be aimed at finding a solution to the dust problem caused by the unloading of Thai cassava in Holland.

INTERVIEW WITH AN ASRCT CASSAVA PELLETIZING TECHNOLOGY PROJECT LEADER

Malee Sundhagul is the director of ASRCT's Industrial Research Department. She also served as the ASRCT project leader for the experiments and analyses carried out jointly with the second European industry mission on cassava pelletizing technology.

In describing how ASRCT became involved in this study, Malee noted that she had been asked to serve on a committee of the Thai Industrial Standards Institute (TISI), which was to gather information to serve as a basis for TISI to set standards for good-quality cassava pellets. It soon became clear that data were insufficient to do this. The TISI committee, however, was aware that a second European industry mission was to visit Thailand to attempt to demonstrate the production of good-quality pellets with Thai equipment. The European mission had requested the assistance of a Thai organization and had suggested AIT.

According to Malee, some disagreement between the Thai and the first European industry mission centered on the mission's main conclusion--that the production capacity of the Thai equipment would be increased when modified to produce hard pellets but that the energy input also would increase. It was the first mission's contention that these two factors would offset each other; therefore, the improvement could be achieved without increasing costs. It was hoped that such disagreements could be avoided by having the second European mission work with a Thai organization. It was questioned, however, whether AIT would be the best organization to work with the European mission. AIT is not a national institution, and the AIT team members who would work with the Europeans also would not be in the position to provide follow-up assistance to producers (because it is mainly an academic institution).

When Malee was asked if ASRCT could handle the assignment, she reflected that the corporation was basically capable and had had some experience in the area. About six years before she had been involved in a project concerned with fungal contamination of processed cassava products and had gone to Rotterdam to discuss the project results. One of her major recommendations, as a result of her investigation, was that a serious study should be conducted on cassava-processing technology. Subsequently, ASRCT had been involved in other cassava-processing projects, including one that concerned spare parts for Thai-manufactured pelletizing mills. While ASRCT's experience in the specific area of cassava pelletizing was somewhat limited, Malee believed that the organization would be technically capable of serving as a counterpart to the European mission.

The European industry mission and ASRCT developed an excellent working relationship before the end of the project. On reflection, however, Malee noted that a little tension may have existed between the two groups at the beginning. For one thing, ASRCT thought it obvious that the hard pellets could not be produced at the same capacity and for the same cost as soft pellets. For a while, some of the ASRCT staff members were under the impression that their European counterparts were out to "prove a point" regardless of the facts. This suspicion soon disappeared as the two groups began working together.

Malee believed that the project had been quite successful. First, it demonstrated to Thai pellet mill owners that hard pellets could be produced, if necessary, at a marginal extra cost that would not price cassava out of the European market. Second, the European mission believed that extra costs would be involved in producing the hard pellets. In February 1979, however, complete agreement on what these costs were had not been reached. Malee thought it natural that in areas of cost uncertainty (due to the limited time available for the study), ASRCT naturally tended to give the Thai producers the benefit of the doubt; on the other hand, the European mission probably would tend to give the benefit of the doubt to the European purchasers of Thai pellets. She indicated that such disagreement, however, was not serious. ASRCT believed the differential cost to be at least 0.15 baht per kilogram.

When the present activity with the European mission has terminated, ASRCT will enter a second phase of the project in which demonstrations, training, and consultancy services will be provided to the processing industry. This phase will be partially subsidized by the Thai government, with some costs being covered by the industry.

Partially as a result of this project, ASRCT has proposed to the Thai National Economic and Social Development Board[6] and the Thai Tapioca Trade Association that a cassava-processing laboratory be set up at ASRCT. The prospects look favorable. The two organizations have verbally agreed to support ASRCT financially in its proposed program. In addition, ASRCT and the concerned EEC organizations have agreed to remain in contact and to consult with each other.

Malee thinks that the future for cassava farming and processing in Thailand is bright. The only possible problem, which she views as short-term in nature, concerns French and Italian pressure to restrict the import of cassava to protect their maize and barley farmers. In fact, she noted that Mr. Geunderlack, the number-two man in the EEC, has proposed to come to Thailand to discuss this particular issue. It is also anticipated that EEC would consider assisting Thailand to strengthen the program on crop diversification with the hope that this will prevent increased production of cassava. Malee believes that any move to drastically restrict cassava imports from other parts of the EEC will probably be resisted because such restrictions would increase the cost of feed and thus meat prices. In the long run, the world will consume increasingly more meat as it becomes more wealthy. This will, in turn, create

---

[6]Thailand's national planning organization.

a growing market for cassava. Malee noted, too, that a compound feed industry in Thailand also was using cassava. This industry already has begun to export to other countries in the Association of South East Asian Nations (ASEAN).

Malee was aware that a Rotterdam city counselor was coming to Bangkok later in the month of February. She said that he was touring port facilities in the Far East, including the one in Bangkok. Because he has important influence concerning the enforcement of pollution laws in Rotterdam, including the city's port, he will meet with representatives of the Thai Ministry of Commerce, the Ministry of Industry, and the Thai cassava-processing industry. Malee noted that although this was not an official visit, it would provide him with background information.

Malee believes that more research and development (R&D) also can give a considerable boost to cassava agriculture and industry. For example, the Department of Agriculture has shown that up to ten tons of cassava per rai can be grown each year. As of 1979, the yield is two tons per rai per year. Indeed, Malee would like to see a cassava R&D center sponsored by a Thai organization that would consider the entire spectrum of cassava problems and opportunities, from agriculture through processing to final marketing.

Finally, Malee said that a previous problem with product quality resulting from adulteration of export products had been reduced drastically. Undoubtedly, she noted, the life prison sentence handed down last year to a prominent cassava processor as a result of deliberate product adulteration has had a sobering effect on the cassava-processing industry. She further noted that the industry was learning in other ways; for example, farmers and chipping and pellet mills are beginning to bypass the middle men in marketing their products, thus insuring better returns on their work and investment.

REFERENCES

N. C. Than, S. Muttamara, and B. N. Lahoni, "Technological Improvement of Thai Tapioca Pellets," Thai Journal of Agricultural Science II (April 1978).

PERSONS INTERVIEWED

Pornsak Pusanguansit
Manager
Chonburi Casting and Machine Work Company, Ltd.
Chonburi, Thailand.

Niyom Chulaserekul, owner of cassava pelletizing factory
Chonburi, Thailand

Peter Hudde
General Factory Manager
Krohn and Company, Bangkok Ltd.
Bangkok, Thailand

T. Tuinenburg
Geraan Elevator Maatschappij
Rotterdam, Holland

Jos A. M. Trum
Memgroeder W. T. Delfia
Maarssen, Holland

Paul E. van de Venne
Wessanen, Holland

Dr. Malee Sundhagul
Director
Industrial Research Department
ASRCT
Bangkok, Thailand

QUESTIONS TO CONSIDER

1. Labor-intensive agricultural export products meet several tests of "appropriateness." What are the particular advantages of the Thai cassava industry in this regard?

2. Pollution standards in industrialized countries are having various effects on developing countries; what is your opinion concerning the way this cassava dust problem is being handled in the case of the European port authorities?

3. The integrated European hard-pellet producers have an apparent market and cost advantage; what should Thai domestic producers do in consideration of this? Or, is the market likely to continue to support both classes of competitors if the domestic firms convert to hard-pellet production?

4. What is the advisability of establishing a specialized cassava technology research and development organization in Thailand? How should the costs of this be met? What role, if any, should government take in establishing it?

5. Is there any additional role that government might take to promote and stabilize the cassava market, realizing the importance of it to the national economy? Should other markets for Thai cassava as animal feed supplement be sought, or should efforts be restricted to Europe?

6. Does there appear to be any latitude for assisting Thai cassava-processing equipment producers to improve their products? Are there any apparent opportunities for innovation in that industry beyond simply adapting foreign technology?

3
Colombia:
The Composite Flour Program

James M. Miller

BACKGROUND

The Instituto de Investigaciones Technológicas (IIT), in Bogotá, Colombia, began work in 1966 on the idea of replacing wheat in commercial food products. This initiative was made by Dr. Norton Young, then the institute director. The technology involved research into the substitution of wheat with locally grown cereals, first in alimentary pastas, then in bread. The first question was to what degree wheat could be replaced: possibly 100 percent in pastas and to a lesser degree in the composite flours used in the two basic breads consumed in Colombia. Breads and pastas were critical because they ranked highest with meat as the food recommended by the institute. The objective was to improve nutrition in products readily acceptable to consumers and was aimed primarily at increasing submarginal nutrition among the poorest group of the nation. Also, since wheat was and is a major food import, the project sought to reduce wheat consumption and thereby decrease the balance of payments problem that Colombia was experiencing.

The Colombian population is divided into five income groups; the lowest group comprises 30 percent of the total population (see Figure 3.1). This group obtains only 62 percent of the protein and 65 percent of the calories recommended by the Colombian Family Welfare Institute (see Figure 3.2). In the most economically depressed areas, 50 percent of the population receives only one-half of the recommended protein and calorie allowances.

Wheat production in Colombia dropped from 142,000 tons in 1960 to 54,000 tons in 1970 and then to 38,000 tons in 1978. Imports increased from 116,000 tons in 1960 to 326,000 tons in 1970, and by 1978 they reached 500,000 tons. The FOB value of imported wheat by 1978 was $63 million in U.S. dollars, or $126 per ton. The National Institute for Marketing of Agricultural Products (IDEMA) projected wheat imports for 1979 of 430,000 tons at a cost of $80 million, or

FIGURE 3.1
PERCENTAGE OF RECOMMENDED ALLOWANCES OF CALORIES AND PROTEINS MET BY FOOD PURCHASED BY THE TWO LOWEST INCOME GROUPS OF COLOMBIAN POPULATION

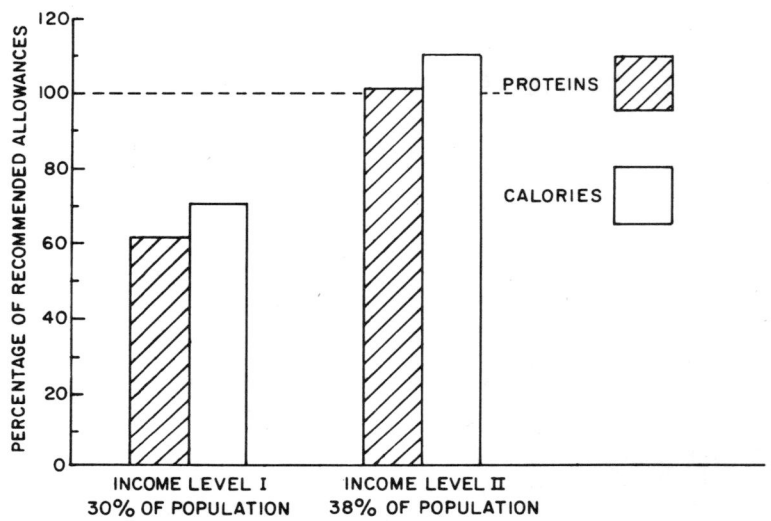

FIGURE 3.2
PERCENTAGE OF RECOMMENDED FOOD CONSUMPTION MET BY FOOD PURCHASED BY THE TWO LOWEST INCOME GROUPS OF COLOMBIAN POPULATION

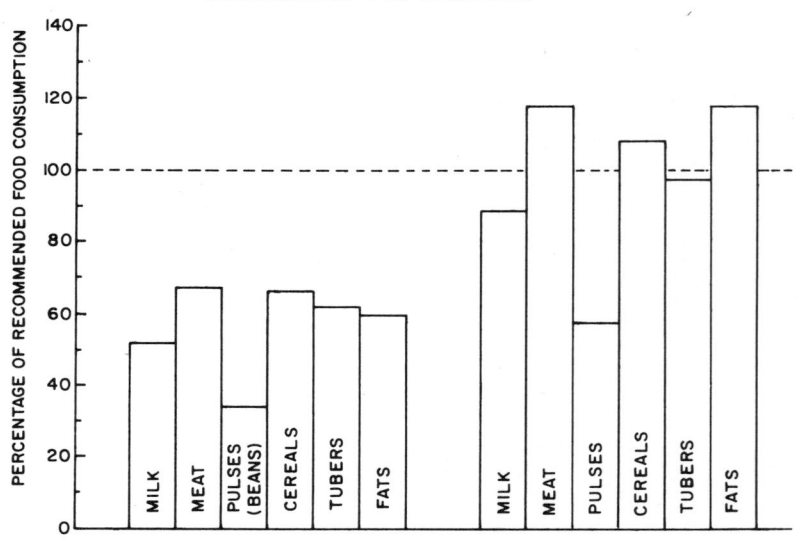

$186 per ton; however, the price had risen by August 1978 to $245 per ton. Wheat is now predicted by Dr. Teresa Salazar de Buckle, project supervisor, to reach $350 per ton by 1980. However, even at the old price of $186 per ton, a 20 percent replacement of imported wheat, using the standard 70 percent wheat, 27 percent rice flour, 3 percent soya and flour formula, taking 70 percent of the imported wheat for breadmaking would equal a balance of payments savings of $11 million.

In addition to obtaining higher protein cereals for better nutrition and an improved balance of foreign trade, another advantage of wheat substitution would be an increase in local crop production with a corresponding increase in the need for labor. The program also would guarantee the availability of raw materials to millers.

TECHNICAL SOLUTIONS

Technical solutions to the problem can be divided between alimentary pastas, the composition of which was undertaken by IIT, and breads. In pastas the composite substitutes were easily achieved and readily accepted. This was not the case in breads.

Eight formula options for pastas were studied by IIT between 1965 and 1975 (see Table 3.1). When first-grade wheat was being imported, IIT concluded that formula VII, composed of 50 percent precooked normal corn, 25 percent soya, and 25 percent semola, was superior. This pasta contained 19 percent protein and had a Protein Efficiency Ratio (PER) of 2.69 compared with semolina pasta, which contained 12 percent protein and had a PER of 0.91 (casein 2.79). With the change to second-grade wheat imports, formula VIII, composed of 15 percent soya and 85 percent semola, temporarily became the preferred product. Each of the two formulas, in turn, was adopted by the Colombian Plan for Food and Nutrition (PAN). In 1978, 750 tons of pasta were distributed to the very poor under a PAN coupon system.

TABLE 3.1
COMPOSITE FLOUR FORMULAS

Flour Formulations	I	II	III	IV	V	VI	VII	VIII
Normal corn		50%						
Precooked white corn				30%			50%	
Precooked opaque corn					30%	50%		
Cassava			50%					
Rice	50%							
Soya	25%	25%	25%				25%	15%
Semola of wheat	25%	25%	25%	70%	70%	50%	25%	85%

It had great acceptance. IIT also showed at the pilot level that substitution of corn or rice flours to levels up to 75 percent was technically feasible. One oil miller[1] currently is producing good-quality defatted soybean flour because of the pasta and composite flours research.

Regarding the precooking process using extruders,[2] tests indicate that scaling down extruders from plants producing 10,000 tons of corn per year at a capital outlay of $1.5 million is very feasible. For example, two inexpensive types of extruders producing 100 kilograms per hour have been compared. The first is one built by the Meals for Millions Foundation, (U.S.), and the second is an extruder designed and built in Colombia originally for corn snacks. The first could be used without design modification; the second requires the building of a prototype that will improve the feeding system and control temperature. Present prices are CIF (cost, insurance, and freight)--between $20,000 and $22,000.

The second solution concerns composite flours for bread. Goals for this research were to have flour produced in commercial mills and the product sold to Bogotá bakers. The first phase, 1970-1972, demonstrated the feasibility of substituting some wheat by adding rice flour, soya flour, and other local products for bread equal to regular commercial bread as judged by consumer acceptance. Rice, corn, soya, cassava, bananas, and sorghum were tested. The standard formula finally determined was composed of 70 percent wheat flour, 27 percent rice flour, 3 percent defatted soybean flour, 0.5 percent calcium steaoryl lactylate (CSL), and 50 parts per million of potassium bromate ($KBrO_3$). Rice flour, cassava starch, cassava flour, and corn starch or corn grits flour were found to be interchangeable in composite flours provided that adjustments were made in corresponding percentages of soybean flour to keep the protein at the same level as wheat flour. Owing to the importation of second-grade wheat, the composite flour recently manufactured was 80 percent wheat flour, 17 percent rice flour, 2.75 percent defatted soybean flour, 0.25 percent CSL, and 50 parts per million of $KBrO_3$. This flour was produced in thirty-ton batches (see Tables 3.2 and 3.3).

In all the composite flour formulas, CSL was added because of the lower gluten content. CSL, in turn, must be imported. Milling of wheat with broken rice was found to

---

[1] An extractor of edible oils.

[2] A process in which materials are heated while mechanically passing through a compression screw and extruded through a die or other restriction.

TABLE 3.2
NUTRITIONAL CHARACTERISTICS OF ONE I.I.T. PASTA
FORMULATION AND A STRAIGHT SEMOLA PASTA

COMPOSITION	PROTEIN (%)	CHEMICAL SCORE FAO 1973	PROTEIN GRAMS 100 CAL.	PROTEIN EFFICIENCY RATIO
Semola 25[1] Soybean 25[2] Corn 50[3]	10.9	73[4]	5.6	2.7[5]
Semola 100	10.7	39[4]	2.9	0.5[5]

[1]Wheat Semola
[2]Defatted Soybean Flour
[3]Pre-cooked Corn Flour
[4]Lysine, limiting amino acid
[5]Protein Efficiency Ratio of reference case in 2.8

TABLE 3.3
NUTRITIONAL CHARACTERISTICS OF RAW MATERIALS
USED IN COLOMBIA FOR COMPOSITE FLOURS

PRODUCT	PROTEIN (%)	CHEMICAL SCORE*	LIMITING AMINO ACID
Wheat Flour	10.9	38	Lysine
Rice Flour	7.5	66	Lysine
Defatted Soybean Flour	46.0	80	Sulfur Content
Composite Flour			
Wheat Flour 70%	11.0	52	
Rice Flour 27%	11.0	52	Lysine
Soy Flour 3%			
Wheat Flour 80%			
Rice Flour 17%	10.8	50	Lysine
Soy Flour 3%	10.8	50	Lysine

*FAO protein and calories requirements, 1973

work well (see Figure 3.3). Only the premix function must be done elsewhere.

The second phase of the technology in breadmaking is to achieve widespread production of composite flours in commercial mills and at a price that will induce bakers to purchase these products. Without public endorsement by the Colombian government and possibly subsidy to achieve the economic advantage, the program regarding this country for the time being is static. Governmental policy could change as wheat prices continue to rise and because the government admits that both importation of wheat as well as consumption of bread are increasing at the rate of 5 percent annually.

FINANCIAL AND ORGANIZATIONAL ASPECTS

The pasta development program originated in IIT with support from the Colombian government and by 1970 was being funded by the Organization of American States (OAS). This project also received funding in its latest stages through the Denver Research Institute Grants Program. Total expense in the pasta portion was estimated at two million pesos or, in terms of 1975 dollars, approximately $67,000. The entire program was managed and operated by IIT. It is now endorsed by PAN.

The composite flour program for breads was and still is more complex. In the early days IIT, encouraged by laboratory findings in 1967, requested funding assistance from the United Nations Development Program (UNDP). The plan included in Phase 1: a technological feasibility study; marketing studies; and economic studies. On the basis of recommendations made by a UNDP/Food and Agriculture Organization (FAO) mission that visited Colombia in 1967, the undertaking proceeded because of promising results achieved in feasibility studies made by the Institute for Cereals, Flour, and Bread, Organization for Applied Scientific Research (TNO), Wageningen, The Netherlands.

The project was initiated with a bilateral agreement between the Colombian and Netherlands governments in 1970, an administrative agreement, and a plan of operations. The Colombian Executive Agency (IFI) assigned the operations to IIT. The Netherlands Executive Agency, Directie Internationale Technische Hulp of the Ministry of Foreign Affairs requested TNO's Institute for Cereals, Flour, and Bread (IGMB) to undertake the technical advisory task.

A separate agreement between TNO and FAO arranged for FAO to provide marketing services related to the testing and introduction of composite flours. TNO's Bureau of International Projects (BIP) prepared an appraisal of the economic aspects. IIT contracted with Interamerican Research Ltd., Bogotá, to assist in the marketing studies.

FIGURE 3.3
FLOWSHEET OF MILLING OPERATIONS IN MAKING COMPOSITE FLOURS

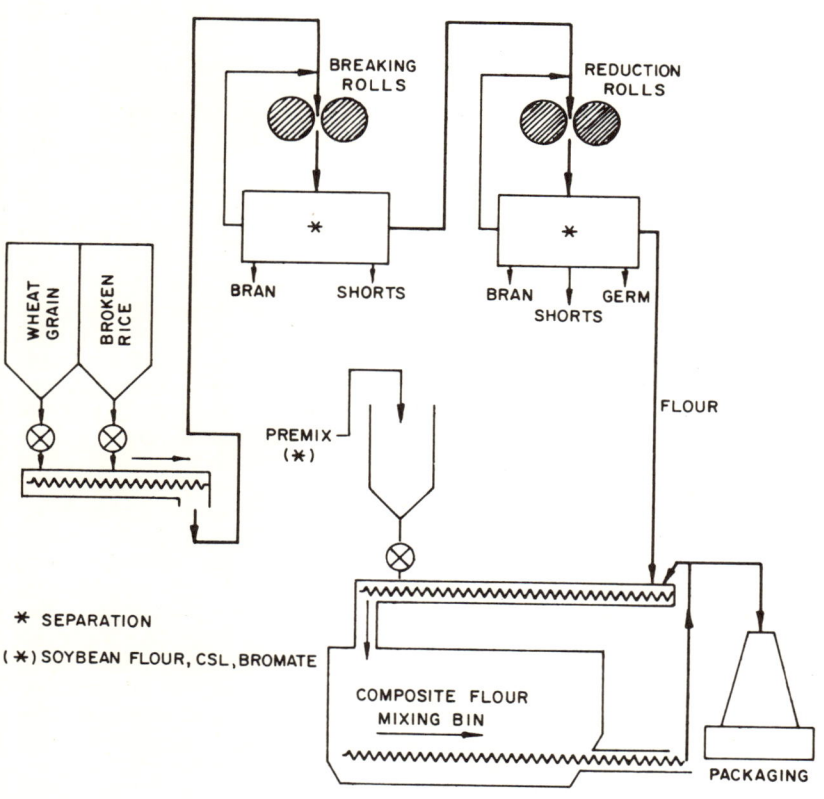

The project was conducted by a managing board, with Jorge Beltran of IIT acting as president, and was assisted by an advisory board that had as its president Dr. Norton Young, director of IIT. Between August 1970 and December 1972, the managing board met twenty times and the advisory board once. The project manager at IIT was Dr. de Buckle, chief of the program division.

In Phase I of this joint program, the Netherlands government appropriated $302,330, which included Dutch experts, raw materials, and equipment. The Colombian government appropriated $338,820, consisting of a physical plant with a bakery, Colombian equipment, and raw materials, plus technical and administrative staff.

Phase 2 under joint agreement between The Netherlands and Colombia began in June 1977 and ended in January 1979. The final report is being written by W. Vander Sluys of TNO, assigned to IIT, Bogotá. This phase involved appropriations of $350,000 by The Netherlands and $90,800 by the Colombian government. The general scope involved, as previously mentioned, the achievement of commercial production of composite flours by a mill and its sale to bakers in Bogotá.

ECONOMIC CONSIDERATIONS

Based on IDEMA projections of wheat imports at $186 per ton, although the cost is now predicted to reach $350 per ton by 1980, a 20 percent composite flour would result in savings in foreign currency of $11 million. Dr. Norton Young, former director of IIT and president of the joint agreement advisory board, states that generalized use of 20 percent composite flour would save $16 million, and 30 percent would save $25 million.

The balance of payments issue presents compelling arguments for composite flours, according to the Netherlands/Colombian joint report. The report also points out a previous Colombian government subsidy of 50 percent to millers on wheat purchases. This stopped in 1974. The report cites other determinants not easily estimated, such as price advantages to bakers regarding composite flour and public support, trade relations, and promotional activities.

Young referred in 1978 to the reluctance of the government to force the use of composite flours in breadmaking, probably because of the uncertainty about composites becoming more costly. PAN already endorses composite flours in pastas, but this is not as complex an issue since it does not involve bakers' economics. Also, the pasta utilization is clearly advantageous nutritionally.

Other criteria mentioned in the joint report but that appear to be met easily are: production and supply of local cereals; attractiveness of the product to the buyer; development of trade channels and storage and mixing facilities; and technological support, e.g., extension services to millers and bakers to guarantee adequate quality of the composite flours.

SOCIAL IMPACT

In terms of nutrition, Young reports pasta composite flour (formula VII) to be 19 percent protein compared to wheat flour, which is 10.7 percent. In breads this difference is less pronounced, namely, a composite flour protein range of 11.7 percent to 11.9 percent against 10.7 percent. Young reports the PER of pasta flour (VII) to be 2.7 versus 0.91 for wheat flour (reference casein 2.75).

The joint report reached the following conclusions: the Bogotan consumer fully accepted the bread; the bread can be made readily in all bakeries subject to some training of bakers and carefully controlled mixing; and the bread can be marketed within the existing trade structure and under prevailing commercial conditions as normal wheat bread.

Mentioned previously is the PAN program, which endorses the IIT composite flour formulas in pastas. Seven hundred and fifty tons went to the very poor under the PAN coupon system.

In terms of bakeries for bread, there are an estimated fifteen thousand in Colombia, most of which are small producers of from ten to one hundred kilograms per day. The millers initially opposed the research in Phase 1, but with the success of pastas, they now readily accept it.

RELATIONSHIP OF COMPOSITE FLOUR TO THE NATIONAL GOAL

This program: (1) provides improved nutrition, especially for the very poor; (2) already is improving the balance of payments situation, at least in the pasta area: all interviewees agreed that expansion into the breadmaking area is inevitable, although at present it is static; (3) provides expansion of local crops (for example, rice and soya), which increases employment and farm income; (4) would guarantee availability of raw materials to the millers of flour; and (5) calls for an absolute minimum of capital outlay, such as small extruders for rural production to meet the needs of the poor people.

REFERENCES

"Interpan--Joint Report of the Colombian-Netherlands Composite Flour Project" carried out in Bogotá 1971-72. This report issued by: IIT, Avenida 30 No. 52-A-77,

Bogotá; Institute for Cereals, Flour and Bread, TNO, P.O. Box 15, Wageningen, The Netherlands; Bureau of International Programs, TNO, P.O. Box 778, The Hague, The Netherlands; FAO, United Nation Food Policy and Nutrition Division, Rome, Italy.

Young, Norton. "The Composite Flour Program for Bread and Pasta Making in Colombia." Presented in Winnipeg, Canada, August 1978.

de Buckle, Teresa, and Carlos Pardo. Annotations to "The Composite Flour Programs in Colombia." IIT (Spanish).

de Buckle, Teresa et al. "Production of Pre-Cooked Corn Flours in Small-Scale Extruders." Presented at the Second International Low-Cost Extrusion Cooker Workshop, Dar es Salaam, Tanzania, 15-18 June 1979.

IIT Technical Report No. 98, November/December 1975.

PERSONS INTERVIEWED

Dr. Teresa Salazar de Buckle, head of programs, IIT, and project supervisor of composite flour program.

Jorge Beltran, head of information services, IIT.

Dr. Norton Young, former director, IIT, and president of joint program advisory board.

Dr. W. Van der Sluys, chemical engineer, IIT; assigned by TNO to Phase 2.

Luis O. Giraldo, Havenera del Valle (flour mill), Cali, Colombia.

QUESTIONS TO CONSIDER

1. Very active foreign assistance is an element in this case; what is felt to be the necessity for this relationship in most circumstances? Is financial and technical support on a high level desirable or necessary? Should there be more international cooperative programs of this nature? What about creation of more research institutions such as IIT; are they viable and cost effective?

2. The government's policy with regard to imports and pricing had a great effect in this instance; is it your

experience that nontechnical factors usually determine whether a technology will be implemented, or is the technology itself (in terms of its effectiveness and efficiency) the principal determinant?

3. Composite flour production is the subject of worldwide study and cooperation through both formal and information networks; is further development of technology utilization networks desirable? Is much to be gained through significantly greater flows of information on appropriate technology? If so, how should this be organized and financed?

4. With the importance of improved nutrition for the poor, why doesn't the government more aggressively pursue the promulgation of IIT's work with composite flour for breadmaking?

4
Haiti:
The Coffee Roads

Laurie Nogg Adler

INTRODUCTION

The design and construction of 110 kilometers of new penetration roads is facilitating the production of coffee in previously barren mountain areas in Haiti. As a result, the Small Farmer Improvement Project is showing signs of increased coffee production, which is affecting not only the small farmer, but Haiti as a whole. The roads also are allowing for increased maneuverability and are opening up communication channels that previously were not available to some Haitian villages.

The roads are unique in that they are being built entirely by hand; no mechanized equipment is utilized on the project, only hand tools. The coffee roads are in great part the responsibility of Chuck Pettis, an engineer under contract to the Agency for International Development (AID). Pettis is convinced that hand-constructed roads fare better than machine-built roads if they are designed carefully and built with close supervision. The project uses indigenous labor and materials and contains a training component, insuring that technicians capable of designing additional projects will remain in Haiti.

Pettis agrees that roads built totally by hand are not always the most cost-effective method of road construction. But in this particular case, with budget constraints, in an area difficult to reach by mechanized equipment, and with an abundant labor supply, the costs ($6,000 per kilometer) are very much lower in line and actually are competitive with other Haitian road projects. Although the roads already have opened access to otherwise remote villages, none of the roads are completely paved; therefore, a thorough evaluation has not been possible.

SETTING

Haiti is one of the most densely populated countries in the world. Sharing the island of Hispaniola with the

Dominican Republic, the 11,000 square miles that comprise Haiti are mostly mountainous; less than 20 percent of the land lies below 600 feet, while 40 percent is at elevations in excess of 1,500 feet. Approximately five million people live in Haiti, creating a population density of about 480 persons per square mile of total land and 1,500 persons per square mile of arable land. Close to two-thirds of Haiti's land is inaccessible and has little potential for cultivation. Over the last fifty years, the quality of agricultural land has deteriorated from erosion caused by an increase in deforestation (charcoal is the most widely used energy source) and poor land use.

Overpopulation, primitive farming techniques in rugged terrain, and climatic disasters (drought and hurricanes) all have contributed to Haiti's inability to provide sufficient food for the health needs of its people.

Yet the economy of Haiti is basically agricultural, with this sector accounting for about 50 percent of the Gross Domestic Product. Approximately three-fourths of the labor force (which comprises 50 percent of the population) is involved in agriculture, utilizing only one-third of the land deemed arable. The economy is somewhat diversified, with the industrial sector (particularly manufacturing and construction), and tourism displaying signs of growth. However, as industry increases, it produces a decline in the agricultural sector. Also, tourism cannot be expected to grow much larger until the supporting infrastructure (roads, public transportation, telecommunications, water supplies) is improved.

With per capita income at about $150 per year and average rural income at $80, Haiti is the only country in the Western Hemisphere included in the United Nations listing of least-developed countries in the world.[1]

Agricultural goods produced include bananas, cocoa, coffee, corn, rice, and sugarcane. Exports consist mainly of coffee, cocoa, sugar, light manufactured goods, and bauxite.

Coffee is the most important commodity to the economy and is considered essential. Its production and export (as of 1976) involved nearly 400,000 farm units and approximately two of every five persons in Haiti. Coffee production provides close to $20 million in foreign exchange annually and additional income through the taxation of a variety of coffee-related activities.[2]

---

[1] JWK International Corporation, Agricultural Policy Studies in Haiti: Coffee (Damien, Haiti: DARNDR, 1976), p. 21.

[2] Ibid., p. 2.

Most coffee in Haiti is produced on small farms throughout the country. The most profitable areas are Cap Haitian, Jeremie, and Les Cayes. Other regions are considered to have potential but are inaccessible due to lack of roads and subsequent market opportunities. In 1976 more than one-half million sixty-kilogram bags of coffee were produced, of which 57 percent was exported. This represented a decline in production over previous years.[3] Coffee obviously is important for the growth of the economy of Haiti, and this has led to recent inquiry and analysis concerning the stability of the crop.

DEVELOPMENT OF THE COFFEE ROADS PROJECT

Background

A 1976 Coffee Policy Study, sponsored by the U.S. Agency for International Development (USAID) and the government of Haiti (GOH), determined that the coffee sector in Haiti was in a crisis situation.[4] Coffee production and exportation had fallen below potential levels, and, in fact, export quotas were far less than those suggested by the International Coffee Organization.[5] The Coffee Policy Study determined that government policy had played a critical role in the reduced cultivation and production of coffee. The coffee sector in Haiti was deemed to be inefficient due to excessive taxation, lack of technical and financial assistance, and underdevelopment of infrastructure--specifically, roads.[6] It was hoped that the recommendations made by the study would increase incomes of small coffee growers, improve coffee productivity and output, and increase foreign exchange earnings from coffee exports.[7]

The major organization affiliated with the production and marketing of coffee in Haiti is the Institute for the Promotion of Coffee and Export Products (IHPCADE). IHPCADE's mandate is to promote the increase of coffee production and assist in the standardization, industrialization, marketing, and control of coffee and other export products.

---

[3]Ibid., p. 7.

[4]Ibid., p. 5.

[5]Ibid., p. 1.

[6]Ibid., p. 2.

[7]Ibid.

Since the early 1970s, the director general of IHPCADE, M. Bertin Dadaille, has encouraged a government-sponsored coffee intensification program. His ideas to increase the quality and quantity of coffee were met with favor by President Duvalier and the Department de l'Agriculture et des Resources Naturelles du Développement (DARNDR), which is the parent organization of IHPCADE. However, lack of funds prohibited the realization of comprehensive intensification programs. Small programs geared toward increased coffee production were implemented, but without full political and financial support from the government, these programs had little effect.

In 1973 Leroy Rasmussen, then agricultural development officer of AID/Haiti, began assisting IHPCADE in designing a program to increase the efficiency of coffee production. By May 1974 AID/Washington had approved the Haiti Small Farmer Improvement Loan, which sought to increase the income and standard of living of the Haitian small farmer through greater production of improved coffee. Such increases were expected to result in significant improvements in the balance of payments and revenue accounts for the GOH. Specifically, the goals of the program were to provide technical assistance to coffee farmers through improved technology, particularly by increasing the use of fertilizer, establishing credit and cooperative facilities, and opening marketing channels by creating coffee centers and developing penetration roads. As a semi-autonomous organization, IHPCADE was to provide technical assistance to the participating coffee farms through regional coffee centers, while the Bureau de Credit Agricole (BCA) was to provide supporting credit.

Economic Aspects

The financial arrangements for the loan were such that USAID provided funds for the design and implementation of eight coffee centers as well as for the procurement and distribution of fertilizer, while DARNDR, specifically IHPCADE, was responsible for road development. However, AID agreed to pay for the services of a U.S. engineer, Chuck Pettis, to assist in the engineering and supervising of the construction of the coffee centers and coffee roads. The AID loan was for $6 million over a five-year period; $650,000 was designated by the GOH specifically for the road-building project.

From the beginning of the project, the coffee roads have been plagued with insufficient monies. Road funds come directly from DARNDR, but projects that are more immediately linked to agricultural output receive priority. It has been suggested by various AID officials that the coffee road project might receive more attention if it came under the Secretariat of Public Works, Transport, and Communications (TPTC), which has more experience with implementing construction and which might better administer funds. Others, however, feel that TPTC already has its hands full and that a small project such as coffee roads might be overshadowed here as well. In any case, because the road project is related to coffee production, it continues to come under the auspices of IHPCADE and must suffer from DARNDR's delinquent administration of funds.

The Coffee Centers

Agronomists from IHPCADE selected sites for coffee centers that had a proper mixture of good soil conditions, adequate rainfall, and favorable climate and that would serve a large population of coffee farmers. Due to the severe deforestation and subsequent erosion in Haiti, many areas that in the past might have been adequate for coffee production were ruled out because trees, which provided necessary shade for the coffee plants, were lacking.

After many months of research by agronomists, eight coffee centers were selected (see Table 4.1 and Figure 4.1). They were designed to be extension centers where small farmers could use credit to purchase fertilizer and receive coffee seedlings. The centers also were to conduct courses in simple coffee-growing technologies, contain nurseries where coffee seeds from other countries would be nutured, and provide a forum for the farmers to exchange informal communications among themselves. The centers were to allow the farmer to be better informed of the market conditions affecting coffee prices. In fact, Haitian farmers are now well aware of the fair market value of their product.

TABLE 4.1

Coffee centers	Roads	Kms
Fond des Negres	Existing road	--
Macary	Perido-Macary	22
Baptiste	Baptiste-Savanette	21
Dondon	Marmalade-Dondon	15
Virgile	L'Asile-Virgile	17
Pilate	Existing road	--
Changieax	Not determined yet	--
Beaumont*	Beaumont-Laurant	35
		110 km

*not yet completed

FIGURE 4.1
MAP OF COFFEE CENTERS, COFFEE ROADS, AND EXISTING ROAD NETWORK IN HAITI

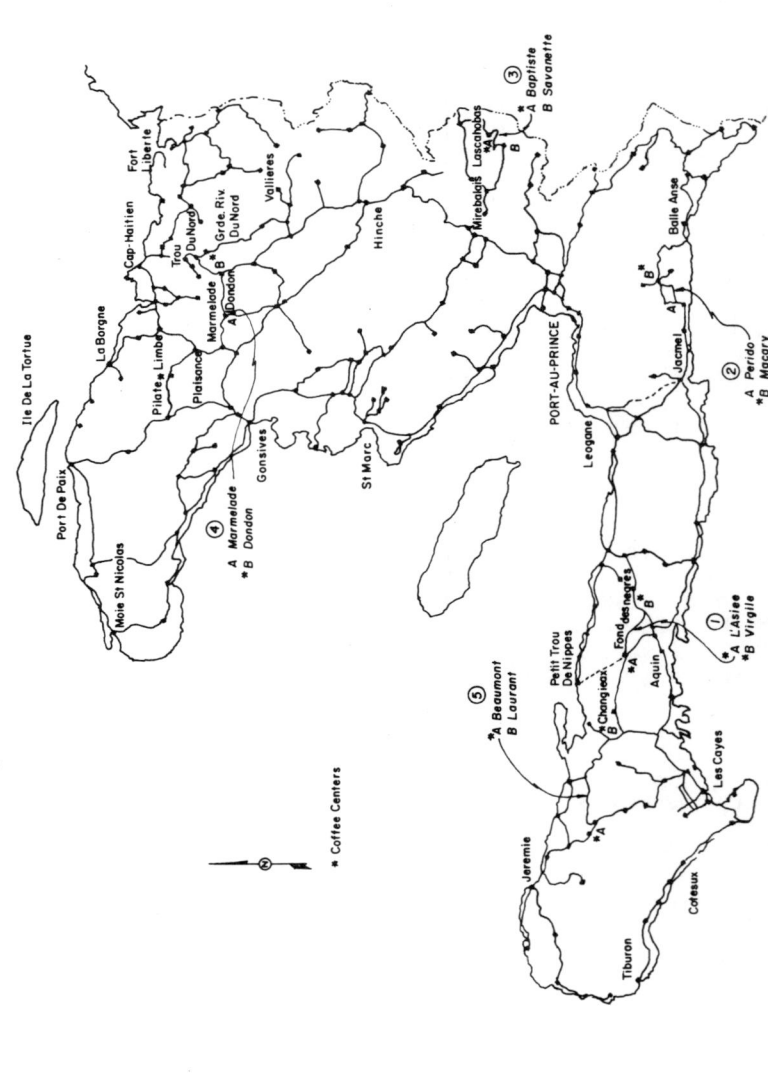

The Coffee Roads

Of particular importance to the success of the coffee project was the design and construction of roads that would open up access to areas previously impossible to reach by vehicle. Coffee grows best in the highlands, but many potential coffee-growing areas had heretofore been ignored due to lack of access by vehicle (and sometimes even by animals). Some of these mountainous areas were so rugged that they could be reached only by foot.

The coffee penetration roads designed to create access to the coffee centers are unique in Haiti because they were built entirely by hand; manpower is the only energy source utilized in their construction. The choice of manual labor was not accidental but, in fact, suggested in the loan agreement. Chuck Pettis lists several reasons to support the use of manual roadbuilding for this project: (1) funding was such that capital intensity, or even a mixture of capital and labor intensity, was prohibitive; (2) there was an abundance of underemployed workers in Haiti; (3) a high skill level was not required to hand-build roads if engineering and supervisory skills were present; (4) some of the mountain areas would not have been possible to reach by machinery at the onset of the project; and (5) the intensive use of labor often creates a community spirit if villagers actually are involved in a development project and can feel responsible for the creation of a product(s). Pettis concurred with the notion that labor-intensive technologies make better use of developing countries' resources, produce higher rates of economic growth, increase employment opportunities, and lead to a more equal distribution of income.[8] Also, Pettis felt that by utilizing labor-intensive methods he actually would create a better road. Were he to utilize equipment on the coffee roads, the $650,000 allotted to the project would cover only the procurement of equipment to satisfy the design of dirt roads. By using manual labor, Pettis could cut costs sufficiently to have paved roads as a finished product.

However, the coffee roads have come under some criticism, partially because they have been constructed only by manual labor. Barney Mosley, an AID engineer and head of AID-funded road projects, feels that the use of manpower is causing major delays since not one road is yet totally completed. Mosley feels that because Pettis still is billing his time to the project, the roads are costing more than just

---

[8]Patrick, et al., <u>Agri-Mechanical Technologies in Latin America. A Survey of Applications in Selected Countries</u> (Washington, D.C.: Inter-American Development Bank, 1978), p. 8.

workers' labor. Pettis, on the other hand, argues that the coffee roads will near completion by September 1979, well within the five-year contract period, and will be completed at a cost of $6,000 per kilometer, which is also within budgeting constraints. Pettis claims further that if red tape had not held up payment for the workers and for the development of the coffee centers in general, the roads would have been completed by now. "If the purpose of the project had been to pave roads as fast as possible, then the project would have to be considered a failure." But Pettis states that the main purpose of the program is to open access to coffee areas, and the temporary roads already have done this. In fact, coffee from these previously cutoff areas can now be marketed.

Another criticism of the coffee roads is that the men must carry rocks great distances in mountainous terrain. Wheelbarrows are sometimes used, but they are at a minimum and are still difficult to pull uphill. Even Pettis agrees that carrying rocks by hand is not an efficient use of time or labor. If he could afford a truck for hauling, a small compressor, and a pavement breaker, Pettis says that he would utilize them.

In contrast to the coffee roads project, the AID-funded Agricultural Feeder Road program, which was created to pave existing roads and increase accessibility of markets to

Women cultivating seedlings in a coffee nursery

farmers, contained a pilot project with a budget of 70 percent for labor, 30 percent capital-funded. This project is expected to cost at least $12,000 per kilometer (estimates were that portions of the road work would run between $2,000 to $7,750 per kilometer) and will complete only twelve kilometers of the original sixty kilometers suggested in the contract. The major portion of the Feeder Road project implemented by TPTC is not designed to employ labor-intensive technology. Mike Rawson, who leads the Tippets-Abbett-McCarthy-Stratton, Inc. (TAMS) consulting team contracted to engineer the feeder roads, also claims that red tape has caused many of the problems in constructing the roads specific to the pilot project. His complaint is that TAMS never claimed to be experts in labor-intensive operations, and they are learning about the problems inherent in a labor/capital mix. He wonders whether the project is "designed to create employment or build roads." Mosley concurs that management is lacking on all Haitian road projects, with inefficiency being the result.

Road supervisor taking measurements; instruments used are tape measures and Abney level

A third road project in Haiti, also funded through AID, is purely a road maintenance job and is run by Frederic R. Harris, Inc., consulting engineers, in conjunction with Haiti's permanent National Highway Maintenance Service (SEPRRN). No new construction methods are used for the SEPRRN project, and it is characterized by the use of maintenance equipment and a minimal level of labor.

However, those interviewed felt that the coffee roads could not fairly be compared to the other road projects in Haiti because (1) the coffee roads are totally hand built; (2) other road projects deal with resurfacing or maintenance of existing roads, not the penetration of new ones; and (3) the coffee roads project is so much smaller that its very nature prohibits comparison. It should be added that the coffee roads project has not received much attention from AID or other road engineers in Haiti. In fact, very few persons interviewed, except those who are intimately involved in the project, have visited the coffee centers or roads.

The importance of road building and maintenance in Haiti cannot be overemphasized. Existing roads in Haiti are at present inadequate. Some of the major highways are closed part of the year due to poor condition caused by heavy rainfall and subsequent erosion. Only recently was a paved highway completed between Port-au-Prince and Jacmel, which are only about sixty miles apart. The North Highway, constructed by a French firm and funded by the World Bank, is considered hazardous at times, and a new, partially completed highway, constructed by a Canadian firm and funded by the Interamerican Development Bank, to Les Cayes after six months is already in need of repair. The insufficient road situation in Haiti is a major deterrent to the economic and social growth of the country and is therefore in the limelight of development assistance. Louis A. Morales, who leads the Small Farmer Improvement Loan at AID, says, "Roads are the best investment you can make in an LDC."

The general consensus among the various engineers and development planners in Haiti is that a careful mix of both labor and machinery is necessary to build good roads. Their thinking is based on the premise that the large number of unemployed persons in Haiti makes the country a likely candidate for labor intensity, while the demands of serviceable roads necessitate workmanship that must include the use of machinery.

Pettis is certain that his roads will be as good as any that have incorporated capital equipment. He states that the TAMS pilot feeder roads already have required maintenance. On the other hand, the "Pettis roads," as they are often called, have not yet been put to a severe test. Although vehicles have penetrated areas that were previously impassable, they have traveled only on temporary roads. Tibor

Nagy, chief engineer, AID/Haiti, says that the coffee roads are fine tertiary roads but that they cannot be compared to or utilized as secondary (feeder) roads. Pettis disputes this claim as he feels the coffee roads are of superior construction and states that roads like his were built by the Romans, and such roads have withstood the test of time. Morales adds, "It would be a crime to do anything but utilize manpower on this project."

Pettis has worked on other labor-intensive projects and is considered by most sources interviewed to be a fine engineer and supervisor. Recently, Panamanian government officials visited the coffee roads and were extremely impressed with the concept and design. A similar contingent from Santo Domingo has heard about Pettis's roads and is scheduled to visit with the hope of planning a similar labor-intensive endeavor.

TECHNICAL SOLUTION TO THE PROBLEM

The Small Farmer Assistance Loan Agreement was signed in 1974. During the following year, a study was initiated to establish the locations of the coffee centers; in 1976 a series of road studies was begun. Pettis personally walked each site three to eight times to determine the best routes for the roads. It was not until 1977, nearly three years after the agreement was made, that actual construction of the coffee roads began. Part of the delay was due to the difficulty in obtaining land clearance for the coffee centers; the roads could not be constructed until such time as coffee centers were determined.

The process by which the coffee roads were built is a relatively simple one and uses fundamental techniques of road technology. The following description of hand-constructed roads deals specifically with mountain roads, as there are few similarities between mountain and level-road construction.

Alignment

Once it was determined by IHPCADE officials and area agronomists where a road was required, technical personnel began to lay out the road alignment. Technicians who traced out roads were taught how to use the Abney level. The maximum road grade was generally established at 6 percent (corresponding to about 3 degrees on the Abney level). Depending on the density of vegetation, a crew of five to ten local laborers was hired to assist the technician. Following the natural movement of the terrain as much as possible, the workers cleared a trail so that Abney level readings were consistent; where the road contour deviated, a wooden stake was driven into the ground. Use of existing roads was encouraged if they followed the general direction and the grade of the proposed road.

Trace

Once the alignment had been staked out, another crew moved in with various hand tools--machetes, matlocks, picks, and shovels. A one-meter trail was constructed to facilitate visual observation of the geological conditions along the proposed route.

Geological Observation

A technician trained to determine soil types and geology was necessary for this phase of construction. A variety of geological conditions were noted: large rock formations; type of rock; rock's degree of durability; types of soil; plasticity of soil; ravines and drainage patterns; and slope steepness. These visual observations are extremely important because they affect the outcome of road construction. In part, the design of the road was based on the context of materials found on site, i.e., rocks are used for pavement and soil is utilized as part of the cement mixture. In addition, a strong familiarity with the contour of the land was necessary if alignment must resort to switchbacks. In such a case, a flat curve must be constructed, with the radius of the curve sufficient to enable the passage of large trucks without danger or difficulty.

Once the visual reconnaissance had been accomplished and changes had been made in the alignment to compensate for the adverse terrain, geology, soil, and drainage conditions, road construction commenced.

Temporary Road

To open access, a road three meters wide was constructed to allow for one-way truck traffic. The first step consisted of clearing and grubbing a six-meter-wide area along the trace by a twenty-person team (nineteen workers and one supervisor). Next, a team of men with shovels and picks concentrated on earth excavation and removal of soft, rocky material. This team did no heavy removal but set the pace for subsequent teams. Two rock-breaking teams with sledgehammers, rock chisels, and crowbars followed. When the workers encountered a large boulder that could not be dislodged, they heated the rock to a high temperature, then doused it with cold water so that the sudden cooling would cause it to crack.

Laborers discussing need for new tools with Pettis

Road construction showing Telford base and macadam

Next, two teams prepared the roadbed. This involved leveling the road and constructing dry-laid rock walls when necessary. These walls were of sound construction, generally about four feet thick. Height varied according to requirements. Constructed on a flat platform but contoured to existing natural slopes, the rock wall consisted of large rock and granular fill. These were constructed where unstable soils might result in slides.

Widening

The temporary roads were then widened to five meters (six meters for a side hill cut to establish drainage ditches) and staked to insure that required width was maintained. Grade was brought to its necessary level, and again, dry walls were constructed when needed. All rocks, the size of cobbles or smaller, that were uncovered during the widening operation were stored on the side of the roadbed to be used for the Telford course (sub-base). Curves, especially at the switchbacks, were opened as much as possible.

Telford Course (Sub-Base)

To insure the best stable grade and level surface, the Telford course was constructed using rock; earth fills as sub-bases were avoided due to lack of adequate compaction equipment.

The existing subgrade was covered with a six- to twelve-inch Telford course, depending on the subgrade of soil. Using a line level (string), the technician set the grade on the side stakes with two rows of edging stones constructed five meters apart. If edging stones were to form a curb, the tops were set three inches above the finished surface. All curves were super-elevated to insure proper drainage into gutters. The road was completed one section at a time after the Telford base was laid down. Voids between cobble-sized rock were filled with rock chips and a mixed granular and plastic material to insure an impervious and stable bond.

Surfacing

The final phase in building the road was the construction of the wearing surface, which consisted of three-inch-thick stabilized macadam. Gravel-size rock (saved from the widening operation) that fell into a size classification of two inches to one-half inch was used for the macadam. A section of chain link fence mounted in a wooden frame was utilized as a screen to remove oversize material.

Hand compacter designed for coffee roads

The road crown was set by the technician using the line level. The two meters of road in the center were set four inches higher than the edges to not only allow for drainage but also for the concentration of traffic in that part of the road.

Insuring compaction of the macadam was difficult. A hand roller consisting of a steel drum with four interior baffles was designed to improve on compaction with hand tampers. Large, rounded rocks were placed within the steel drum; as the drum was rolled, the baffles caused the rock to tumble, introducing the necessary vibrations that would settle the coarse macadam rock. The hand roller was designed by Pettis and his technicians and was used on the TAMS project as well.

In order to seal the course macadam rock on the road, an indigenous mixture of lime (produced from the rock-burning process) and bagasse (sugarcane ash) was created to form a Roman cement. The local farmers use a similar type of Roman cement to stablilize their rock houses. By experimenting with different blends of material from the area such as caliche, satisfactory sealers and binders were

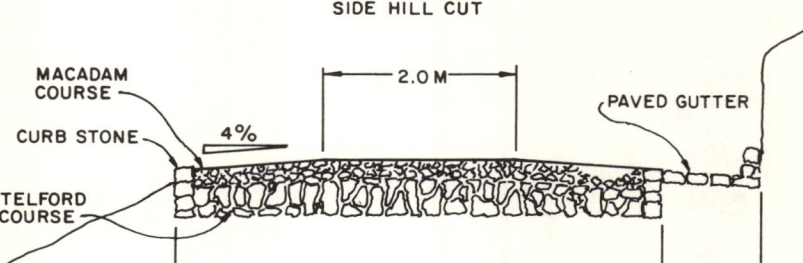

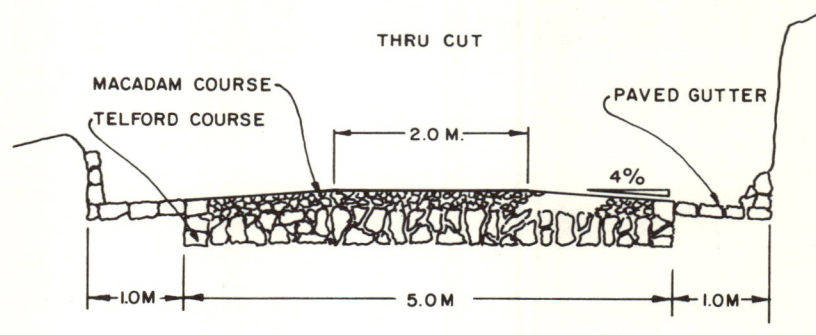

FIGURE 4.2
TYPICAL ROAD SECTIONS

produced. The variety of soils utilized includes lean silt (ML material), sandy silt (SM material), and lean clay (CL material). Six test sections were constructed utilizing the sealer, and after six months the macadam surface has not deteriorated, even after heavy rainfall. However, constant heavy traffic has not passed over the final surface, so the actual durability of the road remains to be determined.

Drainage

The last step in the road design was the construction of drainage structures. When the natural soil consisted of material classified as ML or SM, the drainage gutters were masonry-lined to prevent excessive erosion. Also, curbs were constructed to prevent water from flowing onto the embankment slopes where erosion might develop. As there are no funds for culvert pipe, water was conveyed across the roadway by means of masonry dips. This method also reduces the necessity of maintenance of culverts.

Construction

The construction of each of the five coffee roads followed a similar pattern. Each road project was headed by a technical supervisor who was in charge of between ten to twenty teams, depending on the level of need. Teams of twenty men--nineteen workers and one "chef d'equip" (or team leader)--performed specific tasks. Roads were sectioned off so that a portion was worked on by each team. When a road section was completed (i.e., all teams met), a new section was begun.

The technical supervisors, who have at least a secondary education and are literate, were trained by either Pettis or his Haitian counterpart, Pierre Milord. The supervisors made the design and technical decisions regarding the roads and referred to the senior engineers (Pettis and Milord) for consultation. In this manner, technicians were trained so that they might have responsibility and make decisions themselves. Suggestions were offered by the senior engineers, but one of the purposes of the project was to create a nucleus of men who could take over when Pettis or Milord must leave the project.

Maintenance

It is planned that maintenance of coffee roads will be organized and paid for by IHPCADE. Agronomists from the coffee centers will periodically check the roads to see if they are in need of repair. It is expected that after the initial road work is completed workers interested in becoming part of the maintenance crew will report their availability to village community councils. IHPCADE will then work through the community councils to set up maintenance crews as needed.

SOCIAL AND ECONOMIC IMPACT

Prior to working on the coffee roads, most of the workers were subsistence farmers. To be employed on the coffee roads, they needed to know how to utilize basic hand tools; this was the only skill required. Many of these farmers had never even received wages.

Approximately 500 workers are employed by this project. Technicians receive an average salary of $375 per month. The laborers' wage rate on the coffee roads is $1.30 per day. The minimum wage presently is $1.60, but at the time the original contract was signed, it was $1.30. Pettis hopes to raise the daily rate to the new minimum when a new contract begins in September, but he remembers how difficult it was to obtain the original $1.30.

As is often the case in LDC labor-intensive programs, there was concern about giving workers who had never received wages a minimum payment. It was feared that money would inflate their already strained economy and prove inequitable to those in the community who received no wages. However, food-for-work programs, which substitute food for labor (Church World Services pays at the rate of eighty cents worth of food per day), were thought by many to be a demeaning manner in which to create employment.

Pettis says that he fought with AID to secure the minimum wage. Now that it is accepted (and actually is higher), it has started a precedent: workers on the TAMS project are paid $1.60 per day. Private enterprise in Haiti was unhappy because it was forced to be competitive and pay minimum wage, when it was accustomed to paying only half that amount.

Through a Creole-speaking interpreter, workers on the coffee roads were interviewed at the L'Asile-Virgile construction site. It was discovered that wage payment is always late, sometimes as much as two months. Although hesitant to complain to Pettis about not receiving money (he is instead consulted about the need for new tools), the workers were eager to complain to a newcomer. Apparently the coffee roads project is not the only one in which workers' pay is delinquent due to bureaucratic red tape. Rawson of TAMS complained that his workers on the feeder roads also fall close to two months behind in receiving their wages. To a man who has little or no income, this can be disastrous. However, the workers continue to help each other and work together on the roads, often singing Haitian songs.

The workers believe in the coffee roads project. In fact, in a recent impact evaluation of the Small Farmer Improvement Project, roads were ranked number one in priority in rural projects by farmers.[9] The road provides access that they previously never experienced. Getting the coffee to market is only one of the advantages of the road; equally important is the ability to get sick persons to

---

[9]Samuel R. Daines, Impact Evaluation of the Haiti Small Farmer Improvement Project (Port-au-Prince, Haiti: U.S. Agency for International Development, 1979), p. vii.

hospitals faster. Until recently, someone needing medical attention would have to be carried as far as twenty kilometers to reach a road; the hospital might then have been another sixty kilometers away. Direct access to hospitals by vehicle is now possible because of the coffee roads.

Schools, churches, and new homes have sprung up in villages since temporary roads have brought vehicles into the communities. Women, who were the traditional means of transportation of goods to market (and who represent 50 percent of the labor force in Haiti), are now free to perform other duties. One example is a woman who, now having more time to sew, saved enough money to buy another sewing machine and has started a business. Other women, however, will be displaced by the increase in vehicular traffic unless the stimulus to the community creates new opportunities.

Pettis and other engineers interviewed feel that the roads have had an immediate impact. They have created both primary (road-building) and secondary (coffee farming) employment opportunities for nearly fifty villages. Population in villages is difficult to estimate, but approximately 3,000 to 4,000 persons live in each. Therefore, close to 200,000 people have been affected by the roads. The villagers have been described as better clothed and better fed (eating more meat) since they have acquired easy access to the outside world. They have purchased land, fixed up existing homes, and bought cows and chickens. In effect, Pettis claims, "we are creating a low middle class in Haiti."

Another impact resulting from the development of the penetration roads is the growth of community development organizations. As villagers' access to the marketplace and consequently to the outside world has increased, coordination of community activities also has increased. Animators, or persons charged with motivating village action, are selected. The animators coordinate with government and/or development groups to make certain that growth will be advantageous to the community. It is through the community centers that maintenance of coffee roads will be organized with IHPCADE. However, many community center leaders have been accused of dishonesty and skimming community funds; therefore, relationships with them have not always proved successful.

Mr. Tamari, the AID liaison to Haiti community centers, feels that development due to road access has negative effects as well as positive ones. He claims that land prices have increased substantially; people who were satisfied with simple things now want more. Oftentimes these increased desires lead to frustration and resulting high crime rates (Haiti's crime rate is currently quite low).

The coffee centers are responsible for the increase in revenues that previously were absent from the communities and from Haiti as a whole. Farmers are now receiving ninety cents per pound for their cash crop. The Haiti Small Farmer Improvement Project has achieved substantial increases in the incomes of approximately 70 percent of participating farmers, and it is the poorest farms that are achieving the greatest improvement.[10] Two-thirds of the production impact is attributable to increases in coffee production.[11] Therefore, revenues are generated for the GOH with increased exports of coffee. Monsieur Dadaille of IHPCADE claims that Haiti must rely on coffee exports for at least ten years until the economy diversifies.

The introduction of coffee centers also has improved the soil conditions in Haiti, where continued deforestation previously had caused severe erosion. (The barren wastelands might well become green again.) Because of the roads, the cost of transporting fertilizer has dropped from two to three dollars a bag to one dollar.

The coffee centers also will increase revenues for the private coffee exporters. Some critics claim, in fact, that the coffee program will only make the rich richer and that the exporter should contribute to and participate in the program.

The coffee centers and penetration roads have changed the social and economic status of the Haitian coffee farmer. The far-reaching effects of this program will have to be analyzed over time.

REFERENCES

Allal, M., and Edmonds, G.A. Manual on the Planning of Labor-Intensive Road Construction. Geneva: ILO, 1977.

Daines, Samuel R. Impact Evaluation of the Haiti Small Farmer Improvement Project. Washington, D.C.: Practical Concepts, Inc., 1979.

Hoskins-Western-Sonderegger, Inc. Final Report, Labor Intensive Road Program Small Farmer Development Project--Haiti. Lincoln, Nebraska: HWS, Inc, 1978.

---

[10]Ibid., p. v.

[11]Ibid.

JWK International Corporation. *Agricultural Policy Studies in Haiti: Coffee*. Damien, Haiti: Department of Agriculture, Natural Resources and Rural Development (DARNDR), 1976.

Patrick, Michael J. et al. *Agró-Mechanical Technologies in Latin America: A Survey of Applications in Selected Countries*. Washington, D.C.: Inter-American Development Bank, 1978.

Sud, I.K. et al. *World Bank Scope for the Substitution of Labor and Equipment in Civil Construction, A Progress Report*. Washington, D.C.: World Bank, 1976.

Thomas, John Woodward, and Hook, Richard M. *Creating Rural Employment: A Manual for Organizing Rural Works Programs*. Cambridge: Harvard Institute for International Development, 1977.

Transportation Department, World Bank. *Study of Labor and Capital Substitution in Civil Construction*. Washington, D.C.: World Bank, 1978.

PERSONS INTERVIEWED

Steve Collins, Church World Services

M. Bertin Dadaille, director general, IHPCADE

Louis A. Morales, project officer, AID Small Farmer Improvement

Barney Mosley, acting chief engineer, AID/Haiti

Tibor Nagy, chief engineer, AID/Haiti

Chuck Pettis, chief engineer, Coffee Roads Project

Georgeanne Potter, anthropologist, AID Feeder Roads Project

Ernie Paultre, engineer, AID

James Purcell, program officer, AID/Haiti

Mike Rawson, TAMS consulting engineers, AID Agricultural Feeder Roads

Elias Tamari, AID consultant to community development organizations

Ben Watkins, Frederic R. Harris, Inc., consulting engineers, AID Road Maintenance Project

QUESTIONS TO CONSIDER

1. Given that all concerned (including the project's chief proponent) feel that the roads could have been built more "efficiently," does this case constitute an example of appropriate technology?

2. For example, is the use of heavy manual labor in a hot and humid tropic environment a humane practice, when (if funds had been made available) the strenuous work could have been accomplished mechanically? Is this an instance of "labor intensity" as it is thought of in connection with "appropriate technology?"

3. What is your view of the lasting effect of the temporary creation of a high-scaled labor market? Does this introduce distortions permanently in the economy and social structure?

4. What can one conclude based on the disagreement within the aid agency concerning the appropriateness of the technology that was used?

5. Do you agree with Pettis' view of the benefits of labor intensity in a developing economy?

6. What is the probability that the technology thus demonstrated in Haiti will continue to be utilized? Does the success of this project depend upon the inspiration and drive of one man, or has there been a permanent introduction of a new approach?

5
Pakistan:
Ball-Point Pen Manufacturing—
A Case of Technology Transfer

James W. D. Frasché

INTRODUCTION

    Pakistan became independent in 1947 and had little indigenous capacity in science and technology, because most of the research institutions, universities, and industries were located in areas that constituted India. The Islamic Republic of Pakistan followed a pattern of development, until recently, that focused exclusively on economic aims and on maximizing the growth of the Gross National Product (GNP), resulting in high discrimination against poorer people and backward regions and leading to great income disparities and maldistribution of wealth. Currently, Pakistan has announced new priorities, and policies are now being pursued that will concentrate on: (1) rural infrastructure development; (2) provision of basic needs; (3) removal of imbalance between rural and urban areas; (4) human resources development; (5) improvement of agricultural extension services; (6) encouragement of small-scale industries; and (7) larger involvement of the private sector.
    Pakistan is a country of 796,095 square kilometers with a population of 74.5 million people (as of 1977) and a per capita GNP of $140 per year. The physical regions are varied, the most fertile region and the most densely populated being the Indus Plain.[1]

BACKGROUND

    Having completed his undergraduate studies in mechanical engineering with first-class honors at Loughborough, England, Sayed Ajmal Hussain returned home to Pakistan and compiled a list of possible business ventures that he felt

---
[1]Information obtained from National Paper on Pakistan, Ministry of Science and Technology.

could be accomplished with the new skills afforded by his education and the amount of financial investment he was willing and prepared to make. His initial list of about fifty such prospects was narrowed down to five or six after further serious and more practical considerations. After a thorough appraisal of marketing, financial, skilled manpower, raw materials, and other input requirements, he decided on ball-point pen making as the option that most appealed to him. The technical challenge of the brass tip making for the pen cartridge, or refill, was especially complex, and Hussain felt that this would prevent any competitors from entering the market, giving him time to get his operation in a market entirely dominated, at that time, by poor-quality imports.

It was the intricacy of the tip and the personal challenge involved in its production (the production of the tip requires precision machining and microscopic control of some forty different dimensions) rather than a detailed economic analysis, perceived need of the technology, or contribution to national development, however, that carried the greatest weight in his decision to place an order for his first set of tip-making machines from the foremost manufacturer of such equipment worldwide--Albe, S.A., of Lugano, Switzerland-- which produced more than 90 percent of the tip-making machines on the market. The reply, which Hussain feels both hurt and stimulated him, was that Albe refused to send equipment where it would not work--in short, that the equipment would not work in a developing country and that, therefore, the order would not be filled.

Hussain wrote to his old university, obtained all available literature on tip making, injection molding, and the other processes involved, then wrote again to Albe regarding his qualifications to own, operate, and maintain the equipment. Albe eventually agreed to provide Hussain with the machinery, provided that one of the representatives of the company could remain on-site in Pakistan for six months to assure a smooth start-up. Hussain then went to Switzerland to participate in the selection assembly of the various units to be used in his factory--primary sliding head, automatic tip-making machines from Tornos, in Italy; multi-station tip-finishing machines from Albe, S.A., in Switzerland; an injection molding machine from G.B.F., in Italy; and a plastic extruder from Mapo, S.A., also in Italy. Albe, S.A., gave advice and acted as a consultant in this selection process, taking especially keen interest because it was the first time that this equipment would be sent to a developing country or even operated at such high temperatures. By participating personally in all stages, not only of selection but also of equipment assembly, Hussain learned every detail of Albe's design, operation, maintenance, and repair and even suggested several modifications which were incorporated into the design. He evidently so impressed the engineers at Albe with his innovative approach, particularly in the installation of the safety features of the injection molding equipment for Pakistan, that Albe rewarded him by providing all modifications and additions free of cost. There was, in the case of the primary tip-making machines, a faster, even more sophisticated model available, but in this, as in other equipment choices, he opted for the simpler, albeit still extremely complex, design to minimize his future maintenance and spare parts problems. Hussain was greatly encouraged by his interaction with Albe personnel; he continued to provide them with progress reports, share design modifications; and provide technical assistance to other users of Albe equipment around the world, and, in return, they continued in their advisory capacity to him for several years after his plant started operation.

The ball-point pen manufacturing facility, namely Sayyed Engineers, Ltd., is located in the Small Industries Estates at Gujranwala, approximately forty miles from Lahore. The site was developed by the Pakistani government with foreign assistance to provide a centralized service center for small, self-supporting industries and a better source of bargaining power for the small industries in their relations with the government.

As the machinery was arriving, Hussain also was recruiting technicians from the Pakistani government ordnance factory (where he planned to buy his brass materials for tip making) and drawing on retired Pakistani air force personnel. He trained them and personally supervised the installation of the equipment.

After the plant was installed, limited production was begun in 1968, and for three months Hussain's company produced pens for research and test purposes. Pen behavior was tested and recorded, and Hussain's team of technicians was trained on the job. At the end of 1968, the first pen, named <u>Tempo</u>, went into commercial production.

THE TECHNOLOGY

The manufacture of ball-point pens involves a sophisticated technology. The basic components and major inspection points in the pen-making process follows. The two basic raw materials used in manufacturing ball-point pens are brass and plastic. Storing these materials is not a problem because there are no limitations concerning temperature or humidity. However, the materials used in the ball-point cartridge--brass, steel balls, the plastic used for the ink capillary, and the ink itself--must be chosen carefully to maintain high quality. The ink must be stored in air-conditioned rooms because of the high temperatures at the factory.

Properties of Raw Materials

The brass must be homogenous and free from abrasive impurities, and its lead content has to be controlled within narrow limits. If the amount of lead is too high, the brass will become soft and the consistency in repeating the dimensions of the brass tips will be reduced. If the brass is too hard, it will be difficult to machine, and the limits between which tool settings may be varied will be narrowed. If the material is not homogenous, the dimensions of types produced with the same machine settings will vary, making it impossible to achieve constant writing characteristics.

The steel balls must have good surface finish and high skin hardness. The diameter and roundness must be controlled within close tolerances, and the surface must be free from contamination. Otherwise, corrosion and a chemical reaction with the ink or the atmosphere may be initiated, resulting in irregular writing at best and the jamming of the ball in its socket at worst.

The material of the ink tube has to be selected very carefully. If it does not have low permeability, the oxygen in the atmosphere will react with the ink, and the solvents of the ink will escape through the wall of the ink tube.

Ink is the most important of all of the materials that go into the cartridge. Much research has gone into the formulation of inks that do not dry up in the cartridge and that have very little reaction with the steel ball, the brass, and the plastic tube. The ink should have a high viscosity index so that it does not thin out in the summer and thicken in the winter. This property is particularly important for countries with extremes of climate such as Pakistan. Also, the surface tension of the ink has to be controlled so that it spreads evenly on the steel ball when writing and is absorbed quickly within the tip when it is not writing.

Plastic Extrusion and Molding Units

The following equipment is employed in the manufacture of the pen bodies, tops, and caps: one extruder for polypropylene tubes, complete with a cooling channel and a drawing and cutting device (imported); four injection molding machines (two foreign and two of Pakistani design and manufacture) that form the pen bodies and caps; injection molds that form the cap with clip and top and plastic bodies (simple molds are Pakistani-made, with imported nickel sleeves); three granulating mills to grind rejects and waste from molding machines (imported); three manual and one automatic (imported) and five manual (Pakistani-made) printing machines for pen bodies; three air compressors (imported); and one pigment-mixing machine (Pakistani-made).

An extrusion molding machine is used to make the refill's ink tube. Clear polypropylene granules placed in a hopper at the top of the unit fall downward onto an extrusion screw and are heated by friction created by the turning of the screw and an outside heat source applied at the extruder barrel. As the screw turns, it forces the hot plastic through a die, making a continuous plastic tube with the desired internal and external dimensions. As the tube exits through the die, it is drawn through a cooling channel and cut to the desired length.

In the injection molding step of the pen-making process, the granular plastic is mixed with pigment for pen bodies, tops, and caps. Each batch is fed manually into the injection molding machine (twelve to twenty-four bodies per mold in each unit). According to Hussain, in an injection molding machine, the plastic (usually from the polystyrene or the polypropylene families) is heated in a cylinder to a semi-fluid state; this process is called "plasticization." The plastic is then injected into a mold, and the two halves of the mold are clamped shut in the machine. After injection, the mold is allowed to cool, and the plastic solidifies to take the shape of the cavities. The two halves of the mold then open and the finished pen component is ejected.

Electrical or electronic timers control the duration of the pause between operations of mold cooling and injection. Electronic controllers maintain the temperatures of the various zones in the plasticizing cylinder within preset limits. Mold closing speed, clamp force (the force with which the two halves of the mold are pressed together), injection pressure, and injection speed are set by adjusting various hydraulic valves; the hydraulic circuit then repeats these speeds and pressures automatically in each cycle. The repeatability of the cycle ensures that the molding maintains dimensional accuracy and constant physical properties.

The injection molding machines also have safety circuits to protect the operator, the mold, and the machine. Such safety systems include the following: until full clamp force is built, injection should not take place; if during mold closing the guard door is opened, the moving platen should return and stop in fully open position; once the mold is closed and injection starts, the mold should not open until after the injection operation is completed and the cooling period is over; and, if any plastic material is left in the mold from the previous cycle, the mold clamp force should not be produced and the mold should open.

Two of Sayyed Engineers' injection molding machines were designed by Hussain and built by Pakistan Engineering Company (PECO) under his personal supervision. The areas in which PECO had no experience (such as hydraulic systems and advanced electrical circuits) were handled by Hussain.

Hussain's machines, the MKIII and MKIIIA, contain all of these automatic systems and safety devices. They also have the following outstanding features that are the result of his own innovative electrical and hydraulic circuitry design: smooth, noiseless, and vibration-free operation; "fool-proof" lubrication of all moving parts; hydraulic and electrical circuits that have been laid out for quick and easy maintenance; continuous filtration of the hydraulic oil with a warning system to indicate clogging of the filter; and a hydraulic pump that unloads automatically when pressure is not needed. In this condition, the pump discharges to the tank at zero pressure, thus cooling the oil, conserving power, and increasing its own life.

No attempt was made to manufacture the individual hydraulic and electrical components such as pumps, directional valves, flow controllers, pressure switches, and relays. Just as the European manufacturers buy components from specialist manufacturers, Hussain's company also purchased these parts. In fact, the European manufacturers also buy most of their fabrications, subassemblies, and control panels, while the Pakistanis manufacture these items themselves.

Mechanical Tip Making and Finishing Operations

These processes are accomplished with the following equipment: six sliding head, automatic lathes (imported); two tip-finishing machines (both imported); one centrifuge for removing cutting oil and solvent from ball-point tips (imported); one degreasing and cleaning unit (imported) and three air compressors (imported).

As described by Hussain, the process for making the tips requires very accurate machining and a program of dimensional checks suited to high-speed production, or one tip per second from each tip-making machine.

The tip, made from brass rod, is finally completed after about thirty operations. The most important of these operations include: external profiling of the tip; drilling of 1.2 mm and 0.9 mm holes in the rear of the tip; facing of the top of the tip; center-drilling the top of the tip; drilling a 0.6 mm hole in the top of the tip to meet the 0.9 mm hole drilled from the rear; boring the top of the tip to a diameter of 0.990 mm; broaching ink channels in the top of the tip; flushing out the holes in the tip with oil and air; inserting a steel ball in the top of the tip; forming a ball seat by pushing the steel ball into the brass tip with a hammer tool; and flanging the brass around the steel ball.

The external dimensions of the tip have to be maintained so that the tip can fit into the ink tube and into the body of the pen. Furthermore, external dimensions are

FIGURE 5.1
BALL-POINT PEN PRODUCTION/STORAGE RELATIONSHIP

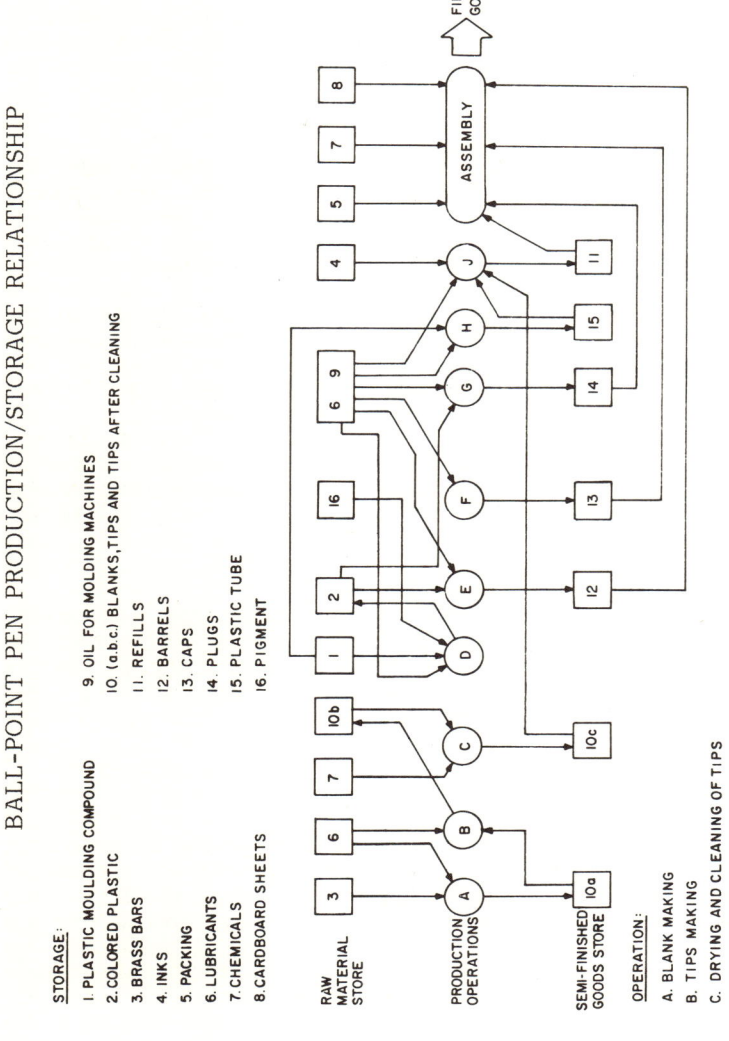

STORAGE:
1. PLASTIC MOULDING COMPOUND
2. COLORED PLASTIC
3. BRASS BARS
4. INKS
5. PACKING
6. LUBRICANTS
7. CHEMICALS
8. CARDBOARD SHEETS
9. OIL FOR MOLDING MACHINES
10. (a,b,c) BLANKS, TIPS AND TIPS AFTER CLEANING
11. REFILLS
12. BARRELS
13. CAPS
14. PLUGS
15. PLASTIC TUBE
16. PIGMENT

OPERATION:
A. BLANK MAKING
B. TIPS MAKING
C. DRYING AND CLEANING OF TIPS
D. COLORING OF PLASTIC COMPOUND
E. BARREL MAKING
F. CAP MAKING
J. INK FILLING AND AUTOMATIC CARTRIDGE ASSEMBLY

important because the tip has to be fed from one machine into another: the feeding and holding devices do not work satisfactorily if the external dimensions are outside the limits of tolerance. The internal dimensions must be maintained even more accurately; some of them must be correct within ± 0.002 mm so that the writing characteristics of the tip will not vary. The machines that work at such high speeds and with so much accuracy are specially made for this job. High-frequency motors running at 30,000 rpm carry out the internal machining.

Ink Filling

Ink filling and refill assembly are accomplished with the help of one automatic machine (imported) for ink injection, assembly crimping, and hot embossing of capillary ball-point pen cartridges; one three-column beaker centrifuge (imported) for vibrationless centrifuging of cartridges; and one air compressor (imported).

According to Ayaz Ahmad Ayaz, physical chemist at the Pakistan Council for Scientific and Industrial Research (PCSIR), a pressure system takes ink from a reservoir and forces it into the ink capillary, and the same machine then fits the finished tip to the filled capillary. The cartridge assembly is then placed manually in a centrifuge, where it stays for approximately five minutes to push the ink into the tip and remove all air bubbles. It is then removed and cleaned by hand.

Assembly and Packaging

All assembly and packaging steps are carried out manually; the only pieces of equipment used are nine printing machines. Refills are manually inserted into the pen body, capped, stamped with the name (<u>Tempo</u>, <u>Allegro</u>, and <u>Moto</u>, or other model name), provided with a top, and hand-packaged in boxes of one dozen.

Inspection and Testing

Product sampling occurs at several points in the process: after tip making but before tip finishing; after tip finishing but before ink filling; after body, capillary, and top molding but before assembly; and after ink filling but before assembly.

By periodic sampling and testing of components at this stage of production, problems can be identified and their sources determined so that raw materials and time are not wasted.

Testing of the pens but, more importantly, of the ink and tips themselves has been a major focus of interest for Hussain. As previously mentioned, the first six months of production at Sayyed Engineers, Ltd., were devoted to producing pens for testing purposes. Unfortunately, after two years, production was halted because of ink and ball problems. The steel balls corroded in the harsh climate of Pakistan. The ink was imported and had a shelf life of only six months, which Hussain now realizes was inadequate. To test for shelf life, Hussain, with the help of the PCSIR, designed a weather simulator that can predict the shelf life of a pen in thirty days of testing. They tested a large number of different imported inks, and, while a suitable ink has been identified and currently is being used, the process also led to the creation of a satisfactory ink-making capability. Plans to have a German company move its ink-production facilities to Pakistan are now being negotiated. At least three European ink-manufacturing companies have set up testing facilities similar to those of Sayyed Engineers, Ltd., and, according to Hussain, their engineers and chemists are sent to him for advice.

Ink bought for production in Hussain's facility must pass the tests now standardized by his weather simulator: to expedite this step, Hussain sends tip samples to European and now U.S. ink manufacturers. Always looking for ways to improve his product, Hussain, with the assistance of PCSIR, also developed a simple evaluation test procedure for the corrosion resistance of steel balls. Using NaCl and a mild acid solution provides an accurate prediction of how the balls will behave in the pen over storage and writing time. His suppliers now use the same methods to test consignments before they are shipped, knowing that they will be returned by Hussain if they do not meet his standards. Several changes have been made in sources of raw materials over the factory's twelve years of operation.

MARKETING AND FINANCE

Before the oil crisis of 1974, Hussain believed that he could not promote his product with massive publicity alone, so he designed his own marketing strategy. He felt that the wholesalers were not friendly to him because they were able to make a higher profit on the other ball-point pens which came to them cheaper. Competitors' pens were accepted even though they dried up, often did not work, broke, and had ink-smearing problems, just because they afforded a higher profit to the middle man.

Hussain divided the consumers into two groups: government and large private organizations--from which his sales team solicited large orders--and the "man on the street."

Hussain initially ignored this latter group because he felt that his efforts and resources would be too greatly dissipated. He designed information-gathering sheets for his sales representatives, distributed samples to organizations, gave demonstrations on how to test a pen by hand, took organization representatives to dinner, and made repeat visits. Wholesalers soon started getting orders for Tempo pens (the first model), and soon their customers would not accept any others. When the wholesalers started substituting European refills in Tempo pens and sold Tempo refills at a premium, Hussain knew that his product would be successful.

After effectively analyzing the pricing structures of his raw materials in 1973 and 1974, Hussain stockpiled enough raw materials before prices rose from the higher cost of petroleum products. He thereby eliminated his competition in Pakistan by calculating his foreign competitors' costs of production and by keeping his own prices constant and low when all others were rising. He invested a substantial portion of his profits in public relations and for eight months in 1975 was the only local supplier of ball-point pens.

Up to this point, in the small villages and poorest areas, Hussain's efforts had had little initial impact because the wholesalers would not distribute a product that brought a lower profit. However, by designing a plan to increase profits to wholesalers as the product's value rose (Hussain offered prizes, fantastic dinners, photos, and contests), demand grew along with production. As a result, the wholesalers slowly moved to "his side." Hussain has an excellent reputation in the industry and attributes a great part of his success to the confidence that wholesalers have in him. He does not use ploys in his distribution practices: one policy exists for everyone, and it is printed. Because of this confidence, Hussain's operation is financed largely by loans paid for in full by wholesalers, who take their reimbursement (without interest) in pens after six months. Each distributor gets a quota of pens and is assured of a continuous, reliable supply. Both the loans and the quotas are very unusual practices for Pakistan.

Hussain started his operation with approximately $50,000 in paid-up capital and $43,000 in Pakistani bank loans. His assets now total over $2 million. He says that operating costs vary monthly but that his profits are "reasonable."

COMPETITION AND LEVELS OF PRODUCTION

Tempo pens, followed by other models (Allegro and Presto), penetrated the small markets, and consumers who had never before used Sayyed products were more or less forced to do so. When they realized that these were superior pens, they requested them, and increasingly more wholesalers and retailers had to carry the Sayyed products. For three years, Hussain had to ration pens, and wholesalers were continually trying to get their quotas increased. By 1977, when all available marketing channels had been filled, he raised his wholesale price from thirty paisa per pen to forty paisa (the pens now sell for fifty to seventy-five paisa retail).

At this point, competitors tried to reenter the market, but they had an inferior product, with the bodies made locally by hand press and the refills imported from Italy. A new, good-quality Chinese pen recently has entered the market and sells for the same price as Tempo, but user acceptance apparently is low in the face of Hussain's introduction of Allegro, Moto, and other pen models. Recently, another Pakistani entrepreneur has installed an injection molding machine with Hussain's assistance, and is currently importing good-quality refills to produce a competing model. Fountain pens (both imported and locally made) also are taking a small part of the market. Now that competing pens are again penetrating the market, Hussain is bringing out two new models. He recently held a competition at an art college in Lahore to design packaging for these two new products. He also has taken on the distributorship of a European-made highlighting pen as a first step toward a joint venture.

Production of ball-points rose from 100,000 units per month in 1970 to 700,000 pens per month in 1975. Currently, over two million pens are manufactured every month. Hussain projects a production of 2.5 million units per month before the end of 1979.

SOCIAL AND ECONOMIC IMPACTS

According to Hussain, his most important accomplishment is that he has "given dignity to this little business world we live in--people are now beginning to rate other people's business methods against his own--this is a small company and well known for honesty and integrity."[2] This perception is apparently widespread, for similar comments were made in Peshawar, Karachi, and Lahore.

In a less abstract way, however, several more obvious benefits have accrued as a result of the introduction of a ball-point pen manufacturing facility in Pakistan. No jobs have been displaced because of automation of an existing

---

[2] Several other sources of information, including PCSIR employees, agree with this comment.

labor-intensive process, as is often the case; on the contrary, jobs have been created. Hussain personally trained employees who work with his equipment. Among the employees are approximately thirty single women from nearby villages who stamp model names on pen bodies and assemble and package the pens. Hussain employs women for these tasks because he feels that they are better with their hands in handling small parts and in rapidly assembly and packaging his product. He also feels that, while an employer must compensate employees according to their needs, this compensation must also be relative to the value of their production. Printing, assembly, and packaging are manual operations, and, therefore, the workers who perform them have low output. They cannot be paid much for this work. A man cannot support his family on this wage, but a single woman can take it as pocket money or contribute it to her dowry. He takes satisfaction in providing jobs to an otherwise unskilled and generally unemployable group and in exposing these women to a progressive and otherwise unapproachable segment of the Pakistani development scheme--a high-technology production facility.

Even though such positions demand a higher wage, women are not employed for the operation or maintenance of equipment in the factory for several reasons. First, in a socioeconomic system where skills are passed from father to son and mother to daughter, women have no technical or mechanical skill or knowledge base to conceptualize the necessary principles and, therefore, cannot work effectively in a job best filled by young men who have been exposed to these principles since their early years. Second, in the "rude and earthy" atmosphere of the shop floor, women would feel out of place and would not be welcomed by their male co-workers. Futhermore, technicians are still looked down on socially as laborers who must work in grease and filth and are generally held in some contempt. This is, therefore, judged as an inappropriate work place for a woman. Their simple presence in the factory in any role at all represents a significant evolution of thought in the country and indicates a broadening role for women in Pakistan. Additional employment in the local area is generated through batch packaging and transportation of the product, in printing and advertising, and in demand for public services on the part of Hussain's operation and its employees.

The aspect of employment generation also enters into the area of technological spin-offs: several types of equipment have been produced locally, and several technological capabilities have been created as the result of the needs of the pen factory. The special-purpose machines manufactured in Pakistan include pigment mixers, printing machines, injection molding machines and molds, and automatic tapping machines.

Many of these items were designed by Hussain, and he takes considerable pride in his utilization of such groups as the PCSIR, PECO, and several individual Pakistani consultants and small-scale shops to produce his needed machinery. Such activities not only have created work but have opened both formal and informal channels of communication and created a problem-solving capability in the fabrication infrastructure where only a machine-building capability previously existed.

According to Hussain, the injection molding machines (MKIII and MKIIIA) and the molds themselves represent particularly interesting achievements for Pakistan. So far, more than fifteen inquiries from plastic-molding companies in Karachi have been entertained by Hussain; he is considering selling his equipment, which is to be made by PECO. So far, about $50,000 have been saved in foreign exchange by fabricating his two machines locally. The successful construction of the first automatic injection molding machine in Pakistan is relevant not only to the plastics industry but also to industries engaged in die casting, die stamping, and deep drawing. With modifications, the MKIII can be made into a press for industry and agriculture, for example, for the pressing or pelletizing of animal feeds. A new twenty-four-unit mold has just gone into operation at Sayyed Engineers, Ltd. It is made by a nickel electroplating process previously not used in Pakistan and therefore represents a "first" for the country. Engineers indicate that this is due entirely to Hussain's effort and that the process should have far-reaching effects in the plastics industry.[3]

Import substitution is another factor to consider. In the past, ball-point pens were all imported. While all the raw materials for Hussain's operation are imported, the value-added component is considerable. Tax revenues generated for the government are substantial; in addition, the pens at one time were exported, earning foreign exchange. All production is now reserved for domestic consumption. Although most of the original equipment was imported, it nevertheless has been instrumental in creating opportunities for the production and export of other machinery made in Pakistan. As Hussain pointed out, "Any time we bring in this type of equipment, we are required to duplicate it;" thus, the export potential is significant.

As previously described, Hussain's efforts in simulating the weathering of inks, brass, and steel have resulted in the transfer of these technologies back to Europe. Hussain

---

[3]The engineers are Hussain and Ayaz.

and his associates give free technical advice and are frequently called upon to design or repair equipment outside their immediate field of interest. It would seem reasonable to describe them as a nucleus for precision engineering that is greatly contributing to the industrial growth of Pakistan.

REFERENCES

Hussain, Ajmal. "Pakistan's First Automatic Plastic Injection Moulding Machine." Engineering News 21 (March 1976): 43-48.

Hussain, Ajmal, and Pirzadeh Zahid Hassan. "Manufacture of Ball-Point Pens." Engineering News 19 (September 1974): 49-56.

PERSONS INTERVIEWED

Dr. Mohammad Aslam

Dr. Amir Khan
PCSIR, Karachi

Dr. Ayaz Ahmad Ayaz

Dr. Eshan Ali
PCSIR, Lahore

Sayed Ajmal Hussain
Director, Sayyed Engineers, Ltd.
Lahore

Chief Accountant
Sayyed Engineers, Ltd.
Lahore

Pirzadeh Zahid Hassan

Three pen vendors in the Bazaar in Lahore

QUESTIONS TO CONSIDER

1. Four elements seem to distinguish this successful technological introduction: sophisticated foreign technology, entrepreneurship, education (technical), and a latent market demand. Is this situation unique, or might this combination exist in other locations? If

so, is there a systematic way by which these ingredients may be brought together in order to initiate similar enterprises in developing country situations?

2. Given the intensive use of imported and very expensive technology, the relatively few jobs created, and a question of the economic utility of the product (except as an import substitution), does this case present an example of "appropriate technology?"

3. Does the employment of women at below subsistence wages constitute an exploitation; or may it properly be viewed as a net addition to the community income and the gainful use of an otherwise unutilized factor?

4. Might the investment required for the creation of this firm have been better expended in some other way that would have resulted in greater economic and social benefit?

5. The demand effect for development of indigenous technology to augment the imported variety appears significant; is this the usual model for domestic technology growth? If so, is there some way of using this method in an organized way to deliberately engender such local innovation. What are the ingredients necessary in the domestic environment in order that such innovation will take place?

6
Java, Indonesia: The Introduction of Rice Processing Technology

Melinda L. Cain

INTRODUCTION

Mechanized rice hullers were introduced in Java, Indonesia, in an attempt to modernize agriculture on the island and thus to increase rice production. In addition, the use of sickles and scales, and the tebasan practice, were employed toward the same goal.[1] However, the introduction of this technology also created labor displacement, particularly among women who traditionally were involved in rice harvesting.

BACKGROUND

Indonesia includes more than 13,000 islands, the largest of which is Java. Java covers only 6 percent of the total land area (about the size of New York State) but contains two-thirds of the country's total population of 135 million. (It is interesting to note that in 1870 Indonesia's population was 1.6 million; by 1965 it had reached 69.2 million. Now, in a little over ten years, it has almost doubled.) More than 80 percent of this population resides in rural areas. In contrast, other islands (Kalimantan, Sulawesi, Sumatra) do not have enough labor for production.

Indonesia is representative of the more than one-half of the world's population that depends on rice for a basic food supply. Approximately 90 percent of the world's rice is produced in Asia.[2] Sixty percent of the total calories and

---

[1]This case history is relevant to one island only--Java. Generalizations suggested here are *not* applicable to other Indonesian islands.

[2]Amir U. Khan, "Present and Future Development of the Mechanization of Rice Production." Paper No. 71-06 (Manila, The Philippines: International Rice Research Institute, 1971).

65 percent of the protein in the Indonesian diet come from rice.[3] Domestic consumption of rice in Indonesia is higher than the country's capacity to produce it, which means that rice must be imported to meet local demand. In fact, in 1978 Indonesia was the world's largest importer of rice.

Obviously, agriculture is crucially important to the Indonesian economy. For example, in 1968 it contributed about 50 percent of the gross domestic product, provided employment for about 70 percent of the total labor force, and produced about 50 percent of all exports.[4] In the second five-year plan of Indonesia, REPELITA II, national efforts were defined as being directed to increase productivity and achieve ultimate self-sufficiency in rice production, to expand agricultural exports, and to reduce rural unemployment.[5] The largest single program in the plan is to increase the production of rice (paddy), of secondary crops (such as corn, sorghum, cassava, soybeans, and peanuts), and of horticulture. The plan outlines increases in rice production from 14.5 million tons in 1973/1974 to 18.2 million tons in 1978/79, or an average annual increase of 4.6 percent, which is about double the rate of population growth. The increase in production is primarily due to the expansion of cultivated land and to increases in per-hectare yields. The governmental program (initially launched in 1968 and revised on several occasions since that time) to support the campaign for increased rice production is called BIMAS, an acronym taken from the Indonesian expression "Bimbingan Massal" ("mass-guidance"). This program is attempting to mobilize the Indonesian peasants in an effort that is involving a massive infusion of fertilizer and high-yield seed varieties (HYV). Some reports indicate that the program has been

---

[3]Personal Communication, Director Tomadias, Social Economic Division, Nutritional Research and Development Center, Department of Health, Bogor, Indonesia.

[4]A. T. Birow, "Pembaugunau Pertanian and Shategi Industrialisasi," Prisma (August 1972): 31.

[5]Government of Indonesia, Department of Information, REPELITA II, p. 19.

far from successful for a variety of political and administrative reasons.⁶ Nonetheless, the national goals of increased rice production and self-sufficiency remain.

Since 1971, when Amir Khan of the International Rice Research Institute (IRRI) wrote that most of the rice production in Asia was done with traditional, nonmechanical methods, a strong effort has been made to mechanize rice cultivation. This effect has not been without problems, however, as noted by Khan:

> Mechanization of rice production in the tropics has many problems which still remain unsolved. Attempts to transfer the highly advanced Western and Japanese mechanization technologies have not produced effective results for the small farm holdings in the tropical regions. The overwhelming need today is to develop an intermediate mechanization technology to suit the prevailing set of agricultural, socioeconomic, and industrial conditions of the tropical regions.⁷

TRADITIONAL RICE HARVESTING

Traditionally, Javanese rice farmers did not restrict anyone who wished to participate in the rice harvest. The harvesters were mostly women from within the village and from neighboring villages. The women used an ani-ani (small finger knife) for harvesting. The ani-ani was suitable for cutting local varieties that matured at different times and had varying stalk lengths. The harvesters carried the rice in sheaves, bound in the field, on shoulder poles to the owner's house. This method of harvesting required large numbers of people, and literally thousands of landless families were involved. In fact, one farm survey showed as many as 500 persons employed per hectare.⁸ The

---

⁶Gary E. Hansen, "Indonesia's Green Revolution: The Abandonment of a Non-Market Strategy Towards Change." SEADAG Papers (New York: The Asia Society, 1971); and Gary E. Hansen, The Politics and Administration of Rural Development in Indonesia. The Case of Agriculture. Research Monograph No. 9 (Berkeley: University of California, Center for South Southeast Asia Studies, 1973).

⁷Amir U. Khan, "Present and Future Development of the Mechanization of Rice Production."

⁸William L. Collier, et al., "Agricultural Technology and Institutional Change in Java," Food Research Institute Studies 13 (2): 174.

harvester's pay was a share of the crop, with a ratio of about seven to ten for the owner and one for the harvester. The division was made by bundles and not by weight.

THE TEBASAN HARVESTING SYSTEM AND THE USE OF SICKLES AND SCALES

Traditional methods of rice harvesting in Java have changed significantly partly because of the increased population pressures on land. Individual farm sizes have become smaller as farms have been subdivided from generation to generation, and it has become more difficult for farmers to run a profitable business. The population increase also has meant that larger numbers of landless laborers are looking for harvesting work. As the amount received by harvesters has grown smaller, they have tried to obtain larger shares than custom dictates. Furthermore, farmers customarily have felt a social obligation to let all the harvesters participate. Therefore, farmers have found their share of the harvest diminishing.

One way of improving the farmers' share is to limit harvesters. This can be done by the adoption of the tebasan system. Tebasan is a harvesting system that enables the farmer to sell his crop to the penebas (middleman) before harvest. This limits the number of harvesters and avoids the problems of supervising the harvest and dividing the shares. About one week before the harvest, the farmer sells his crop to a buyer, who then arranges for the harvest and sells the rice. The penebas may be from the same village or from outside the village. The farmer usually is paid within one week after the harvest if not at harvest time.

The penebas is recognized as a trader, and his right to a profit is accepted. Individual harvesters may benefit from this system, especially when the penebas can control the number of participants, thereby ensuring larger returns for each harvester. According to village surveys, some rice always has been purchased by the tebasan method. However, the system has become more important with the use of HYV because there are now two harvesting seasons and thus more rice to harvest.

A comparison of costs of harvesting with the penebas were estimated from a sample of village surveys. With the ani-ani and the traditional system, the estimated harvesting costs were about $39 per hectare. Comparing those costs with about $15 per hectare that it costs the penebas to harvest, it is evident that the harvest costs can be reduced about 50 percent by using the tebasan system.[9]

---

[9]Ibid., p. 185.

Tebasan and the introduction of HYV have caused an important technical change in the method of harvesting rice: this is the use of the sickle. The ani-ani is more suitable for cutting traditional varieties of rice; the sickle is preferable for cutting the HYV. When the sickle is used, the rice is threshed in the field, then carried in sacks to the penebas's house, where harvesters are paid in cash according to the weight, not according to bundles. Thus, when the penebas began to use sickles, scales became necessary to weigh the shares for the harvesters. Furthermore, harvesters must provide their own sickles, threshing mats, and sacks to carry the rice. With sickles, only about 75 person days are required to harvest one hectare, while with the ani-ani, 200 or more person days may be needed.

THE INTRODUCTION OF RICE HULLERS

Due to governmental initiative, mechanized rice hullers were introduced in 1970 to 1971. The diffusion of hullers occurred very rapidly after 1970 (see Table 6.1). By 1978, only about 10 percent was being hand pounded, mostly for family consumption.[10]

An English model by Engleberg is widely used in the Philippines. This machine has few moving parts and is very durable. However, a Japanese model that uses rubber rollers is more common in Indonesia. Pasawahan, a village in west Java, has three milling centers that use the Japanese hullers and polishers.

Rice must be processed through the machine four to eight times. It is first poured into the top of the huller; the hulls and excess material (bran) then travel through a pipe and are discarded outside the building. The hulled rice is then run through the polisher three or four times.

CHOICE OF TECHNOLOGY AND ECONOMIC ASPECTS

Timmer has discussed the choice of the rice hulling technology in Indonesia by analyzing the four alternative milling/storage/drying facilities that were considered by USAID/Jakarta and the Indonesian government in order to "modernize" the rice marketing sector. He mentions four efficient alternatives, of which the most capital-intensive required $65,000 investment per worker and the most labor-intensive required only $700 per worker. Timmer also points out that beneath the decision to modernize lay a deep-felt

---

[10]Personal Communication, Pudjiwati Sajogyo, Center for Rural Sociological Research, and Soesarno, Food Technology Development Center.

TABLE 6.1
NUMBER OF SAMPLE FARMERS PROCESSING RICE WITH HULLERS,
AND NUMBER OF HULLERS IN SAMPLE VILLAGES, 1970 AND 1973

	No. of Farmers in Sample	No. of Farmers Processing Rice with Hullers		No. of Hullers in the Village	
		1970	1973	1970	1971
West Java					
Kab. Serang					
Sentul	27	0	0	0	0
Warungjaud	24	0	17	0	*
Kab. Cianjur					
Jati	29	15	29	3	5
Gekbrong	27	0	0	0	0
Central Java					
Kab. Banyumas					
Kebanggan	30	0	29	0	1
Sukaraja Lor	30	0	22	0	*
Kab. Kebumen					
Bulus Pesantren	30	0	27	0	*
Patemon	30	2	25	0	1
East Java					
Kab. Ngawi					
Geneng	29	0	26	0	3
Gemarang	30	-	21	0	2
Kab. Jember					
Sukosari	30	0	26	0	8
Tanggulwetan	28	8	27	-	3

*Farmers from these villages used hullers in neighbouring villages.

Source: 1972/3 wet season surveys (6th round); 1970 data based on 3rd round.

Source: Collier, "Rice Milling," p. 109.

Drying of rice before milling

bias on the part of Western and Western-trained technicians that "identified capital-intensive with modern, and modern with good.[11] Such value judgments may play an important role in determining technological choice, as shown by the widespread introduction of the Japanese rice mill.

In part due to Timmer and other work, the Indonesian government chose the mechanical but less-high-technology alternative because it was economically preferable. Also, loans to buy hullers were available at 1 percent per month interest, whereas regular village credit runs about 5 to 10 percent per month. Therefore, the machines were well subsidized and available to those who could afford them.

Collier estimates the average investment costs of a hulling center to be $3,111 for machinery, buildings, and land. Such a hulling center would have an average capacity of .58 tons per hour. This figure is based on the combined

---

[11]Peter Timmer, "The Choice of Techniques in Rice Hulling In Indonesia." <u>The Choice of Technology in Developing Countries</u> (Howard University: Center for International Affairs, 1975), p. 25.

Rice huller in Kepala Desa village

use of old and new equipment. Timmer estimates $8,049 as the initial cost of a hulling center with a capacity of .42 tons per hour.[12]

In Pasa wahan village the initial cost of a huller in 1976 was Rp 2.5 million (or about U.S. $4,000). The owner of the huller said that operating costs were low except for repair, the need for which did not occur very often. Both the huller and polisher were diesel-powered, using a crude kerosene fuel that cost Rp 30 per liter (five cents per liter, or less than twenty cents per gallon). Ten liters would run the huller for five hours or about one ton of rice. Repair costs so far had been low. The owner pointed out a small part that had recently been replaced for Rp 40,000 ($66).

---

[12]Collier, "Rice Milling," pp. 113-115.

At this particular hulling center or mill, about two tons of rice could be hulled per day. This compared to the hand-pounding of forty kilograms per day by one woman. Two men who operated the huller and polisher could process about 100 kilograms of rice in twenty minutes and were paid Rp 45 for every 100 kilograms. Labor use as estimated by Collier was four to five hand laborers to hull 92 tons per month (average). Timmer estimated twelve laborers to hull 1,000 tons per year.[13]

It was found in one survey that the average cost of hand pounding was $1.45 per 100 kilograms. In comparison, the average cost to the farmer of using a huller was $.54 per kilogram. In addition, the by-products were kept by the miller while in the traditional harvest, women were able to keep the by-products to use as animal feed.[14]

Hand pounding of rice by women in Java

---

[13]Collier, "Rice Milling," p. 115.

[14]Ibid.

IMPACT OF THE TECHNOLOGY

During the last five years, the mill has taken over work traditionally done by women. Two examples illustrate these changes: "A former rice trader, now turned mill owner, stated that he used to employ eight women to hand pound his rice. Four women working five hours could hand pound 100 kilograms of gabah. This rice trader could buy 200 kilograms per day of gabah. The women's wages were 10 percent of the rice they provided, which amounted to just under two liters of milled rice per day. Thus, over the harvest season these eight women earned perhaps sixty liters of milled rice each or enough to feed themselves for four months." "In Kendal, Central Java, a farmer said that in the past there were more than 100 women 'hand-pounder' laborers in his village. But now they have no work."[15]

Estimates of jobs lost ranged as high as 1.2 million in Java alone and as high as 7.7 million in all of Indonesia as a result of the introduction of the new technology. Collier estimated that the loss to laborers in earnings due to the use of hullers was $50 million annually in Java, representing 125 million woman days of labor.

The rice farmer pays less to the mill for threshing and the process is much quicker, but the women have lost a highly remunerative source of income. They are now forced to work longer hours at other jobs, if such can even be found. The shift from a traditional technology to a more modern one has eliminated one of the more important sources of income for landless villagers.

Thus, the adoption of the use of HYV, tebasan, sickles, scales, and rice hullers has not helped to solve the problems of unemployment and income distribution in Java. Rather, these problems have been exacerbated. There is little evidence to indicate that the rural unemployed are being taken up by work opportunities in the cities.

REFERENCES

Abdullah, J. A. and Sondra Feidenstein. "Socioeconomic Implications of High Yielding Varieties Rue Production on Rural Women of Bangladesh." Dacca, April 1975.

Collier, William. "Recent Changes in Rice Harvesting Methods: Some Serious Social Implications." <u>Bulletin of Indonesian Economic Studies</u> 9 (2):36-45.

---

[15]Ibid., p. 108.

Collier, William L., Soentoro, Gunawan Wiradi, and Makali. "Agricultural Technology and Institutional Change in Java." Food Research Institute Studies 13 (2):169-194.

Collier, W. L. "Choice of Technique in Rice Milling: A Comment." Bulletin of Indonesian Economic Studies 10 (March 1974): 106-120.

Collier, William. "Food Problems, Unemployment, and the Green Revolution in Rural Java." Prisma: Indonesian Journal of Social and Economic Affairs: 20-37.

Government of Indonesia, Department of Information, REPELITA II.

Hansen, Gary E. "Indonesia's Green Revolution: The Abandonment of a Non-Market Strategy Towards Change." SEADAG Papers (New York: Asia Society, 1971).

Hansen, Gary E. The Politics and Administration of Rural Development in Indonesia. The Case of Agriculture. Research Monograph No. 9 (Berkeley: University of California, Center for South Southeast Asia Studies, 1973).

Khan, Amir U. "Present and Future Development of the Mechanization of Rice Production." Paper No. 71-06 (Manila, the Philippines: International Rice Research Institute, 1971).

New York Times, Sunday, November 30, 1975.

Sinaga, Rudolf and William Collier. "Social and Regional Implications of Agricultural Development Policy." Prisma: Indonesian Journal of Social and Economic Affairs.

Stoler, Ann. "Class Structure and Female Autonomy in Rural Java." In Women and National Development: The Complexities of Change (Chicago: University of Chicago Press, 1977).

Timmer, C. Peter. "The Choice of Technique in Rice Hulling in Indonesia." In The Choice of Technology in Developing Countries. Some Cautionary Tales (Center for International Affairs, Harvard University, 1975).

PERSONS INTERVIEWED

Dr. Benjamin White
Agricultural Development Council, Inc.
P.O. Box 62
Bogor

Dr. William Collier
Social Economics Department
Bogor Agricultural University (Institute Pevtanian Bogor)
Jalan Raja Pejajaran
Bogor

Professor Rudjiwati Sajogyo
Center for Rural Sociological Research
Bogor Agricultural University
Jalan Oyo Iskandardinaka
Bogor

Walter C. Tappan
Chief Agricultural Officer
US AID
American Embassy
Medan Merdeka Selatain 3
Jakarta Pusat

Sidney R. Jones
Ford Foundation
Taman Kebon Sirch I/4
P.O. Box 2030
Jakarta

Tomadias, Director
Social Economic Division
Nutritional Research and Development Center
Department of Health, Bogor

Soesarsono, Director
Food Technology Development Center
Bogor Agricultural University

QUESTIONS TO CONSIDER

1. In this situation, was the increased productivity of rice production in food-short Java worth the displacement of large numbers of women agricultural workers? Is this an example of appropriate technology?

2. Is there anything the government might have done to provide alternative employment to those who were displaced?

3. Is there sufficient information provided to conclude whether, having started the mechanization process for rice hulling and processing, it would have been better to go all the way and use the most modern and efficient equipment.

4. Does it appear that there was sufficient study of the economic and sociological aspects of the Java situation? If Timmer's study had an effect, as indicated, what more might have been done by authorities who instituted the policies that resulted in the use of the new technology?

5. What additional information do you feel is needed in analyzing this situation, and how would you propose developing it?

7
Central America: The Lorena Cookstove

Suellen Sebald Edwards

INTRODUCTION

The Lorena cookstove is in use throughout Central America, especially in Guatemala, by people in the highlands. With its efficient design and durable construction, the Lorena cookstove offers a low-cost, fuel-saving alternative to those who have to live in a wood-scarce environment. The stove uses 50 percent less wood than the conventional open fire. It is hoped that the use of the stove will eventually cause a significant change in fuel demands of the highlanders and contribute to the reversal of the severe deforestation that is plaguing Guatemala.

BACKGROUND

Guatemala, where 50 percent of the population lives in the rural highlands, is a small, mountainous country about the size of the state of Kentucky in the United States. As the population increases, so does the demand for firewood, both of which contribute to the problem of deforestation that besets not only Guatemala but other Central American countries as well.

The shortage of firewood means that people must travel long distances in search of their cooking fuel source--wood. The backbreaking labor not only consumes considerable time, but also up to 25 percent of the monthly family income. The time is ripe for reconsidering cooking habits, or else a family may face spending 100 percent of its income on fuel.

This was the situation faced by Ianto Evans and Donald Wharton when they began their work to design a cookstove that would save wood. They also accepted the challenge of designing a stove that would fit into the cultural life-style of the highlanders and would eliminate other problems plaguing the cooks (e.g., smoke, open fire hazards for children, intolerable heat generation in the summer, and backbreaking labor). With $1,000 and six months of experimentation time, Evans and Wharton went to the Choqui

Experiment Station in Quezaltenango, Guatemala, to develop a stove. During 1976 and 1977, the stove was designed and tested, after which the experiment station started to promote it.

Several problems had to be considered in the planning and designing of the Lorena cookstove; all of them were related to the existing cooking methods. First, many poorer women cook over an open fire, using three rocks or "tituates" on which to place the cook pot. This method is very inefficient in terms of fuel requirements: only prime-quality wood can be used, excluding sawdust, cornstalks, and so on; also, the combustion is poor due to the high altitude, causing smoke to permeate the room and house and finally escape through cracks in the roof. Second, in the conventional cooking method, food pots are placed on the ground, all within reach of rats, children, dogs, and cats. The practice results in burn accidents, food loss, back breaking labor for the cook, and unsanitary conditions. Third, most Guatemalan highlanders have very little money and no access to money sources. This means that even the most inexpensive stoves (cast-iron plate on a brick platform) are beyond the reach of most families. Fourth, few tools beyond the machete and cooking utensils are available. Any construction of stoves would have to employ these tools, for the people have no financial resources to rent or buy other items.

With these problems in mind, Evans and Wharton began to experiment. The solution was based on knowledge of other stoves in operation throughout the world, e.g., Ghanaian and Indian stoves. These were rejected due to their nondurable construction and inability to contain heat very well. The result was the Lorena cookstove, a smokeless, heat-conserving, fuel-saving, modern-looking device that is spreading rapidly throughout Guatemala. Its success is credited to the fuel-saving design.[1]

The final design was the result of trial-and-error prototypes. The designers worked closely with rural Guatemalans, who patiently gave suggestions for revisions. One such suggestion was to reject the low stoves, even though sociologists stressed the cultural importance of families crouching around the open fire in cold mornings and evenings.[2] The people wanted a modern-looking stove of "proper" height.

---

[1]Choqui Experiment Station. "Preliminary Progress Report of the Lorena Stove Program," February 1979, p. 9.

[2]Ianto Evans, "Using Firewood More Efficiently." Paper submitted to the 8th World Forestry Congress, Jakarta, Indonesia. N.D.

Test stoves were built at the Choqui Experiment Station in Quetzaltenango. Tests of local clay and sand mixtures were made to find a combination that would be resistant to cracking and deterioration. Also, the designers tried to build the stove exclusively with tools available to all Guatemalans. Simplicity in terms of money, technical design, tool requirements, and materials was the underlying philosophy of the project.

THE STOVE

The Lorena cookstove offers unique design characteristics that make it very desirable to the Guatemalans. The stoves are carved from a high monolithic block of clay and sand, resulting in great structural strength and heat retention. The fire is totally enclosed, preventing burns and inefficient heat transfer as are common with open, "three-rock" fires. Also, the Lorena stove has a high proportion of sand to clay, thus preventing shrinkage, the common cause of cracking. The Spanish name "Lorena" comes from the sandy mixture used to construct the stove--"lodo" meaning "mud" and "arena" meaning "sand."

The Lorena stove is wood-burning, with each stove being designed to fit the location and needs of its users. Because the stove is custom-built, various designs and shapes ranging from bakery stoves to small family stoves exist throughout commercial Guatemala, Honduras, and Mexico. Three basic shapes are promoted: square, round, and L-shaped (for corners).

The materials required to build each stove cost four to five dollars, and the average labor time is three to four days. The simplicity of design and minimum tool requirements are no accident; they are the result of calculated design revisions by Evans and Wharton. The basic materials needed to build a Lorena stove include clay, sand, water, scrap sheet metal, chimney pipe, twenty to thirty adobe bricks, a machete, and a spoon. Construction proceeds more quickly with the use of a sifting screen, a shovel, a large hoe, a bucket, and a mason's trowel. These are not essential, however.

A stove is constructed in the following manner:

(1) Determine the size, shape, and design of the stove (e.g., number of pot holes, room for eating space, cutting board, location, etc.).

(2) Mark the full size of the stove on the ground.

(3) Construct the base support with bricks or stones ten centimeters in from the mark, leaving the center empty. The indentation creates a foot space for the cook.

(4) Add the second layer of bricks so that it protrudes five centimeters farther than the first layer. Add a third and fourth layer of bricks and rocks. (Additional layers can be added depending upon the desired height of the finished stove). Forty to forty-five centimeters is standard for the base height.

(5) Fill the center space with stones, debris, soil, or broken adobe; pack it down. Allow the base to dry overnight.

(6) To make the body of the stove:

   a. Mix sand and clay. A high proportion of sand prevents shrinkage of clay, the cause of most stove cracking (a good ratio is 60 to 80 percent sand to 20 to 40 percent clay). It is difficult to prescribe the exact clay/sand mixture because of the differences in clay from region to region. The one important detail is that the clay must fire to prevent crumbling and subsequent firebox collapse.

   b. Sift the sand and soil to remove rocks.

   c. Mix twelve to twenty buckets at a time. This is enough for one layer of the stove. Do not add too much water, only enough to hold the mixture together.

(7) Build the body of the stove layer by layer. Do not add the next layer until the first stops glistening and is firm. Work from the outside to the inside; the drying of each layer may take as long as one day, depending on the location of the stove.

(8) When the layers measure forty centimeters, work a board across the top to level the stove. A machete or trowel can be used to even up the sides or to shape them.

(9) Allow the body to dry until a finger pressed into the side leaves only a slight depression.

(10) To design the layout of the cooking surface, it is necessary to have all the pots that will be used in cooking. Place the pots on the surface of the body. The heat "tunnels" will be carved out of this monolithic block. The idea is to have two tunnels from the firebox meet beneath the final pot before exiting to the chimney (see Figure 7.1). The placement of dampers on either side of the final pot and at the opening of the firebox allows for heat control.

(11) To carve out the inside, a machete and spoon are needed. (The metal chimney can be used to start the holes for the pots.) Cut the damper slots with a wet machete, making a groove that slopes into the slot.

(12) After making the initial holes for the pots and the firebox opening, allow the body to dry. (The length of time varies from several hours to one day.) If it is soft inside, allow a longer drying time.

(13) Use a spoon and the pots that will be used to make the final pot holes. Wet the pot sides and twist with slight downward pressure; enlarge the hole with the spoon.

(14) It is necessary to get a good fit between the pot and the pot hole; the farther down the pot sits in the stove, the better the heat transfer.

(15) Excavate the firebox and the tunnels using a machete, spoon, or trowel. The entrance to the firebox is an arch higher than it is wide. The water box hole should be about the same depth as the cook's height. The tunnels circulate the hot air and therefore should be wide enough to accommodate a hand holding three eggs.

(16) The chimney should be fitted with the tapered end down. The hole for the chimney should be deeper than the tunnel inside the stove (to avoid clogged passageways).

FIGURE 7.1
INTERIOR VIEW OF CORNER-BUILT LORENA COOKSTOVE

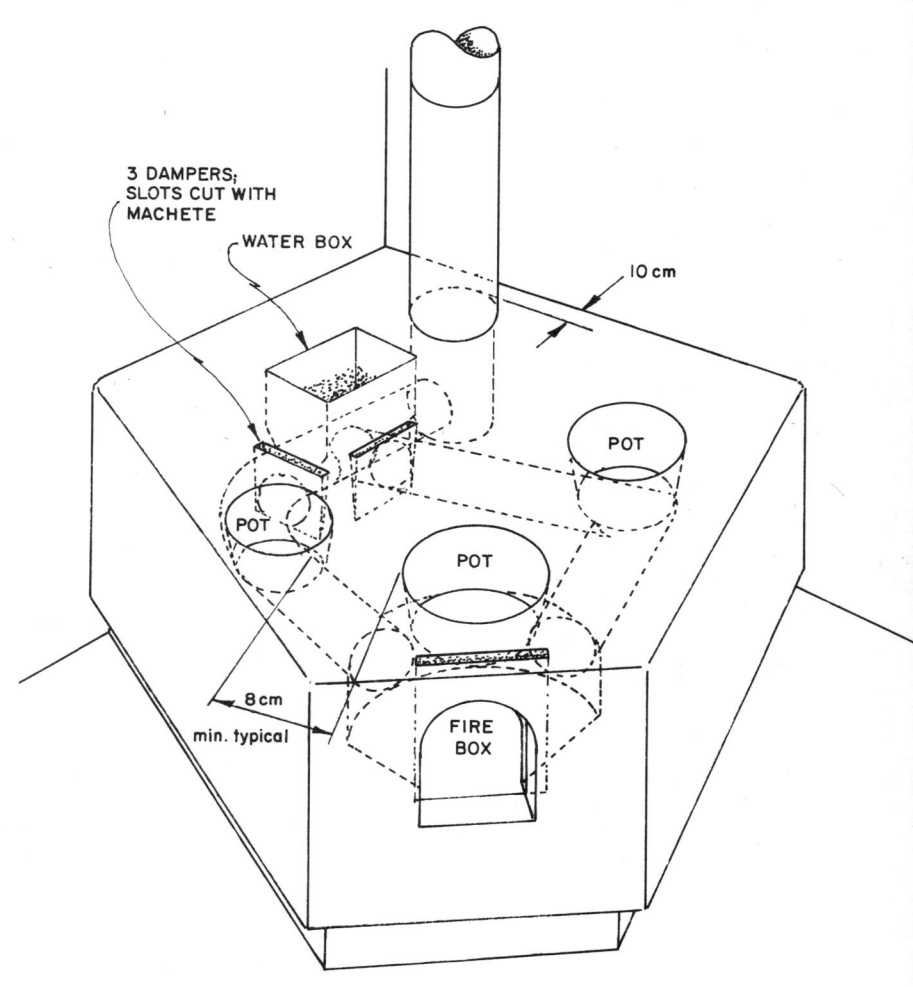

(17) The final details of damper doors and finishing the outer surface of the stove are not difficult but are necessary to insure proper operation and optimization of the stove. The outside of the stove should be whitewashed or varnished to prevent deterioration.

STOVE EFFICIENCY AND ADVANTAGES

The Choqui Experiment Station has conducted efficiency studies under different cooking conditions. The stove consumed 50 percent less wood than the traditional open fire system.[3] Tests by other groups record the wood consumption at 20 to 75 percent less than conventional stoves.[4] The wood consumption figure does not reflect the total savings to the owner, for the Lorena also can burn cornstalks, bagasse, sawdust, and organic wastes. This allows the owner to further reduce the time and money spent on finding or buying wood. The open-fire system requires prime wood because of poor combustion at the altitude of the Guatemalan highlands. The average fuel bill now is five to eight dollars, a 50 percent reduction.[5] For a family whose total monthly income is fifty to sixty dollars, the savings are significant.

Not only does the stove save money for the owner in initial capital investment and fuel bills, it also provides the cook with advantages. The stove is smokeless. With open-fire cooking there is constant exposure to smoke, creating eye, lung, and sinus irritations. Many older women in Guatemala have bronchitis, emphysema, and lung defects, and some women are blind due to a lifetime of smoke in the eyes. Another aspect of the stove that contributes to its attractiveness is its height. The stoves are built to an eighty-centimeter height, eliminating backbreaking work for the cook. This height also keeps the cook pot out of the reach of children, dogs, cats, and rats. The Lorena stove keeps the heat inside the stove, providing the cook with an exterior work surface and eliminating the suffocatingly hot kitchens with open fires.

---

[3]Ibid.

[4]Volunteers in Technical Assistance (VITA), "Wood Conserving Stoves, Draft" (21 November 1978). Submitted to Action/Peace Corps under Contract No. 78-043-1032, Mount Rainer, Md.

[5]Centro Mesoamericano de Estudios Sobre Tecnología Apropriada. Conversation with project director, March 1979.

Lastly, the surface design, i.e., sinking the pots in the surface of the stove, increases the efficiency of the stove. Little heat loss to the cool atmosphere occurs since the stove sits directly in the stream of hot air. The stove also allows for slow, overnight cooking without additional stoking. This is accomplished by closing the dampers all night; enough heat remains to simmer food and heat coffee.

The airflow in the Lorena stove is notably different from the Ghanaian and Indian stoves in which the air flows under successive pot holes in a direct path to the chimney. The Lorena circulates the air through tunnels under each pot and finally through the chimney. This type of routing has two important effects: maximum dispersion of heat within the stove and reduction of draft (which draws much of the heat out through the tunnel in the Ghanaian stove).

PROMOTION

The stove is being promoted by various local, national, and international groups throughout Guatemala and Honduras. Training classes are the major vehicle for promotion, supplemented by training manuals and demonstration stoves. By far, the classes are the most effective.

The training class is given to eight to ten people at one time at the request of an organization or group. The course lasts for three to five days, depending on the trainer and the drying conditions. The group provides the materials to make a prototype stove and chooses the location of the stove. Travel expenses and per diem are paid to some trainers (Centro Mesoamericano de Estudios Sobre Tecnología Apropriada [CEMAT], for example), and a donation to the organization (such as the Choqui Experiment Station) acts as payment in other cases. An entry fee of three dollars is charged for each student. It is not known exactly how many people have been trained, for there is no record of the "spin-off" trainers; however, CEMAT over the past fourteen months (as of February 1979) has trained 160 persons, and the Choqui Experiment Station has trained 640. Whether all of these students built stoves after the course is not known, but it is estimated that 800 stoves exist in Guatemala.[6]

The training programs are advertised in several ways: word-of-mouth, seminars (lecture visits to groups describing the range of activities that CEMAT offers), and promotional flyers. The best advertisement, however, is having actual stoves in operation.

---

[6]Interview, Larry Jacobs of the Choqui Experiment Station, Guatemala City, Guatemala, February 1979.

Lorena cookstoves at the Choqui Experiment Station
in Quetzaltenango, Guatemala

Cooking surface of stove; note the way the pots "fit in"
the stove and the position of the dampers

During the training program, the instructor often visits each trainee's house or clay/sand source and tests the clay for firing qualities. The clay/sand mix is critical to the durability of the stove, yet the "proper" ratio is hard to prescribe due to the variation of clays throughout Guatemala. Therefore, the trainer tries to analyze each clay source and "guesstimate" the ratio for the builder.

The following groups are involved in training courses throughout Central America:[7]

Local	National
CEMAT	Alliance for Community Development
Instituto Guataliko	Ministry of Education Extramural Development
International	Ministry of Agriculture
Peace Corps	National Guatemalan Reforestation Group (INAFOR)
Save the Children Federation	

FEEDBACK

The Choqui Experiment Station personnel have collected data on stove dissemination by visiting places where courses have been given. They visited sixty-four stove locations and interviewed the head of the household. Ninety-one percent of these stoves were working daily; 64 percent had been built by people who had taken the Choqui training course; 90 percent used less firewood; and 58 percent of these used 50 percent less firewood than before the Lorena.

Informal follow-ups also have been conducted by CEMAT. The data of CEMAT and Choqui personnel have raised several problems that all agree should be resolved before the training programs continue.

The major complaint expressed by those involved with the promotion project and by those using the stoves concerns the cracking problem. Seventy-three percent of the stoves visited had cracked in some way.[8] As previously

---

[7]The list is the result of conversations and interviews with the people in Guatemala who are involved in the dissemination of the Lorena stove.

[8]Preliminary Progress Report, p. 4

stated, the usual source of cracking is improper mixture of clay and sand. However, the cracking usually does not affect the operation of the stove.

Another problem voiced was the difficulty in trying to build a stove from instruction manuals, without personal instruction. The manuals just do not adequately explain the concepts of shape and design, making the visual conceptualization difficult.

The follow-ups also discovered that people wanted to change the proven design. This is attributed to the people's lack of knowledge of how to work the damper doors or the chimney. Also, stove builders are not putting a protective sealer on the stove body; thus, spills contribute to the deterioration of the stove.

It has been discovered that some of the stoves are cracking from too hot a fire in the firebox. This may be because women, who do most of the cooking, usually do not attend the training courses and do not know how the stove should be operated. Both of the trainers interviewed admitted that women probably would be excellent stove builders and that more efforts to include them in the training program should be pursued. Also, the builder sometimes becomes impatient and builds a fire before the stove is completely dry, causing the stove to crack.

IMPACT

Local promoters have a distinct effect on the "success" of the dissemination of the Lorena stove. A local promoter understands the people and their design needs and is available for consultation when builders run into problems. Thus, communities with promoters usually have more good, functioning stoves than communities without a local promoter.

Evidence indicates that stoves are being built by a local entrepreneur in one section of Guatemala--San Pedro la Laguna. He buys and assembles all materials, then builds stoves by contract, charging fifteen dollars for the finished product. This one incidence of local entrepreneurship does not mean that factories for Lorena stoves will be springing up. The finished stove is so heavy that transporting it would be impossible or at least too expensive to make it feasible. Also, the appeal of the stove is based partially on its flexibility--custom stoves to meet the needs of each situation. Traveling stove builders may multiply, but it is not believed that mass production will occur.

The dramatic reduction in needed firewood will affect the rate of acceptance of the stove and eventually will affect the rate of deforestation. Complete diffusion of the Lorena stove would put natural growth back ahead of the cutting rate, since it appears that the stove can save one-half to three-quarters of the wood normally used in cooking.[9]

The stove already has had an impact on the individual owners, who now do not have to spend countless hours and expend energy collecting wood.

Efforts to get the stove accepted by urban and rural planners involved in housing design also are being encouraged. After the 1976 earthquake, significant changes in construction of houses, e.g., compartmentalization, different roofs, and materials, were made. Part of the acceptance of the stove is credited to the change in attitude of the people--they were awakened to the fact that change is necessary. The Lorena stove is part of that transformation.

THE LORENA STOVE AS AN APPROPRIATE TECHNOLOGY TOOL

The Lorena cookstove exemplifies characteristics that lend themselves to wide dissemination. The stove is easy to build, utilizing simple tools and locally available materials. It can be adapted to meet the individual needs of the user, i.e., it can be small or large, can be made to fit a corner, can be placed out in the open for community use, and so on. Also, most importantly, the stove saves the user money. It cuts the monthly fuel bill in half. This factor alone is credited with the most word-of-mouth dissemination of the stove.

Another characteristic associated with the dissemination of the stove is the fact that local Guatemalans are doing the training. There are no language barriers, nor any feelings of a technology being brought in from the outside. The stove was developed with direct input and feedback from Guatemalans, and this has contributed to the steady dissemination of the stove. The beneficial characteristics of the stove are "selling" it; neighbors tell neighbors of the fuel-saving advantages, the smokeless kitchen, and the comfortable work space design, and this convinces the listener to build the stove.

Another factor that may contribute to the success of the stove is the tuition for the training course. Because the course costs three to four dollars, the skills and knowledge gained are viewed as worthwhile, or at least costly enough to utilize. No one is giving away anything.

Problems with the stove are presented to local promoters or to trainers when they pass through the community

---

[9]Ianto Evans, Lorena Owner-Built Stoves, eds. Jim Kalin and Ken Darrow (The Appropriate Technology Project of Volunteers in Asia, January 1979).

to check the stoves. The opportunity for improving or correcting a badly built unit is important. The stoves are not abandoned as a result of the people becoming discouraged with the design or with the dissemination programs.

Finally, the low cost of the stove contributes to the success of the technology. It is within the reach of all Guatemalans.

REFERENCES

Evans, Ianto. <u>Lorena Owner-Built Stoves: A Construction Manual for Highly Efficient Low-Cost Stoves That Can Save at Least Half the Firewood Normally Used in Cooking</u>. Jim Kalin and Ken Darrow, eds. (San Francisco: Appropriate Technology Project of Volunteers in Asia, January 1979).

<u>The Lorena Cookstove</u>. Estacion Experimental Choqui, Apartado Postal 159, Quezaltenango, Guatemala CA (no date) (Author's experiment station)

"Wood Conserving Stoves Draft." VITA, 21 November 1978.

<u>Lorena Owner-built Stoves</u>: a construction manual for highly efficient low-cost stoves that can save at least half the firewood normally used in cooking. by Ianto Evans, Aprovecho Institute. Edited by Jim Kalin and Ken Darrow, January 1979. A publication of The Appropriate Technology Project of Volunteers in Asia, San Francisco, California.

Preliminary Progress Report of the Lorena Stove Program. The Choqui Experiment Station. February, 1979.

PERSONS INTERVIEWED

Rodolfo Castillo of Cemat, Guatemala City, Guatemala, 20 February 1979

Phone interview with Larry Jacobs, Choqui Experiment Station, Guatemala City, Guatemala, February 1977

QUESTIONS TO CONSIDER

1. This is an example of a very simple "backyard technology." Is it possible and useful to try to formalize such simple technology-producing activities by creating

local institutions for this purpose? Are there enough situations present in the underdeveloped environments to warrant a formal institutional approach to providing research and development at these levels?

2. The development and use of more efficient cookstoves has been widespread throughout the world for thousands of years; what would explain the lack of adoption of technological improvements in similar situations?

3. In this and other situations described in these cases, the intervention of "outsiders" has seemed to be important in the innovations that have taken place; to effect change in the way things have traditionally been done may require such intervention; do you agree? What other elements are necessary or useful in effecting change of this sort?

4. Would it be feasible to systematically survey developing country situations, looking for needs for improved technology? Is this, in fact, done in any instances?

5. In what way did dissemination play a role in the success of technology? Should this practice be a more active part of technology transfer?

8
Honduras:
An Experimental Lime Kiln

Judith Evans Blum

BACKGROUND

In the latter part of 1974, Rigoberto Arrevalo, a rural entrepreneur, stimulated the search for a new lime kiln design in Honduras. As president of the Los Pinos lime cooperative in the Talanga region,[1] Arrevalo was acquainted with the technicians of the quasi-governmental Centro Cooperative Técnico Industrial (CCTI), which had helped the cooperative organize that year. Arrevalo discussed problems of the lime industry in Talanga with CCTI. In his view, the industry was facing several obstacles related to the fuel utilized in the technological process for calcining lime: (1) wood was becoming more scarce; (2) the governmental forestry commission, CODHEFOR, was beginning to regulate cutting; and (3) the price of wood was rising. Arrevalo was interested in developing a kiln that would operate more efficiently and require less wood than the traditional pot kilns used in Talanga and other parts of Honduras for decades.

CCTI was very interested in helping Arrevalo because the institute was just beginning to expand into rural development activities. Although CCTI technicians had involved themselves primarily in stimulating artisan industries, their goals included developing intermediate technology projects that would raise the standard of living of Honduras' poor. Their guidelines were to introduce modified production techniques that could be efficiently adapted to labor-intensive, low-cost situations.

Because CCTI's engineers lacked specific experience with kiln design, the institute wrote to Volunteers in Technical Assistance (VITA) in Washington, D.C. VITA is a private, nonprofit association that has provided technical

---

[1]Talanga is located approximately twenty miles north of Tegucigalpa, the capital of Honduras.

information and assistance to developing areas since 1960. CCTI and VITA had maintained contact with each other since 1969, and in 1974 VITA had a full-time representative in Honduras.

VITA sent a Washington representative to Honduras to confer with CCTI and VITA/Honduras and ascertain the level of commitment to the lime kiln project. After learning that local interest and willingness to work were substantial, VITA concentrated on developing a close cooperative relationship with CCTI; it was VITA's viewpoint that CCTI and VITA should work together on the project. In this spirit, VITA requested background materials from CCTI, including a 1957 study on the Honduran lime industry by the Arthur D. Little Corporation that had been contracted by the Honduran government. This study had recommended the installation of kilns that used oil for fuel, but no follow-up had been done. It was postulated that this was due to the high cost of oil in Honduras. At first, CCTI was reluctant to deliver informational materials, but this reticency disappeared as the trust relationship between the two organizations grew.

VITA also began to gather information in the United States, including standard texts on lime production and kiln design. During this period, the VITA/Washington representative stopped in Honduras again to determine the lime cooperative's commitment to the project. Although other members were interested, it appeared that Arrevalo was the one actively concerned individual.

Through a subsequent exchange of correspondence between VITA, CCTI, and VITA/Honduras, a number of questions were raised and answered concerning the best fuel for the kiln, the present uses of the lime produced in Talanga, and local procedures for quality control. Although alternative fuels such as charcoal, sawdust, and agricultural residues were considered, wood was still chosen as the best option. It was felt that charcoal was not feasible because no charcoal industry had developed in Honduras and because the use of charcoal would require a kiln technology too complex for the Talanga cooperative; sawdust and agricultural residues presented problems of combustibility; and all three alternative fuels posed the major issue of not being accepted by lime producers who were accustomed to using wood.

During this first stage of communication between VITA, CCTI, and VITA/Honduras, the following criteria were established: (1) the kiln should be locally constructed; (2) it should improve the fuel efficiency; and (3) it should produce a better quality of lime. In early 1975, VITA communicated with several volunteer consultants who were kiln experts and civil engineers to discuss the background materials and to ask the consultants to apply their expertise in designing a kiln that could meet the above criteria.

As of February 1979, Talanga was the only lime region in Honduras where the producers had formed a cooperative. According to Arrevalo, there were 700 active members. Work was a joint effort, with the owners of the kilns sharing laborers. Six types of lime workers were employed in Las Quebradas (a small town in the Talanga region), the site chosen for the experimental kiln: (1) stone drillers/dynamiters; (2) stone cutters/pilers; (3) stone haulers (who usually used an oxcart or wheelbarrow to bring the stone to the kiln); (4) kiln chargers or stone stackers; (5) fire stokers; and (6) lime unloaders.

Lime was quarried by drilling a seven-foot-deep hole in the limestone using a four-foot-long bar with extensions. The bar was tipped with a chisel point and was pounded with a sledgehammer. Customarily, one or two sticks of dynamite were placed in the holes. After blasting, the stone was broken up by sledgehammers.

The traditional method for calcining the lime involved the use of a pot kiln built into the side of a hill. The pot kiln was made of earth and stone and had a beehive shape. A large oven was typically twenty feet high with walls about two feet thick at the bottom. The inside diameter was approximately six feet. The kiln chargers would pile limestone within, first using large, flat pieces to construct a dome with a space below for a fire. When the dome reached a certain height and stability, stones would be added freely from the top. The construction of the dome required considerable skill gained over time.

When the kiln was full, it was left uncovered except in the event of rain, and the firewood was inserted from below. The fire was lighted, and the rising heat cooked the rocks. A large oven typically was fired for four days and nights. The kiln was loaded about once every thirty minutes with four or five pieces of wood ranging in size from twenty to ninety pounds. Since limestone shrinks when it has been calcined, the lime workers had determined from experience that the rock was sufficiently cooked when the pile of rocks had dropped approximately four inches below the top of the kiln.

After the calcined lime had been left to cool for a couple of days, it was prodded with a stick and loosened. At this point it consisted of small stones. These were sold to truckers who then transported them and resold them at higher prices in the marketplace. The kiln unloaders checked the lime for quality as they removed it from the kiln, pulled out underburned pieces, and loaded it in the truck. The truck driver usually helped also because his reputation in the marketplace was at stake. Unburned pieces were determined by their weight and color. The president of the Los Pinos lime cooperative reported that, on the average, 3 percent would be underburned.

The lime was measured in cargas, with one carga equaling one-half of a fifty-gallon oil drum, or approximately one-tenth of a ton. Wood was measured in carretadas; one carretada equals forty pieces of different sizes weighing twenty to ninety pounds each. According to CCTI, an average traditional pot kiln would use one carretada of firewood for every ten cargas of lime.

Lime from Talanga was used primarily for mortar and plaster in the construction industry, and the major marketplace was Tegucigalpa and the central region of the country. Selling the lime to a new sugar mill in nearby Canta Ranas also was discussed. For a while, Honduras was importing substantial quantities of lime from Germany for the sugar industry. However, it was evident that a lime plant in the northern region of the country near San Pedro Sula had replaced German imports for surrounding sugar mills, since lime imports from Germany had declined from 692,000 kilos in 1974 to zero kilos in 1976.[2]

In Honduras, lime also could be used as fertilizer for growing grapes, in road stabilization, and for the stabilization of soil blocks that would be adequate for modest load-bearing blockwork or as infill to concrete frames.

PROJECT DESIGN AND IMPLEMENTATION

In March 1975, VITA's Washington representative procured samples of lime and limestone from the Talanga region for analysis by the consultants designing the kiln. After considering all factors, one of the volunteer consultants suggested the design of a vertical kiln constructed of refractory brick. The limestone would be stacked in the kiln from the top so that the stones could be wedged against the sides of the brick wall. A firebox would be constructed underneath, allowing heat to pass upward to cook the stones. As the limestone burned, lime would descend to a receptacle at the bottom.

Although such a kiln was much more sophisticated than existing Honduran kilns, it was considered to be the most simple one that would produce a higher-quality lime and that could be adapted later to other fuel sources, such as bagasse. It would cost approximately five thousand U.S. dollars, which included $1,000 for the refractory brick. The justification for the cost was that savings would be realized over time with the production of better lime at lower costs.

---

[2]Ernie Burg, with modifications by Jeff Brown. Lime Kiln Report, VITA Volunteer On-Site Visit to Honduras. Mt. Ranier, Maryland: Volunteers in Technical Assistance, 1977.

The design was sent to CCTI and approved, although the Honduran institute had reservations about the need to use refractory brick that would have to be imported. CCTI, therefore, started to gather data on Honduran bricks to ascertain whether they could be used instead of imported ones. The criteria was that they be capable of standing up under very high temperatures, the minimum temperature for calcination being 830°C.

At CCTI's request, VITA planned to send the consultant who had suggested the vertical kiln to Honduras to help with construction. This proved impossible for personal reasons, and instead, in October 1975, VITA sent a consultant with experience in low-cost housing projects and kiln construction. Upon on-site review of the local situation, this consultant decided that the vertical kiln was probably too large and complex for the lime cooperative's requirements. The consultant proceeded to study the local production techniques, the local market demand, and techniques being employed in neighboring El Salvador and Guatemala. During the investigation, an abandoned lime operation was discovered in Comayagua, Honduras, approximately thirty-five miles away from Talanga. Expensive miner cars and hammer mills lay rusting away. The consultant learned that the equipment was owned by a local lawyer and that the kiln had collapsed; this supported the conclusion that the technology should be kept as simple as possible.

During his study, the consultant considered three options: (1) import an expensive, sophisticated technology; (2) build upon the existing traditional technology; or (3) adapt an imported technology to meet requirements for a low-cost, fairly unsophisticated technology.

The consultant decided against option one and instead recommended that the existing pot kilns be modified by reducing the diameter of the mouth from 3.5 meters to 2.0 meters, by building a kiln chimney 4.5 meters high to improve the draft, by fabricating a lid for the mouth of the kiln to decrease energy loss, and by constructing a door in the kiln to regulate the influx of air. Regarding option three, the consultant suggested using an inclined chimney kiln based on kilns that had been used for hundreds of years in Korea and other eastern countries for firing ceramics and bricks.

Both CCTI and the lime cooperative were interested in the inclined chimney kiln, although plans were also made to adapt one of the existing kilns owned by the co-op president in the manner suggested. VITA decided to convene a panel of the volunteer consultants who had been considering the Honduran situation to discuss and evaluate the new kiln design. The following major issues were debated: (1) that locally made brick might not fire sufficiently to withstand the high temperatures required for lime calcination; (2) that the 20 to 25 percent slope used in the Korean kiln might be dangerous and cause the limestone stacked within the kiln to collapse; and (3) that the mortar needed to be refractory.

To solve these problems, the panel decided that the slope should be changed to 12°, that the bricks to be used should be fired at the highest possible temperature (at least 850-900°C) and for a longer period (one and one-half to two times as long), and that the mortar should be made by crushing some of the bricks produced in the manner just described. The panel agreed upon a final design for a kiln that would be 20 meters in length and 1.8 meters in height and width, with the cross section in the form of a catenary arch. A door would be built at each end for loading and unloading. Firing holes would be included every two meters on each side of the kiln for adding wood fuel, and a chimney would be constructed at the upper end.

Meanwhile, CCTI had been negotiating an agreement with the lime cooperative in Honduras. In April this resulted in the following mutually agreed upon points:

(1) The cost of the new kiln would be U.S. $1,755, which would be shared equally by CCTI and the cooperative.

(2) The kiln would be located on a site owned by a member of the cooperative and, once constructed, would be the property of the cooperative and CCTI.

(3) Because of the experimental nature of the undertaking, losses, should they occur, would be sustained mutually.

(4) If the kiln were to prove successful, others like it would be constructed in the cooperative.

(5) CCTI would help to analyze the quality of the lime and to search for markets.

(6) The kiln would be maintained and operated by both CCTI and the lime cooperative during the experimental tests.

CCTI also prepared a work plan for construction that delineated tasks and outlined delivery schedules. Construction began in June 1976. CCTI requested that the VITA consultant be present to supervise. However, since this was not possible for financial reasons, CCTI and the VITA/

FIGURE 8.1
CUTAWAY VIEW OF INCLINED CHIMNEY KILN SHOWING
LIMESTONE STACKS AND FIRE CHAMBERS - NOTE
LOWER PLACEMENT OF SQUARE FIRE HOLES

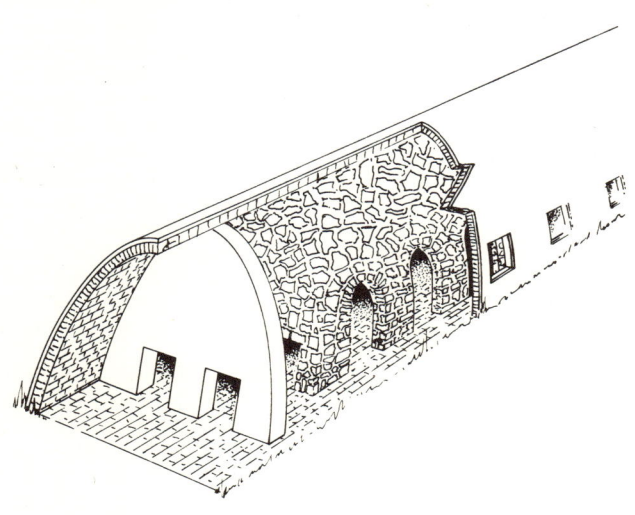

FIGURE 8.2
ORIGINAL DESIGN OF INCLINED CHIMNEY
KILN BEFORE MODIFICATIONS

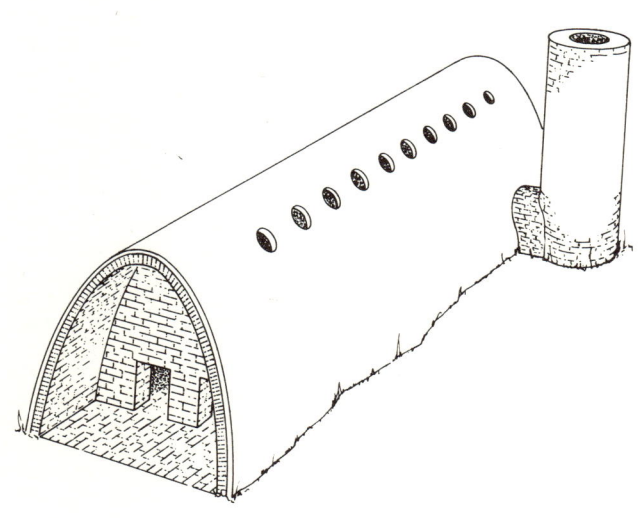

Honduras representative handled problems as they arose at the site. Whenever possible, they communicated by telephone and mail with the consultant to obtain his suggestions.

A major problem arose in the first month of construction. In order to build the arch, the workers used a wooden form over which they laid the bricks. Once one section was finished, they were supposed to push this form along and start another. However, when they removed the form from the first section, the section collapsed. The VITA consultant suggested that they build a roof over the kiln during construction to facilitate drying and that they employ different masonry techniques. The VITA/Honduras representative recommended the design of a new form that used less wood and modified the arch slightly. These suggestions helped and construction continued. As other problems arose, CCTI engineers solved them by recommending minor modifications in design and construction. The kiln was finished in September 1976.

The first test firing, conducted shortly thereafter, caused several cracks to appear in the side of the kiln. It was postulated that these may have been caused by rainwater leaking through the thatch roof during the burn, by expansion of the limestone during the burn (on this first burn the limestone was packed all the way to the ceiling and sides), and/or by uneven construction and insulation. CCTI wrote to VITA suggesting that the kiln walls be buttressed and, not hearing anything to the contrary, tried this. The VITA consultant later found out that the walls should be free to expand and contract as they heat and cool, and the buttresses were removed.

By May 1977, four test firings had been completed. Various problems had become apparent and, when possible, corrected. It was especially difficult to educate the kiln workers about how the type of wood used and the manner of fueling the fire affected the ability of the kiln to maintain a high temperature and hence affected the efficiency and quality of the firing process. Water, as present in green and/or wet wood, would reduce the temperature of the kiln. It was estimated that wet wood would lose approximately 40 percent of its value with respect to calcining limestone. However, the lime workers would pay the same price for all wood, dry or green, then leave it exposed to the weather until it was used. Thus, in the rainy season, very wet wood entered the kiln. The VITA consultant tried to get the workers to sort the wood but was relatively unsuccessful. He also suggested that the roof over the kiln be enlarged to shelter the woodpile and that a rack on which the wood could dry be placed over the kiln. However, these ideas were not implemented.

The traditional manner of fueling the fire was to load the kiln about every one-half hour with four or five pieces of wood. However, this method tended to cool the fire and cause incomplete combustion. From experience it was determined that hazy smoke indicated the best fire for calcining lime. The VITA consultant decided to have one or two pieces of wood added to the fire every five minutes, which resulted in a significant increase in the time that the smoke was hazy. However, he reported that the traditional fire stokers were unwilling to change their ways, and he felt that this caused an excessive consumption of wood and a high risk of poor lime. It was postulated that one possible reason why it was difficult to change the workers' habits was because the laborers (as opposed to the owners) were given no incentive. Neither a higher salary nor easier work were to be personally derived by changing old routines.

Another problem involved the placement of the fire holes. These were at first constructed quite high (125 centimeters) because more fuel could be added. However, the higher the hole, the more the flame shot out when the door was open. This made it harder to stoke the fire. Also, the burners would bang the kiln considerably as they struggled to get heavy pieces of wood through the high holes. To combat these problems, twenty extra holes were cut into the kiln at seventy centimeters above ground level. The old holes were covered up. This made the work easier for the fire stokers but caused wood to shoot across the kiln and bang the opposite wall forcefully.

It was questioned whether several fires should have been started at one time. Because the fire hole acted as a gas producer for the succeeding hole, the VITA consultant advised the use of one fire hole at a time. By the time the last, or tenth, hole was reached, it was not necessary to add wood because the gas fire was sufficient to calcine the stone.

Other aspects concerning the use of the new kiln that still remained in the experimental stage after the fourth burn included the location of draft tunnels, the size of the limestone piles, the need for thicker insulation for the walls, and the development of a method for determining when the limestone had been calcined. Compared to the traditional kilns, the new kiln had not yet proved consistently more fuel-efficient, nor had the new process been sufficiently worked out to demonstrate a higher yield of quality lime. The length of time required for a burn in the new kiln also was still greater than that required for the traditional ones. VITA had sent test instruments to aid in the analysis of the kiln during the experimental burns but, unfortunately, these were delayed in the air freight shipment and did not arrive until the fifth burn.

The objectives of this burn were to gather data using the test instruments, continue to develop criteria that would indicate when the limestone was calcined, produce a burn that yielded limestone of good quality, and start the fire with the kiln sealed. To the disappointment of all concerned, the middle of the kiln collapsed when the fire reached the ninth hole.

When the kiln was unloaded, the piles in the first nine chambers were found to be completely calcined, and 174 cargas of saleable lime were produced. In relation to the amount of firewood burned, CCTI determined that in the fifth burn the new kiln had produced 7.56 cargas of lime per carretada of firewood. This represented the best ratio of lime to firewood so far produced. However, according to CCTI, the new kiln still was operating at only 75.6 percent of the efficiency of the traditional pot kilns.

PROJECT FOLLOW-UP

The project participants from Honduras were dismayed at both the collapse of the new kiln and at the lack of conclusive evidence that during the period of its operation the new kiln was better than the traditional pot kilns. VITA also was disappointed but maintained the perspective that from the beginning the new kiln had been an experiment. VITA had attempted to encourage the experimental attitude throughout the project, but many, especially co-op members, continually expected that the demonstration kiln would be a solution to their problems. In the first planning stages, one of the VITA volunteer consultants had proposed building and testing the kiln in a U.S. laboratory. However, it was decided that field experimentation in Honduras was necessary for facilitating and determining the appropriateness of the technology.

Structurally, the following conditions were considered by VITA to be possible causes of the collapse of the kiln:

(1) The firing temperatures may have been too high for the bricks to withstand.

(2) The kiln was built by masons not familiar with the rigorous requirements of kiln construction.

(3) Work was supervised by several different people, and on-site supervision by CCTI was not possible on a continual basis.

(4) Some critical dimensions were wrong, resulting in stress (i.e., width: narrowness

allowed firewood to bang against opposite wall; and height of fire holes wood was banged against holes in the loading effort).

(5) On the first burn, the tightly packed limestone may have expanded and weakened the structure.

(6) Twenty extra fire holes were cut, weakening the structure.

(7) Buttresses built for the walls may have hindered the expansion and contraction of the walls during heating and cooling. CCTI, on the other hand, believed that the buttresses were necessary and that their removal contributed to the collapse of the kiln.

(8) Parts of the top of the kiln were wedged with small bits of brick because the arch was not laid up quite right.

(9) The insulating layer varied in thickness from zero to four inches, causing unequal expansion.

(10) Several leaks in the roof over the kiln allowed water to drip onto the hot kiln during the rainy season, causing considerable stress.

After the structural analysis, questions still remained about whether the kiln would have proved viable had it not collapsed. VITA recognized a number of problems, including the relatively high initial capital outlay and the subsequent skepticism on the part of co-op members that the return on the investment might not be worth the risk. If the fuel savings and improved marketability of the lime justified the added cost in the long run, VITA realized that considerable education about the technology would have been necessary and that credit possibilities would have had to be explored. The acceptance of the new technology by the local labor force also was considered a possible problem. VITA realized that the established hierarchy and pattern of work, especially concerning the use of fuel, did not match the requirements of the new technology and that education and incentives would have been needed.

CCTI believed that a new kiln costing ten times more than a traditional one should significantly improve fuel efficiency and provide unique capabilities for producing high-quality lime. CCTI concluded that the inclined chimney kiln could not do this but that certain improvements, based on the technology of the new kiln, could be made in the traditional pot kilns. For example, wood could be used more efficiently, and the walls of the kiln could be lined with brick so as not to contaminate the lime.

CCTI currently is exploring methods for modifying the traditional pot kilns. The president of the lime co-op maintains close contact with CCTI regarding the research. He has modified one of his pot kilns in the manner earlier suggested by the VITA consultant to try to increase the efficiency and quality of production. A chimney for the kiln has been constructed, and a cover for the kiln reportedly is in use. Arrevalo stated that his modified kiln was 40 percent more efficient than the traditional ones and yielded only 1 percent uncalcined lime compared to 3 percent in the traditional kilns.

PROJECT SPIN-OFF

Seven modified inclined chimney kilns have been constructed in a brick factory in nearby Valle de Los Angeles. Coincidentally, the brick factory was the original brick supplier to the Las Quebradas experiment. During the course of the experiment, the owner obtained the plans from CCTI, modified them to meet his needs, constructed the kilns, and has used them successfully in his business.

It was learned that the impetus for building the kilns was that the owner's Hoffman ring kiln was slated for revision and that alternate kilns were needed in the meantime. Wood shavings were used for fuel. Lower temperatures (maximum 800 to 900°C) were sufficient for firing the bricks, and the duration of the burn was relatively short. Moisture control was important to keep the upper bricks from melting caused by condensation of the steam generated from the drying of the lower bricks. Therefore, three fires were used at one time, with the fire holes spaced about one meter apart. The design modifications made by the brick factory owner included the construction of stepped shelves in the inclined shaft so that the bricks would not slide down, the construction of sixteen-inch-thick walls consisting of one layer of bricks on edge with a second layer on end, the use of a semicircular arch built on top of short walls, and the connecting of as many as three kilns to one chimney. Therefore, although the inclined chimney kiln in Las Quebradas is no longer operating, the concept has not left the region.

REFERENCES

Discussions with persons involved in or knowledgeable about the lime kiln project from VITA, the lime cooperative, CCTI, and AID/Honduras.

Bury, Ernie, with modifications by Jeff Brown. Lime Kiln Report, VITA Volunteer On-Site Visit to Honduras, May 1977. Mt. Rainier, Maryland: Volunteers in Technical Assistance, 1977.

Lime Production in Honduras. Mt. Rainer, Maryland: Volunteers in Technical Assistance, 1977.

Lopez, Maris, and Lardizabal, Ernesto. Informe de las Actividades Realizadas en el Proyecto de Instalación del Horno de Cal en Talanga, F.M. Tegucigalpa, Honduras: Centro Cooperativo Técnico Industrial, 1977.

Parry, J.P.M. Proposals for Building Materials Manufacture in Honduras Using Light Capital Technologies. London: Intermediate Technology Development Group, 1978. pp. 28-33.

Parry, J.P.M. Proposals for Method Improvements and New Technologies for Small-Scale Building Materials Manufacture in Honduras. London: Intermediate Technology Development Group, 1978. pp. 3-8.

QUESTIONS TO CONSIDER

1. What can be said about the nature of research and development, with regard to the ratios of success, that might apply to the case situation?

2. In terms of project organization, what improvements could be made based on the experience in this situation?

3. Would it have been a good idea to build a laboratory prototype before attempting the full-scale experiment? Would the timely arrival of the test instrumentation have made any difference in the outcome?

4. What type of incentive program might have been designed for the kiln workers?

5. If the project were to be revived, what should be the order of procedure? What would be done about the modified pot kiln experience? Should the project be revived?

6. What are the implications of the adoption of the technology in another setting, at the brick works?

9
Malaysia:
Small-Scale Brick Manufacturing

Ronald P. Black, A. Rahim Bidin, Woo Seng Khee, and Nik Ahmad Kamil

INTRODUCTION

Malaysian brick manufacturing was initiated in the early part of the twentieth century. From the outset, old rubber trees were used as a source of fuel for firing bricks. While this is still the practice, one small-scale brick manufacturer recently raised a question as to whether this practice would remain viable much longer.

BACKGROUND

Small-Scale Brick Manufacturing Firms in Malaysia

These firms provide bricks for partitioning in Malaysia's home and building construction industry, which has been growing rapidly. Based on data provided by the Malaysian Department of Statistics for 1974, the authors estimate the current brickmaking industry for Peninsular Malaysia[1] to be a M$ 50 million-per-year business.[2]

In Malaysia, companies that produce more sophisticated decorative brick, as compared to the small-scale brick manufacturers, do exist. According to a preliminary survey of the Standards and Industrial Research Institute of Malaysia (SIRIM), the decorative brick companies provide about 2 percent of the total number of bricks produced in Malaysia. The decorative brick manufacturing companies are much more capital intensive than the "common" brick firms. According to the SIRIM survey, a typical decorative brick factory would cost ten times more than a common brick

---

[1]Peninsular Malaysia excludes primarily Sarawak and Sabah on Borneo.

[2]One U.S. dollar is equal to approximately M$ 2.2.

factory. The decorative brick factories are more automated and quality conscious than the common brick factories and use diesel fuel for firing bricks.

Rubber Trees in Malaysia

Rubber trees were introduced in Malaysia during the nineteenth century. Rubber is now the country's number one foreign exchange earner--currently representing 23 percent of Malaysia's export revenues. Malaysia now produces 47 percent of the world's natural rubber.

Rubber trees are replanted every twenty to twenty-five years. This is encouraged by the Malaysian government through a subsidy to planters for acreage replanted. This subsidy remains in effect until the new trees begin to produce. In 1975, Peninsular Malaysia had 1.4 million acres planted in rubber trees. Reportedly, 3 percent of all Malaysian rubber trees are felled each year.[3]

Over the last several years, the number of acres planted in rubber trees has decreased slightly as a result of shifts to palm oil production. Recently, however, the government has begun to explore tax incentives and other measures that may halt these shifts.

Wood from old rubber trees in Malaysia originally was used for fuel. At present, it also is used for producing charcoal, wood chips, and blockboard (see Table 9.1 for amounts used in 1974).

TABLE 9.1
USES FOR OLD RUBBER TREES

Use	Amount (tons)
Cooking	Unknown
Brick kilns and smokehouses	2,000,000
Charcoal	1,000,000
Wood chips	300,000
Blockboard	10,000-20,000

Source: "Utilization of Rubber Wood," pp. 1-2.

A SMALL-SCALE BRICK MANUFACTURER

Eng Huat and Company is typical of small-scale brick manufacturers in Malaysia. It is family owned and managed

---

[3]Tan Ah Goh and Chang Wai Pong, "Utilization of Rubber Wood for SMR Pallet Construction" (Shah Alam Selangar, Malaysia: Sirim, 1974), p. 1.

and has been in existence for about twenty years. Raw materials must now be brought to the company site, located in suburban Kuala Lumpur. In its early days, however, the site was surrounded by rubber trees; the clay used for the bricks was obtained from nearby deposits. These materials have long since been depleted.

When the plant was visited by the authors in early 1979, Mr. Yap, a young company manager, was pleased with the current success and future prospects of the business. While he envisioned developments that would require changes for Eng Huat, he seemed confident that these could be met successfully.

Facilities

Eng Huat and Company, together with the family residence, is situated on three acres of land. Office activities are conducted in the home.

For transporting clay from the pits to Eng Huat, the company has two five-ton trucks, one a 1974 Bedford, the other a 1970 Mercedes. Each cost about M$ 40,000. The company also has a 1960 Michigan bulldozer that is used for moving and mixing the different types of clay. The bulldozer cost approximately M$ 100,000. Fifteen wheelbarrows are used for moving the bricks; each wheelbarrow cost M$ 100.

The company also owns a locally made, screw-type extruder that is more than ten years old. The extruder can process clay for approximately 20,000 bricks in an eight-hour shift. Its present power source is a forty-horsepower Lister diesel engine of about the same age as the extruder. It is located in a room near the extruder and uses eight gallons of diesel fuel per eight-hour shift. Attached to the extruder is a manual wire cutter that cuts the clay into bricks. The cutter produces ten bricks at one time.

The extruder, cutter, and diesel engine are located under a shed that measures approximately 35 by 150 feet. The shed also is used for drying and storing the "green," or unfired, bricks. The shed is constructed of local materials. It has a roof of galvanized zinc sheets, and its sides are open.

The two kilns used by the company are ten years old, which is the normal lifetime of a kiln (see Figure 9.1 for a sketch of one of the kilns). The kilns are constructed of the type of bricks produced by Eng Huat. The kilns measure fifteen feet in height, forty feet in length, and twenty feet in width, and the wall thickness is three feet. The top of the kiln is open. Along the length of the kiln, on both sides, are eight firing holes, two feet by two feet, into which rubber tree wood is fed for fuel. The kilns are located under an open-sided shed that has a galvanized zinc roof.

Process

Eng Huat and Company produces approximately 300,000 bricks per month of a quality acceptable to Malaysian building contractors for partitioning walls in houses and other buildings (see Figure 9.2 for a schematic diagram of the process used for producing these bricks).

The two company trucks obtain clays from pits located within five miles of the business. The clays are dumped about two hundred feet from the shed containing the extruder. The front-lift bulldozer then does the preliminary mixing of the clays and brings them, as needed, to the extruder shed. These subsequently are shoveled into a box approximately ten feet square located over the extruder. Here the clays are mixed, if necessary, to give the bricks their desired characteristics. Water is added to give the mixed clay a suitable moisture content for shaping by the extruder.

In the center of the box is an opening for the extruder's hopper, into which the mixed, moistened clay is shoveled. The extruder shapes and forces a rectangular column of clay from its mouth. The clay column then moves across a table to three persons who operate the manual wire cutter.

The green bricks are then manually carried by three persons from the cutting table to another part of the shed, where they are stacked for drying by two other workers. The green bricks are allowed to dry for three days and are then ready for loading in the kiln.

The green bricks are hand carried to the kiln, where they must be stacked in a pattern that allows for easy transmission of heat throughout the kiln. Although many stacking patterns are possible, Yap uses the one that his family has always employed. The bricks are stacked to a height of approximately twelve feet, a process that requires about four days and eight workers and that produces approximately 100,000 bricks.

Firing the bricks requires about seven days. Approximately six days of low-intensity firing are needed to gradually remove moisture from the bricks, followed by one day of intense firing. Yap can tell when the intense firing should begin by observing the smoke from the kiln. When the smoke ceases, intense firing should commence. The time required for low-intensity firing is determined by the moisture content of the green bricks.

During the low-intensity firing, rubber tree wood is fed into each of the kiln's firing holes every six hours. This is

FIGURE 9.1
ENG HUAT KILN

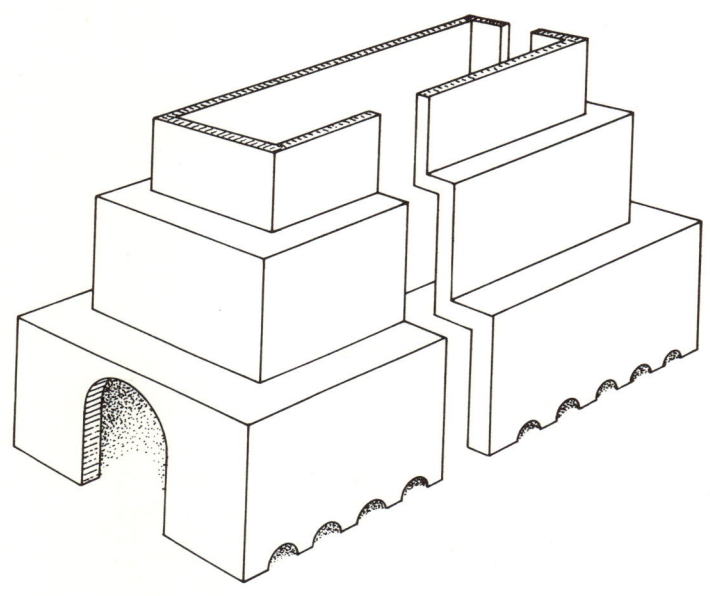

FIGURE 9.2
BRICKMAKING PROCESS AT ENG HUAT

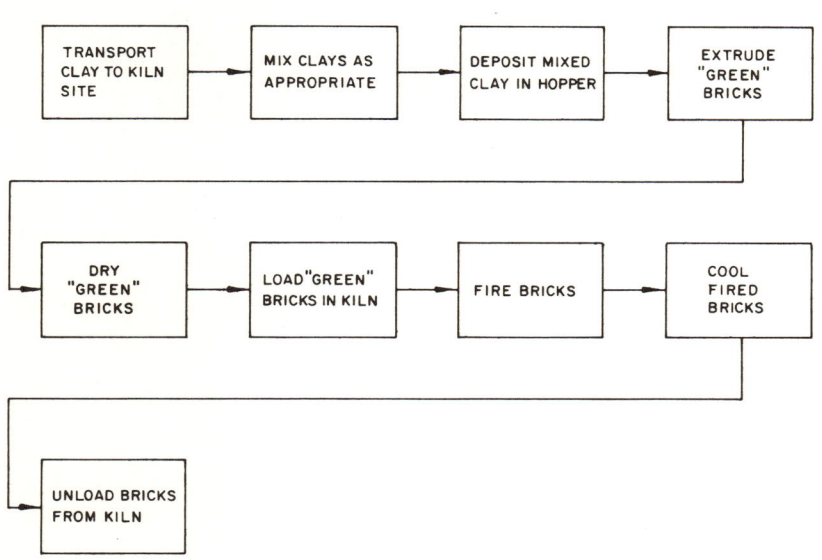

Shed covering brick kiln

normally done by two persons. When intense firing begins, fuel is added every thirty minutes, requiring a team of four persons.

Finally, the kiln must cool for two days before the workers can begin to unload the bricks. This usually is done directly into a purchaser's truck by about five persons. Unloading takes one to five days depending on the number of workers. Five persons can unload a kiln in about two days.

Stacking bricks after firing

Fuel

Eng Huat uses primarily old rubber tree wood as fuel. Occasionally other scrap woods are used, but Yap prefers rubber tree wood because it provides a more consistent heat.

The rubber tree wood is purchased from dealers who collect it within a thirty- to forty-mile radius of Eng Huat, then deliver it to the company. For one kiln firing, approximately twenty-five truckloads of the wood are required. Each load weighs about three tons and costs about M$ 60.

Although old rubber tree wood has been the traditional fuel for small-scale brick manufacturers in Malaysia, Yap says that it is getting increasingly difficult to obtain. He believes that within five years he will have to shift to a petroleum-based fuel that will cost more and that will require a kiln of a different and more expensive design.

Personnel

Yap and his brother manage the company's operations. Eng Huat hires two truck drivers, one bulldozer operator, two persons for shoveling the clay, three persons for cutting the clay into bricks, three workers for carrying the green bricks to where they are to be stacked for drying, and two persons for stacking the green bricks. The two truck drivers and bulldozer operator receive between M$ 400 and M$ 500 per month. The other workers are paid on a piece basis. The ten-person team is paid M$ 50 for each 10,000 stacked green bricks.

Eng Huat contracts for the work associated with the firing operation. Yap pays M$ 80 for 10,000 fired bricks. The staffing pattern for brick firing seems to vary considerably depending on the availability of workers. However, based on discussions with Yap and his brother, a typical pattern is estimated to be as follows: (1) loading the kiln--eight persons working one shift for four days; (2) low-intensity firing--two persons working three shifts for six days; (3) intense firing--four persons working three shifts for one day; (4) cooling--none; and (5) unloading the kiln--five persons working one shift for two days. Therefore, the firing operation is estimated to provide 13.5 person-months of employment each month. This figure is based on obtaining an average of three firings per month from the two kilns.

Yap noted that it is getting increasingly hard to find workers; therefore, he has thought about automating parts of his operation. He said that his first step probably would be to purchase a fork lift.

Yap expressed the opinion that the increasing difficulty in finding workers was due to the growing number of work opportunities in the Kuala Lumpur area. Not only are more places hiring workers, but a wider variety of types of businesses are offering employment opportunities. According to Yap, many of these jobs are more attractive than working as a manual laborer in a brick factory.

Market

Eng Huat sells bricks directly to building contractors. Because of the current construction boom in Malaysia as a whole and in Kuala Lumpur in particular, Yap notes that he easily can sell all the bricks he can produce. He does not advertise; contractors come directly to him to buy the bricks.

Yap said that several years ago Eng Huat exported bricks to Singapore because a better price could be obtained there. However, in the mid-1970s the Malaysian government passed a regulation preventing the exportation of bricks because of the brick shortage in the Malaysian construction industry.

At present, Eng Huat receives M$ 0.08 to 0.10 for each brick depending on its quality. Since the company sells about 300,000 bricks per month, its annual sales volume reaches approximately M$ 324,000.

Finances

According to Yap, capital costs for setting up a brick factory similar to Eng Huat would total M$ 1,268,500 (see Table 9.2). Monthly operating expenses for Eng Huat are M$ 12,654 (see Table 9.3).

Because Eng Huat and Company owns all of its facilities and consequently has no loan to pay off, its monthly profits can be estimated at approximately M$ 14,000.

To calculate the funds that an investor would need to create a job place in a small-scale brickmaking factory in Malaysia, several assumptions were made. First, such a factory would not be established on property as valuable as the land on which Eng Huat is located. For approximately M$ 90,000, three acres of cleared land may be obtained twenty-five kilometers out of Kuala Lumpur. This would be

TABLE 9.2
CAPITAL COSTS FOR BRICK FACTORY SIMILAR TO ENG HUAT

Requirement	Number	Price (M$)
Land	3 acres	1,000,000
Office/residence	1	15,000
Trucks	2	80,000
Bulldozer	1	100,000
Extruder	1	15,000
Diesel engine	1	15,000
Shed	1	2,000
Kiln	2	40,000
Wheelbarrows	15	1,500
	TOTAL	1,268,500

TABLE 9.3
MONTHLY OPERATING EXPENSES FOR ENG HUAT

Requirement*	Cost (M$)
300 truckloads of clay @ M$ 8 per load	2,400
300 gallons of fuel for trucks @ M$ 1.20 per gallon	360
Wages for two truck drivers	900
Wage for bulldozer operator	450
Fuel for operating extruder	144
Wages for workers producing and stacking green bricks	1,500
Contract for firing bricks	2,400
Rubber tree wood	4,500
TOTAL	12,654

*Based on a production output of 300,000 bricks.

a more likely location for a new brickmaking factory than inside metropolitan Kuala Lumpur.

Second, a used bulldozer, costing approximately one-half the price of a new one, would be purchased to move the clay. Third, it is assumed that a loan could be obtained to cover 60 percent of the capital investment costs and that payments on the loan would not commence until the factory was making an income.[4] Finally, it is assumed that the new factory would need the equivalent of two months'. operating expenses.

Based on the preceding assumptions, an investor in a small-scale brick manufacturing operation in Malaysia would need about M$ 100,000. Assuming the staffing pattern described earlier, one job place could be created by an investor for about M$ 8,000.

A SIRIM PROJECTION[5]

The building industry in Malaysia is growing at 8.4 percent per annum, and this trend is expected to continue into the 1990s. Brick, being a major component of building construction in the country, is expected to be produced with an output growth rate approximating that of the building industry. Other materials, such as concrete blocks, boards, wooden planks, and concrete panels, can compete with bricks, but they are not expected to pose a serious threat. The relative costs and convenience, as well as the availability of clay throughout Malaysia, favor the continued use of clay bricks. Only concrete panels and cement blocks can completely substitute for bricks. Wooden planks and boards can be substituted for bricks only in certain applications, such as partitioning walls that are constructed internally and where soundproofing and other properties possessed by bricks are not required.

The use of cement blocks and concrete panels depends on the availability and accessibility of cement. The current cement supply in Malaysia just about meets the demand. The cost of manufacturing Portland cement is dependent upon the cost of fossil fuel, namely oil. And the cost of oil is expected to continue to climb due to the depletion of the world's oil reserves. Furthermore, investments for cement

---

[4]According to a loan officer at a local bank, these are both reasonable assumptions.

[5]For the purposes of this case study, the SIRIM staff projected the outlook for the brick industry and the use of rubber tree wood as a fuel source in Malaysia. The SIRIM projection is contained in this section.

plants are high. As a result, it is not expected that the supply of cement will increase dramatically in the near future. Switching from bricks to cement blocks in building would cause a severe shortage of cement in Malaysia and thus constrain the increased production of cement blocks. The continuing use of bricks therefore seems assured.

Most of the bricks produced in Malaysia are fired in kilns that use rubber tree wood as a fuel. A surplus of rubber tree wood exists in Malaysia today, even though the supply is restricted in parts of the more developed areas of the country. Old rubber trees are continually cut down to make way for new rubber trees; this is expected to continue. The government recently has tried to encourage a greater use of land for the planting of rubber trees; this should help to ensure a continuous wood supply.

A small percentage of bricks produced in Malaysia today is being fired by oil-fueled kilns. These bricks are used mostly for decorative purposes and cost a great deal more than the common bricks. The price of decorative brick is expected to increase as the price of oil rises. It is also expected that the price of bricks fired by rubber tree wood will increase but at a slower rate than the cost of oil-fired bricks. The price of common bricks will rise gradually because of inflation and also because rubber tree wood is being used increasingly as a fuel by many industries that do not want to pay the escalating costs of oil.

CONCLUDING REMARKS

The common brick manufacturing industry in Peninsular Malaysia, 80 percent of which employs fewer than fifty persons at each factory, accounts for an annual M$ 50 million contribution to the gross domestic product of the country.[6] Assuming that Eng Huat and Company is typical (which SIRIM surveys support), then the industry provides jobs directly to approximately 4,500 Malaysians, most of whom are unskilled.

Bricks provide approximately a 5 percent value input to the construction industry. According to the SIRIM projection, if bricks were not available, the country would have to substitute cement. This, in turn, would cause a severe cement shortage.

Since old trees are a waste product of the rubber industry, their use as a fuel in firing bricks is valuable.

---

[6]See Survey of Construction Industries, Malaysian Government Department of Statistics, Kuala Lumpur, 1974. The figure for the annual contribution to the country's gross domestic product has been extrapolated from this report.

In addition, jobs are created and income is generated for the persons who cut and transport the 140,000 tons[7] of old rubber tree wood consumed by the brickmaking industry in 1979.

The use of rubber tree wood as a fuel for the brickmaking industry in Peninsular Malaysia represents the replacement of 6.16 million gallons of heavy fuel oil. The difference in the cost of fuel alone is M$ 4,501,000.

As a result of this study, the SIRIM staff began to question whether the "typical" common brick factory in Malaysia is indeed the most "appropriate." Much of what Mr. Yap described to the team about his operation and facilities was based on "handed down" knowledge. The SIRIM staff began to question whether the facilities, particularly the kiln design, and the operations (for example, the pattern in which the bricks are stacked in the kiln), are optimum. As a result, SIRIM may initiate a new program to determine, scientifically, what is the most "appropriate" technology for common brick manufacturing in Malaysia.

REFERENCES

Goh, Tan Ah, and Pong, Chang Wai. Utilization of Rubber Wood for SMR Pallet Construction. Shah Alam, Selangar, Malaysia: SIRIM, 1974.

Malaysian Government Department of Statistics.

Far Eastern Economic Review: Asia 1979 Yearbook. Hong Kong: Far Eastern Economic Review Limited, 1979.

PERSONS INTERVIEWED

Mr. Yap, manager, Eng Huat and Company

QUESTIONS TO CONSIDER

1. "More modern" obviously does not equate with "appropriate" in this case. Does this use of a renewable energy resource represent a unique situation, unlikely to be found in any but a few other places, or is it likely that waste organic material could be used rather

---

[7]This 1979 figure (estimated) is based on the amount of wood required for one kiln firing and the annual output of brick in Peninsular Malaysia.

than fossil fuel in many instances? Would this be a good subject for a general survey in a developing country situation, or may it be assumed that such practices are employed already where "natural economics" suggest it?

2. To what extent have countries conducted energy surveys that are comprehensive quantitative evaluations of energy use? Would this not be a sound idea?

3. What is the most effective role for an in-country research institution such as SIRIM, with regard to studying unsophisticated small industry like this? Is the subject of small-scale, dispersed, and traditional industry best handled by specialized institutions? Or is it more efficient to include these "appropriate" or "intermediate" technology subjects within the purview of general industrial research institutes such as are found in most countries? Or, put another way, is the "small is beautiful" idea something around which specialized research institutions should be organized?

4. Should government take an active role in starting and aiding appropriate small-scale industry as this brick manufacturing activity, including finance, technology, and market development, or would this be inefficient and wasteful of public funds?

10
Colombia:
An Innovative Approach to Rural Development

James M. Miller

BACKGROUND

The Foundation for the Application and Teaching of the Sciences (FUNDAEC) is a private, nonprofit organization that was granted legal status by Resolution #4747 of the government of Cauca Valley, Colombia, on 25 October 1974.
FUNDAEC actually had begun several years earlier at the University of Valle (funded by the Rockefeller Foundation), where questions had been asked about the role of higher education in the field of development. Many years of involvement in community projects, primarily in the area of health, had led several of the professors to question the very concepts of development. It was clear that traditional indicators such as Gross National Product and exportation indices were not applicable in terms of describing the conditions of the poor. In spite of economic growth, health conditions, nutrition, living conditions, and educational opportunities for the majority of the people had not improved and, in many cases, had actually worsened. This is attributable to the large gap that exists between the established institutions--governmental or private, national or multinational--that have know-how, human and financial resources, and infrastructure, and the small farmers, who each year abandon their villages in large numbers to migrate to the slums of urban areas.
One of the often-mentioned solutions to needed improvements in development projects is to formulate interdisciplinary groups and multi-institutional programs. One such group consisting of professors at the university met regularly for almost one year and began to consider models for social development. However, problems related to the integration of their efforts at the level of community action arose. An element seemed to be missing: someone or a group of persons who could coordinate actions or needs at the family and community levels. Since no one with this type of skill

could be found from among the group of university experts, issues were raised about the effectiveness of the various academic disciplines themselves.

To begin with a specific problem, that of rural development, a new approach was taken by asking the question, "What would a person need to know in order to help the small farmers improve their living conditions?" Many fields of knowledge were seen as indispensable, especially agriculture, health and nutrition, sanitary engineering, civil engineering, formal education, and elements of the social sciences.

THE SMALL FARMER IN COLOMBIA

The size of landholdings in itself is a somewhat variable parameter since in regions such as the western plains of Colombia, twenty hectares is a small expanse, while in the fertile land near Bogotá, one to three hectares cultivated in flowers or vegetables are sufficient for a good income. Traditionally, in Colombia a small farmer is defined by the government as one who possesses less than ten hectares of land, who consumes most of what he produces, and who utilizes primarily his own manual labor. However, the majority of the small farmers have plots much smaller than ten hectares, consume little of what they produce, and are dedicating an increasing portion of their time to working in larger enterprises. These farmers, who do not have access to the services offered by the Colombian Social Security Institute, the Agriculture and Livestock Financial Fund, or other public institutions, grow approximately 60 percent of the food commodities produced in Colombia. This same rural population, according to data from the Survey of Human Resources for Health and Medical Education, has the least access to income, health care, sanitation facilities, and education of any group in the country.

The amount of land dedicated to agriculture in Colombia during 1970 was 31.3 million hectares, of which 76 percent was owned by 7 percent of the landowners in individual holdings generally far in excess of 50 hectares. In the decade of 1960-1970, the number of holdings with less than ten hectares decreased by approximately 5.5 percent due to abandonment or sale of the property. Of these, the farms most affected were those consisting of less than five hectares. This represents a total loss of land for more than 90,000 small farmers and their families, or an equivalent of 540,000 persons.

Twenty percent of persons employed in the rural sector work only six months of the year; 30 percent work up to eight months. Only 50 percent work all year. Although the national average income is $162[1], small landowners receive only between 21 percent and 60 percent of this figure, or from $34 to $97. Therefore, in most instances, only a farmer with holdings between ten and fifty hectares can begin to have the possibility of an acceptable standard of living. As of April 1979, only 22 percent of the farmers are in this category.

Consequently, the rural population has the worst conditions in housing and sanitation and the lowest levels of literacy and primary education in the country. Nutritional levels are lower than those found in urban areas. The number of malnutrition cases, which in urban areas is around 30 percent, reaches levels of almost 90 percent in some rural regions.

A particular example of these conditions is present in northern Cauca Department. This region comprises various rural communities that are characterized by an unequal distribution and inadequate utilization of the land. With a population of 65,000, of which 25,000 live in the small town of Puerto Tejada, the region is inhabited primarily by small farmers who have partially abandoned their own agricultural labors in order to work for the nearby large sugar industries. Data show that 60 percent of the rural population are involved in this type of labor. Some 80 percent of the rural population have plots smaller than three hectares, of which 27 percent have only one-half hectare. Fifty-six percent of the houses lack sanitary facilities and, in some places, malnutrition reaches levels of between 65 and 70 percent.

The principal causes of these problems in the rural sector traditionally have been identified as the scarcity of credit, the poor quality of technical assistance, and difficulties in marketing. As mentioned previously, while accepting the indisputable importance of these factors, it seems in general that other aspects of possibly equal importance are not fully considered in the life of the small farmer. These include the need for new and adequate systems of agricultural production and animal husbandry and new technology that would permit the small farmer to live with a certain degree of independence.

The majority of development agencies have investigated and promoted solutions that have been successful for the larger agricultural industries. Such solutions have then been reduced in scale to be applicable to the small farmer. Much experience indicates, however, that a mere reduction in scale is not always effective and may not produce improvements in rural living conditions.

---

[1]U.S. dollars.

In the Cauca Valley the credit guide published by the Regional Corporation of Cauca Valley (CVC) shows an average profit of between $30 to $40 per hectare per month for commonly cultivated crops such as corn, soybeans, cassava, and red beans--a sum that represents an appreciable income for the owner of fifty or more hectares. These figures assume optimal conditions in the accessibility of modern machinery, credit, and marketing. In other words, assuming the perfection of existing national agencies and the accessibility of these services to the small farmer, the owner of two hectares (which is considerably more than the average in northern Cauca) would have a monthly income of sixty dollars. Considering that the prices of basic foodstuffs are relatively high ($1.25 per pound of meat, $.60 per pound of red beans, and $.30 per pound of rice), it would be extremely difficult to feed a family.

Confronted with this situation, the majority of those who have not already sold their lands to the large agricultural industries maintain their traditional farms with permanent cash crops such as coffee, cocoa, plantain, or fruit trees. Although the lack of techniques such as crop rotation and the use of fertilizer has diminished the productivity of these landholdings, the crops nevertheless contribute to the low, but fixed, income of the rural family. The other part of the income derives from occasional low salaries from the agricultural industries.

Historically, and even today in many parts of the world, small farming is not as bad an occupation as the foregoing discussions show. Where, then, lies the contradiction between the actual situation of the farmers in northern Cauca (and in many similar parts of the world) and the potential for small farming? The quality of land in the Cauca Valley is so good that one hectare could produce abundant food for a large family. Why, once working within the economic system of a society that even in its most inefficient form is designed to render at least some benefits to its members and is seemingly preferable to a system in which individual families work their own farms in isolation from society, do the small farmer and his family suffer from severe malnutrition?

Part of the answer is found by examining modern agricultural production and animal husbandry, which, because of their mechanized and technical nature, have high requirements for investments of capital, energy, and natural resources. The question arises, then, whether science, which has permitted the development of a system applicable to a certain life-style, could not also improve the life of the small farmer. Wouldn't this contribute to the solution of the worldwide food problem given that, for example, in many countries in Latin America more than 60 percent of all food already is produced by the small farmer under present adverse conditions?

Other questions examined are: (1) what experiences are there concerning making science and appropriate technology available to the small farmer? (2) what are the proper channels? (3) what methods are showing promising results? (4) where is the manpower for such a vast program of community education? (5) how is this manpower in itself to be trained? (6) what portion of the existing knowledge of appropriate agricultural practice actually needs to be propagated and to what extent? (7) how should the recipients of these services be organized? (8) what are the methods to bring about community participation? (9) what is really meant by it? (10) who should make the decisions about priorities? (11) are these organizational models at the village level? (12) what is the role of the economically advantaged section of the community? (13) how should the private sector participate in rural development and through what institutions?[2]

TECHNICAL SOLUTIONS

Farzam Arbab, who received his Ph. D. in elementary particle physics from the University of California, Berkeley, and currently is a professor of physics at Valle University in Cali, concluded that solutions extended beyond traditional academic disciplines and indeed beyond the traditional educator. A concept of supplying a person with knowledge about his/her environment was required. Arbab gathered a small group of professionals from the natural sciences, social sciences, humanities, medicine, agriculture, and engineering fields. He set up what is now known as the Rural University (FUNDAEC), but which is not accredited in the usual academic manner.

The staff of FUNDAEC totals twenty-three full- and part-time professionals in the following fields: agricultural systems, mechanical and civil engineering, sociology, physiology, education, biology, medicine and public health, physics, anthropology, and veterinary medicine.

With the objectives of educating rural candidates in their own environment and having them remain in the rural scene rather than migrating to urban areas, selection teams went into rural areas discarding aptitude testing and concentrating instead on ability to learn. They gave courses to volunteers, then tested the volunteers for potential achievement in an accelerated program.

---

[2]All of the foregoing was extracted from papers by Dr. Farzam Arbab, head of FUNDAEC, and Ms. Christine Tucker of FUNDAEC.

The first group of candidates, twenty-six in number between the ages of sixteen and twenty-three, started in May 1975. The lowest level of schooling was through the fourth grade in grammar school (three in this group remain), and the highest level of schooling was the third year in high school. Six were young women. After the first two years, the group had dropped to sixteen, three being young women. The class then stayed this size for the remaining four years. Training is for six years and, for lack of a better name, is called "Engineering for Rural Welfare."

The second class began two years later with twenty-two students. Four were women. The class still retains fourteen, four of whom are women.

All classes are held in FUNDAEC facilities built in the community of Arrobleda, forty-five minutes from Cali. The students are completely subsidized, and they stay in dormitories for the three-month periods, some returning home for a one-month vacation following each trimester.

Textbooks are all originals written by FUNDAEC staff. The Colombian Ministry of Education has approved the curriculum as experimental for Arrobleda, in the Cauca Valley, making two years of FUNDAEC equivalent to four years of the Colombian high school program.

The objective of the "engineering" program has been to educate a new professional who can work within the rural milieu and look for answers to problems concerning health and nutrition, sanitation, housing, agricultural production, income, and community organization. The six-year training program has been designed in three stages, each consisting of two years. The curriculum fits into the following scheme:

Preparatory Stage (First two years)	University Studies Stage (Second two years)	Professional Stage (Third two years)
Language	Science Courses	Integral Courses
Manual Arts & Technology	(special texts)	
	Community Service	Community Service
Natural Science		
Math		
Community Service[3]		

The five general areas of the first two-year phase are designed to develop within the student scientific and investigative abilities. They do not follow the usual system dedicated almost totally to information delivery. Development of these fundamental capacities enables the students to learn in university-type courses of the second phase. The students

---
[3]Ibid.

learn to study texts, extract information, express themselves, and communicate with others. They are given the opportunity to utilize their creativity and become familiar with the concepts of appropriate technology for translation into community service.

Looking at the first of three "text" examples--language--the objective is to regain and even strengthen the student's ability to communicate in terms of his rural language. Tapes and lecture material are built around this concept of retaining idiomatic usage as well as proceeding with developing language ability to cope with university-level education. The text begins with concepts like forms, size, geometry, substances and their properties, and color and expands into more complex concepts such as systems, changes in the state of a system, and interactions. True and False tests are given every five days.

In the area of mathematics, although abilities are taught systematically, immediate applications to real situations or to other areas of knowledge contribute to the process of integration. For example, in the first text, after a review of the concepts related to sets, a few lessons are dedicated to the classification of the plant and animal kingdoms. The text on geometry includes a number of activities in simple topography that are applied immediately by the students to their work in the community. The text on fractions includes many pages of material on epidemiology, a subject that generally is introduced in advanced university courses. The first concepts of the topic, however, only use rates and ratios, which are applications of the mathematical concepts well within the grasp of the students. The student uses these tools immediately to examine the health indicators for the Cauca Valley, the rates of mortality, and the prevalence and incidence of the most important diseases. This reorganization of knowledge is an important task for educators, and what FUNDAEC has done is indeed a very small step toward a far more ambitious program of changing the content of the text for primary and secondary education.[4]

During the first deliberation about possible methods and content in the areas of natural sciences, the educational team of FUNDAEC became aware of the inherent limitation of teaching the sciences as distinct disciplines. According to Arbab, the student, when introduced to the field for the first time through highly structured disciplines, with their definitions and formulas, never learns the fundamental task of observing, discovering, and describing that which surrounds him. The student mentally divorces science from nature, which, paradoxically, is what science tries to study.

---

[4]Dr. Farzam Arbab.

It therefore was decided to teach the sciences of systems and natural processes related to the future work of the students. A list was made of the themes that are covered in pre-university science courses. These themes were separated concept by concept and reorganized into units that examine systems and processes such as the cultivation of beans and corn, the biosphere, the digestive system and nutrition, and the betterment of varieties of crops and animal breeds. In addition to developing the basic scientific capacities of the student, this approach has led to increased motivation and greater clarity of the material presented.[5]

The second phase of science courses (third and fourth years) is designed around University of Valle courses but with special texts. Professors from the university conduct many of these courses. Throughout all six years, emphasis remains on community services. This emphasis involves observation and description of the environment and its needs, which is carried out in cycles through specific packages developed by the facility of the Rural University around "research-action-learning." The fifth and sixth years of professional integration are almost entirely devoted to community services and are organized around these packages.

Currently, packages under development are child development, production in small farms, appropriate technology, environmental sanitation, and community organization. The following sample details one of the packages selected at random.

Concepts of Sanitation: A Three-year Program.[6]

The first year includes lessons in diagnosis:

> General sanitation
>
> Problems of intestinal parasites relating to gastrointestinal systems
>
> Taking fecal samples
>
> Taking data on incidence of diarrhea
>
> Laboratory diagnosis at the University of Valle of intestinal parasites
>
> Diagnosis of water supply and disposal systems
>
> Diagnosis of nutritional systems

---

[5] Dr. Farzam Arbab.

[6] One of the authors of this program is Dr. Gabriel Carrasquillas, M.D., who also holds an M.A. in public health.

The second year includes action/interventions based on diagnoses of the first year:

Water supply and disposal systems

Continue measuring indicators

Deeper theory and practice

The third year:

Analyze what has been accomplished.

As the training activities progressed, it became increasingly evident that more was required than placing new professionals as agents for rural change. The need for activities in the field of appropriate technology led to the establishment of a new department within FUNDAEC called "Technology for Rural Welfare." This effort seeks to adapt technology that is more suited to the economic capacities of the small farmer.

Two full-time engineers are dedicated to research in the field, supported by additional staff when necessary. They are seeking viable alternatives for production systems for the small farm. Work is underway in minimal irrigation, fertilizers, land preparation, credit, and availability of seeds.

This work on appropriate technology opens the opportunity of educating technicians and para-professionals in shorter training programs within the six-year training of Engineers for Rural Welfare. This program has two levels: (1) the practitioner for rural welfare, who undergoes one year of study; and (2) the technician for rural welfare, who has a three-year period of study and is classified as a middle-level professional. These two programs will become available during the coming year.

ECONOMIC ASPECTS

A large number of people (on staff and otherwise) plus organizations and industry have continuously contributed services and materials to this program. No donated value appears in the following fundings:

Appropriations: (all in U.S. dollars)

1974	$32,000	Rockefeller Foundation (design programs and curricula)
	4,000	Colombian Foundation for Higher Education (Arrobleda land)
Total	$36,000	
1975	$29,000	Rockefeller Foundation
1976	53,000	Rockefeller Foundation
	2,000	Colombian Foundation for Higher Education (FES)
Total	55,000	
1977	75,000	Rockefeller Foundation
	10,000	FES
	10,000	Carvajal Foundation of Colombia
Total	95,000	
1978	50,000	Rockefeller Foundation
	30,000	UNICEF
	10,000	Colombian National Plan/Foods & Nutrition
	23,000	Fundo Nacional para el Desarrollo (FONADE)
	7,000	Eder Foundation of Colombia
	27,500	U.S. Inter-American Foundation
Total	147,500	
1979	70,000	Rockefeller Foundation
	55,000	PACT of the U.S.A.
	15,000	United Nations University
	*27,500	Inter-American Foundation
	7,000	Eder Foundation of Colombia
	**45,000	Colombian CIID (estimates pending)
	24,000	Ford Foundation
	50,000	PAN of Colombia
Total	293,500	

* Divided into 1978 and 1979.
**To cover two years.

SOCIAL AND ECONOMIC IMPACT

The community services "packages" vary in length throughout the six years. The sanitation package, for example, is three years. They then are to be repeated in subsequent classes, which will provide the basis for longer-term evaluations that will indicate the success or failure of specific interventions.

At present, each of the students in the first class (1974) has done about eighty family case descriptions that have led to assessments of need, e.g., water supply, animal feed, waste utilization, agricultural needs, and social/ economic associations between families. The latter, for example, has resulted in separate cooperative systems for chicken production, processing, and marketing as individual small business enterprises.

The success of the program, according to Arbab, can only be measured by the welfare of the families and communities in which the rural engineer works. Evaluation plans are drawn up to yield the educational program's improvement. Immediate conclusions are premature, but the small projects now being implemented show excellent relationships between students and communities.

An obvious indicator of success is the fact that all sixteen students in the first class have been offered positions in various Colombian government agencies, even two years prior to graduation. They are to remain in the rural environment. Another strong indication of acceptance is the funding growth from $36,000 in 1974 by two agencies to $293,500 by eight agencies, national and international.

In terms of potential social impact, all aspects of health, education, nutrition, housing, social and community development, public utilities, recreation, and retention of professional leadership in the rural scene are areas of great promise for the future. Arbab advised that the best student with the most potential is one who finished only the fourth grade. According to Arbab, it is inspiring to see intellects saved for productive use for their communities.

Favorable economic impact, increased agricultural production and utilization, higher income levels, strengthened family relationships, and local and regional development with reduced urban migration all appear to be quite possible.

RELATIONSHIP TO NATIONAL GOALS

The Colombian PAN Food and Nutrition program plus the Integrated Rural Development program call for increased food production, better nutrition, improved sanitation, and improved rural infrastructure. The National Health System is extended to small towns and inaccessible areas. The National Education System aims at the application of knowledge to local conditions. FUNDAEC, with its engineers and technology for rural welfare through its university and community service, intends to fulfill all of these objectives.

PERSONS INTERVIEWED

Dr. Farzam Arbab
President of the Board of Trustees, FUNDAEC
Apartado Aereo 6555
Cali, Colombia

Dr. Gabriel Carrasquillas, staff member, FUNDAEC

Ms. Christine Tucker, staff member, FUNDAEC

QUESTIONS TO CONSIDER

1. The more direct application of academic education to the problems of the rural areas is a sought-after objective. Do the students in this program have sufficient opportunity to learn the basics of their fields of interest, or are they too dispersed across their rural activities to develop a real specialty? If not, should there be an institution to accomplish these sorts of development activities, in addition to the more traditional forms that concentrate on academic preparation of students?

2. The minimum land area necessary for self-sufficiency in this region of Colombia appears to be below that which is typically owned. Is there a clear need for land redistribution policies in this instance? If so, how might they be accomplished?

3. Is the assistance program introduced by the university faculty likely to be perpetuated, or is it a phenomenon that depends on the motivation of a few individuals for its success? Is this concept replicable in other situations?

4. Other than the limited testimony of a few interviewees, is there any evidence that the program was having a material impact on the rural economy? In view of the land tenure situation and the economics of agricultural self-sufficiency, is there any real prospect for the long-term viability of the small farmer system?

11
Brazil:
The Role of CRUTAC in Community Development

James P. Blackledge

INTRODUCTION

The migration of people from rural areas to urban centers causes similar problems in nearly all developing countries. Such migration tends to reduce agricultural productivity on the one hand and at the same time to greatly increase congestion in the cities with measurable impact on transportation, housing, health, employment, and a number of other socially and economically related factors that are nearly impossible for city administrators to resolve.

The desire to migrate to urban centers is fairly easy to understand. The environment in rural areas frequently is not conducive to a meaningful way of life. Unemployment or underemployment, extreme poverty, malnutrition, lack of educational opportunities and adequate health care facilities, i.e., a lack of minimal living conditions in general, are prevalent and persistent. People arrive at the inevitable conclusion that there is no foreseeable solution to this dilemma.

Thus, as various means of communication have improved in the rural area, the attention of rural inhabitants has focused on descriptions of a better way of life in the urban centers, more employment opportunities, and improved education for their children. Under these circumstances, even though such opportunities are seldom verified, people leave rural areas in large numbers to journey to the cities. As is well known, such migration does not significantly change the life-styles of the majority of people so involved. Rather, it exacerbates their existing problems.

Many attempts have been made to change living and working conditions in rural areas with the dual objectives of improving the quality of rural life and diminishing rural migration to urban centers. Intense effort is currently being directed to an integrated approach to rural development, encompassing a variety of social, economic, and technological considerations. Usually, assistance is provided through international and bilateral entities, although many developing countries are increasingly determined to take the initiative in establishing local or national plans and procedures to implement their own action-oriented programs for improving rural life.

During the past twelve years, attempts have been made by the Federal University of the state of Rio Grande do Norte (UFRN) in northeast Brazil to ameliorate living conditions in selected rural areas within the state. The program, called Rural Center for Training University Students and for Community Action (CRUTAC), has, as dual objectives: the training of students to better understand the rural environment and to find solutions to the problems of rural life; and to provide direct social, medical, legal, educational, and technological services to rural areas through use of students in situ.

This pilot experiment, initiated in August 1966 in the state of Rio Grande do Norte, has been the basis for similar programs now established in several Brazilian universities. While it must be recognized that there is no universal panacea for improving the quality of rural life, it appears that the CRUTAC approach is a mechanism worthy of consideration by universities in the developing countries.

BACKGROUND

The state of Rio Grande do Sol in northeastern Brazil provides a contrast in terms of economic growth and development on the one hand and continuing poverty and minimal living conditions on the other. One of the principal Brazilian space facilities is located near Natal, the state capital. Industrialization is increasing around the capital at a reasonable rate. The land, to a depth of approximately fifteen kilometers inland from the state's extensive coastline, is well suited for agriculture and has accounted for most of the state's agricultural production. At the same time, agricultural production has declined from 50 percent of the state's domestic product in 1950 to 23 percent in 1971.

The contrast between the coastal areas and the interior of Rio Grande do Norte is startling. The interior region experiences a shortage of water and energy. Drought is endemic. Health and sanitation conditions are marginal to poor. Unemployment is high. Agricultural productivity is low and usually consists of subsistence farming. Opportunities for self-improvement are limited, and migration by rural families to urban centers continues. This situation is by no means unique to Rio Grande do Norte but is generally experienced by most of the states in northeastern Brazil.

The federal government of Brazil has been attempting, for a number of years and with massive inputs of funding, to ameliorate this situation. Understandably, such efforts involve a long-term process of education, creation of adequate health services, and meeting community needs. These programs have been conducted on a regional basis (where the region is very large), and measurable results are still meager.

The UFRN, located in Natal, receives principal financial support and administrative direction from the Federal Ministry of Education and Culture. UFRN currently has a student body of around 8,000, projected to grow to around 15,000 in the next five years. As is typical in Brazil, the students, for the most part, come from rural areas and small urban centers within the state. As is also typical, most of the students desire to find employment within the state upon graduation. Academic departments include medicine, pharmacy, dentistry, public health, social services, sociology, economics, education, law, home economics, agronomy, and engineering. The faculty is composed largely of natives of the state.

For a number of years UFRN struggled to develop a clear perception of its role in the three functions of a university, namely, teaching, research, and community service. This perception was codified in 1965 by the then rector of UFRN, Dr. Onofre Lopes, who realized that UFRN tended to be structured in a rather rigid mold of teaching and research, with minimal provision of community service. He reasoned that, since many UFRN students came from rural and small urban centers, a feedback mechanism should exist to encourage students to transfer their knowledge back to their respective communities.

He further reasoned that the university had an obligation to utilize its resources--students and faculty--in focusing on specific community problems where a definite impact could be made. He foresaw a structured approach whereby the resources of the university could, in an integrated manner, attempt to solve community problems in various regions of the state. This structural approach he called the Rural Center for Training University Students and for Community Action (CRUTAC). Dr. Lopes persuaded the Federal University of Education and Culture that CRUTAC was not only clearly within the objectives of UFRN, but also was a viable mechanism for providing appropriate and useful community service. The necessary legislation, bylaws, and regulations were completed, and CRUTAC became an operating entity of UFRN on 28 December 1965.

CRUTAC ORGANIZATION AND OPERATIONS

As already mentioned, CRUTAC uses a multidisciplinary approach to community action that involves UFRN faculty and

FIGURE 11.1
ORGANIZATION OF CRUTAC

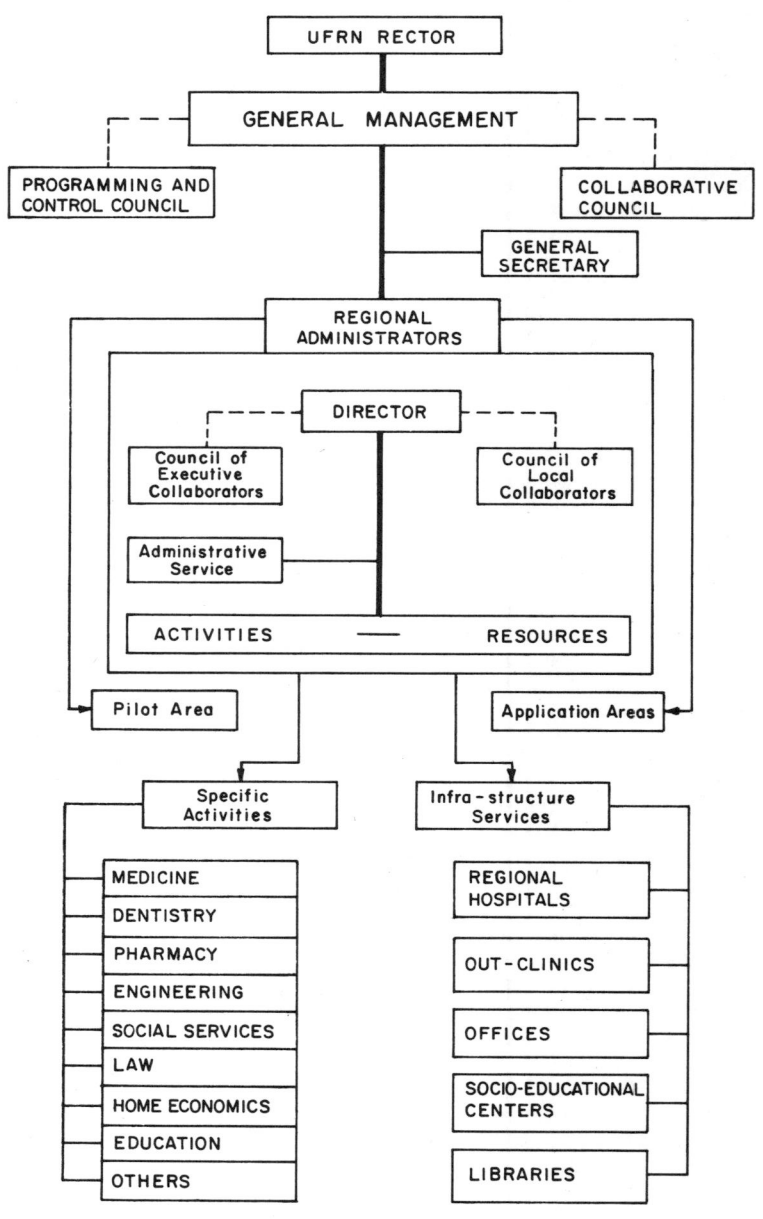

students in the final year of university training. The decree by the Federal Ministry of Education and Culture, which established CRUTAC, states that the university must create, in the geo-economic region in which it is situated, training programs and the application of different methods of teaching, through the rendering of services coordinated with the programs and needs of the local governments and the interests of the respective communities.

The budget for CRUTAC's operation is derived from federal, state, municipal, and university funds and is controlled by the University Governing Council (see Figure 11.1). Two important committees assist the CRUTAC director: (1) the <u>Collaborative Council</u>, formed by chairmen of the academic departments, the presidents of student associations, the directors of public services of state and federal governments, and civic leaders, which provides advice and counsel to the CRUTAC director in terms of cultural and administrative needs of society; and (2) the <u>Programming and Control Council</u>, formed by representatives of the departments involved, a program specialist, a communications specialist, and a pedagogical advisor, which establishes the general program of activities, coordinates the integration of the various disciplines, and evaluates the effectiveness of the program.

CRUTAC does not attempt to compete with existing public and private organizations that provide social and other assistance to areas of the state. CRUTAC maintains liaison with entities of the state and federal governments, with private organizations such as the Rotary and Lions clubs, and with the press, radio, television, and other mass communication media in order to avoid duplication of effort and to provide services where none exist or are deemed to be insufficient.

The state of Rio Grande do Norte has been divided into sixteen microregions of ten to twelve municipalities each, with the objective of impacting to the maximum extent the greatest number of people within a maximum radius of sixty miles from CRUTAC headquarters. Two such microregions, with populations of 73,000 and 90,000, currently are being provided service by CRUTAC. Each microregion being considered is subjected to a socioeconomic analysis that includes a number of factors to help determine the extent of assistance and the available resources, both existing and latent. These factors are:

(1) <u>General information</u>: land characteristics; urban and rural population; population density; number of families (using an index of six people per family); number of municipalities; and maps of the municipalities.

(2) Climatical and physical aspects: general data; type of zone; higher above sea-level; highest and lowest areas; and precipitation.

(3) Infrastructure: highways--distance from Natal to the region center; railroads; telegraph; telephone; mail service; and energy.

(4) Economic conditions: agricultural situation--exploitation of farming; animal raising--herds or flocks; mills; meat; by-products; ownership of rural land--number of owners; system of working land; and payment of workers--salaries, crop sharing, rental of land.

(5) Organizations assisting in agricultural production: federal; state; municipality; individuals; agro-associations; and general conditions for credit--banks, cooperative, individual.

(6) Industries: number and types; persons employed in industry; principal raw materials and possibilities for use; and crafts and artisans--type and number of employees.

(7) Commerce: number and types of stores, business, etc.; persons employed; and existing banks.

(8) Aspects of health: registrations of births--archives, churches; deaths--causes; hospitals, maternity centers, first aid clinics, pharmacies; doctors, dentists, pharmacists; midwives; and healers and unlicensed practitioners.

(9) Educational education: number of children of school age; number of children in schools; levy the attendance in schools; number of schools in community; number of professors, nonprofessionals, trainees; adult education--individual; elementary; secondary; libraries--numbers, public, or private, use factor; and museums--numbers, public, or private.

(10) Social situations: social centers--numbers, types, use, location; clubs and associations--dancing, sports, syndicates, music, etc.; cooperatives--number, types of credit, etc.

(11) <u>Churches</u>: Catholic, Protestant, others.

(12) <u>Public services</u>: police; municipal offices; army cantonments; civil registry; and judiciary.

(13) <u>Communications</u>: method of transportation--rail, highway, etc.; and newspapers, radios, television.

(14) <u>Reasons for communal concentration</u>: public holidays--dates; civil holidays--dates; and religious holidays--dates.

(15) <u>Folklore</u>: dances; regional customs; musical instrument players; and poets and authors.

Such a socioeconomic analysis, to the extent that data are available, provides a good picture of the resources as well as the needs of the region in which CRUTAC wishes to operate.

For each microregion a director is appointed and advised by two committees. These are: (1) the <u>Council of Local Collaborators</u>, composed of people of the region, including municipal officials, a judge, minister, or priest, directors of public services, and community leaders, who provide guidance and counsel to CRUTAC in consonance with the region's needs; and (2) the <u>Executive Council</u>, composed of team leaders, university faculty, and representatives of local services, who conduct meetings with involved students, analyze projects underway or contemplated, and advise on redirection or reemphasis to meet community needs.

As new regions are added to CRUTAC operations, the regional management structure is expanded so that each region has its own management team.

UFRN DISCIPLINES AND NATURE OF CRUTAC TEAMS

CRUTAC can draw on a broad range of capabilities from the university, particularly where the need for a certain service is critical or where the impact of the service can be expected to be significant.

The composition of a typical CRUTAC student team is as follows: (1) one medical surgeon student for urgent and selective surgery, trauma, and obstetrics; (2) two clinical medicine students with capability for outpatient treatment, auxiliary surgery, and anesthesia; (3) other medical students in various subdisciplines where necessary; (4) one dental student for preventive odontology and restoration; (5) one pharmacy-biochemical student for dispensing medicines, laboratory tests, etc.; (6) one education student to evaluate the system of local education, present teacher training courses, teach in outlying municipalities, assist in cultural diffusion, organize libraries, etc.; (7) one law student for various types of assistance, including civil and criminal laws, rural and worker rights and legislation, legal rights of individuals, families, etc.; (8) one social work student for community training, organization of community groups, artisan and craft training, domestic education, recreation, etc.; and (9) one engineering student, one economics student, one agricultural student, and one veterinary student, as needed, to solve specific problems or assist the above.

The numbers and mix of students can be changed according to specific regional requirements as determined by the local collaborative and executive councils and the regional CRUTAC directors. The previous list is regarded as the core or minimum team for assignment to a region. Every attempt is made to overlap student assignments to avoid breakdown in communications and to transfer information regarding specific tasks or projects in various stages of progress. A full-time professional team leader resides in the region to ease transitional problems and to keep CRUTAC activities at an optimum level of performance.

As stated earlier, the students are assigned to CRUTAC during their final year of academic training so that they are close to a professional level of experience (see Table 11.1). In the majority of disciplines, students receive academic credit at the university for their participation in CRUTAC projects and are graded similarly to classroom participation. Engineering students, who usually are not in residence in regions, may receive credit either for inputs or for making

TABLE 11.1
PARTICIPATION OF UFRN STUDENTS IN CRUTAC (1975)

Academic Discipline	Students Per Year	Hours Per Student
medicine	120	260
dentistry	70	260
pharmacy	70	260
social work	40	130
engineering	50	80
education	100	80
law	50	80
economics	50	80
accounting	50	80
administration	50	80
sociology	40	80
agriculture	30	80

up lost work in classes. Participation of students in UFRN departments of medicine, dentistry, pharmacy, social work, and law, is obligatory. For example, students in social work must participate two days per week for fifteen months; medical students must participate for thirty-five days continuously, and the integrated teams are required to participate for two days per week over a period of eighteen months. Engineering students usually spend only a short time in the field to gather the data and information necessary to design a building, a park, a sewage system, and so on.

Some UFRN professors also are in residence, particularly in the area of medicine, as team leaders. They usually have three to five years of experience and view their assignments as a form of internship that gives them more practical experience than can be gained in larger urban hospitals. Similar data were not obtained for 1978; however, in one of the three microregions where CRUTAC currently is operating, the following number of students are involved each month: 104 in medicine, 86 in dentistry, 46 in sociology, plus 20 professors in health services and 5 professors in other areas. Since CRUTAC is now concentrating on an integrated team approach, the number of students involved has increased considerably, although in some areas, such as engineering, student involvement has declined as the obvious projects such as the design of a sewage system, or plans for a park, have been completed.

STUDENT REACTION TO CRUTAC

A number of the students currently involved in the CRUTAC program have been interviewed relative to their perception of the beneficial aspects of CRUTAC as well as the constraints that may have impeded full effectiveness of the program in providing community services.

First, the students feel that the experiences gained during their tour of duty have been beneficial both to themselves and to the microregion where they have served. The students have commented that (1) the training and in-residence periods should be of longer duration; and (2) students should receive exposure to the problems of community service in the field, perhaps at the end of the second year of academic training at the university, which would help to prepare them for service with CRUTAC during their final year at the university. It appears that the students who have been involved regard their CRUTAC experience as very meaningful in terms of their own professional development. Since most of them will stay in the state upon completion of their academic training, they recognize the importance of impacting, even in a small way, on their state's economic and social development.

The students are particularly proud that their involvement in CRUTAC has promoted and strengthened the concept of self-help activities in the community from the municipality, civic leaders, and the churches to direct involvement in the process by the people themselves. The students have expressed the opinion that it is virtually impossible for the university to provide sufficient numbers of students to solve all community problems unilaterally. People in the community must have the desire to help themselves. Then the student inputs can be meaningful.

The students' professors regard the CRUTAC program as a meaningful, practical experience to augment theoretical academic training. While the two opinions do not necessarily conflict, it appears that CRUTAC management views community services as most important, while the faculty and students view the training and practical experience gained as the most important element of the program.

Several students have commented that they feel the need for more interaction with their professors during their field assignments. They have said that sometimes they feel inadequate to undertake assigned tasks and could benefit from more experienced leadership and better communications. On the other hand, the students recognize that they are given more opportunities to gain practical experience than would ever be possible in a totally academic environment.

The students also recognize that the demand for CRUTAC assistance exceeds the number of students available to participate. They continue to raise questions as to how the state and federal governments can be motivated to provide more support to CRUTAC and to expand the program.

CRUTAC ACCOMPLISHMENTS

CRUTAC operations are diverse, as demonstrated by the following examples. Medical students operate a hospital in each of the two microregional capitals. Emphasis is on surgery and treatment of illnesses. An emergency service is maintained for accidents. The hospital is staffed twenty-four hours per day. As roads have been built into the interior of each microregion, the use of hospital services by people from remote regions has increased markedly. Clinics are being established in a number of smaller towns to provide first aid, treatment of minor illnesses, and emergency services.

Considerable emphasis has been placed on prenatal care and on postnatal care for six months. The children born under this service receive medical treatment for six years following birth. In one microregion, 850 children have been so treated. While the number is small in terms of total regional population (73,000), continuation of this service should have a long-term beneficial impact.

Children between the ages of seven and fourteen in the same region are being trained in fluoridation of teeth through the use of a simple mouthwash. This treatment appears to have reduced cavities by 30 to 40 percent.

Students have provided assistance in municipal administration, accounting, and other public services. In particular, law students have been active in advising rural people of their rights under Brazilian law with respect to land use and ownership, employment, legal matters, and public assistance programs.

Engineering students have designed public buildings, roads and streets, drainage and sewage systems, and parks. In one municipality a system was established for installing 400 septic tanks, with materials provided by the Federal Ministry of Health and the state government and labor supplied by the local people. Another plan that was implemented involved the distribution of ceramic water filters for home use. This activity is maintained by monthly payments from users, which provide funds to supply additional filters. In both of these cases, the provision of technical inputs was accomplished by training and guidance in the use of sanitation and water treatment.

Students provide training in production of handicrafts from sisal, palm fronds, leather, ceramics, and other locally available materials. These crafts are sold at competitive prices through a retail outlet maintained by CRUTAC in Natal. Profits are returned to the producers of the articles.

Students in education, social work, and home economics provide training in preparation and preservation of foods, basic homemaking, reading and writing for adults, and improved educational techniques for local school teachers.

COMMUNITY RESPONSE TO CRUTAC

Community response to CRUTAC can best be adjudged by the fact that there are more requests for assistance and projects that CRUTAC has resources and students to handle. The state of Rio Grande do Norte has been subdivided into sixteen microregions, but to date CRUTAC has been able to provide services in only three of these. At the same time, 150 municipalities have submitted requests to CRUTAC for assistance.

Twenty-one other universities in Brazil have established similar programs in their states, and other universities are contemplating establishing such programs.

The response to CRUTAC by community leaders in the three operational microregions has been excellent. People are more aware than ever before of their rights, opportunities, conditions for better health, education, and so on.

Community leaders have viable evidence of improvement in community affairs, relations with the people, an improved standard of living, and small increases in prosperity.

REFERENCES

Lopes, Onofre. <u>CRUTAC and CINCRUTAC: University Rural Training Service Rendered to the Community</u> (Natal, Brazil: UFRN Press, 1973).

Lopes, Onofre. <u>ESTA Extensão Chamada CRUTAC, Uma Inovacao Educacional.</u> (Natal, Brazil: UFRN Press, 1975).

Lopes, Onofre. <u>CRUTAC: Estudio Piloto Sobre Medicina e Farmacopéia Popular Na Região do Trairi</u> (Natal, Brazil: UFRN Press, 1976).

Lopes, Onofre. <u>CRUTAC: Manual Do Supervisor</u> (Natal, Brazil: UFRN Press, 1978).

Lopes, Onofre. Presentation to General João Batista de Figueiredo, President of Brazil, Natal, Brazil, September 1978.

PERSONS INTERVIEWED

Professor Onofre Lopes da Silva, former rector of UFRN and founder of CRUTAC

Professor Domingas Gomes da Lima, rector of UFRN

Professor Clovis Concalves, vice-rector of UFRN

Professor Marco Antonio Calvacanti da Rocha, pro-rector for Research and Graduate Studies, UFRN

Professor Jose Pereira da Silva, director, Center of Technology, UFRN

Professor Maria da Nascimento Bezerra, supervisor, CRUTAC

Professor Marshall Jamison, Department of Educational Technology, UFRN

Two regional CRUTAC directors

Approximately fifty students working in the two microregions visited

QUESTIONS TO CONSIDER

1. What does this case tell the reader about the motivations of university students to assist community development in pragmatic ways?

2. Does this system provide a means by which younger persons may be productive at an earlier stage of their development than would otherwise occur?

3. Given the concepts of contemporary Western higher education, can these students spend the apparent amount of time on these development projects and still receive the necessary degree of education in their chosen fields?

4. Should academic credits be allowed for this type of community service activity, irrespective of the fields of study of the involved students?

5. Is there anything unique about the Brazilian situation that makes this program successful, whereas it might not be so in another environment?

6. Would it be useful to survey students who have been active in the program to determine its effect upon them some years after the exposure to the concept? Has it had any lasting effect on their careers?

7. Conversely, what has been the measurable long-term effect of the program on the microregions involved? Has there been any measurable effect of the program? What are the problems of measurement in this type of situation?

8. Recognizing the emphasis on education and pragmatic training of the appropriate technology concept, how does this activity correspond to the ideal of integration of community service activity and the provision of special education programs to abet development?

12
Western Samoa: The Samoa Methodist Land Development Program

Ruth Lechte

BACKGROUND

Forty years ago church land at Levaula, located near the capital of Apia, was established as a farm. All pastors were required to work at the farm for one year during their general training program. The church had originally intended the farm to provide funds for the support of church schools, since the Samoan government, and South Pacific governments as a whole, could afford to provide education for only a small percentage of the population. For most of the intervening years the farm was losing money and certainly was without profits. In 1973 it was incorporated by a decision of the Methodist Conference into the Samoa Methodist Land Development scheme (the SMLD). The board of directors was composed of local people, overseas agricultural or educational experts, and representatives of overseas funding groups or church mission boards. The SMLD now is run by a board whose members all are local Samoans.

In 1975 the Reverend Lalomilo Kamu and his wife, Donna, took charge of the SMLD, not as a source of church revenue, but with the aim of developing it as a training, development, and community education activity. Prior to that time, the farm project had had seven directors in as many years. In 1975, while there had been considerable investment and improvement in the previous three years, only T108 was in the bank account (one Tala Samoan or 100 senes equals $1.40 in U.S. dollars), and most early activity centered on rejuvenation. Fences were repaired, fields and tree crops were cleaned, water was piped, and animals were restocked. A second farm, Afiamalu, located high in the hills above Apia, also was included as part of the project. Other goals were developed as the possibilities were assessed: cash from the farm would be used to develop training in agriculture and other areas; appropriate technology would be applied to village development; nutrition programs using farm produce would be implemented; and an "outreach" program would be developed. These programs were set in motion along with the basic rejuvenation project.

Farm activities are basic to the program, and many other goals are being attempted simultaneously. The programs are still developing and evolving. Farm activities are intended to be labor-intensive so that as many individuals as possible will benefit from the SMLD activities. High levels of unemployment and underemployment exist in Samoa, as in most South Pacific countries. Out-migration also is common--probably as many Samoans live in New Zealand, Hawaii, and on the U.S. west coast as in Samoa. The following statement vividly describes these circumstances: "A high percentage of our food is imported at high cost. Unemployment is high and our people migrate to low-paying jobs around the world. Coconut and cocoa are the ruling crops which cover most of our arable land and form the basis of our economy. These colonial crops were introduced into Samoa to be sold overseas. Their prices are set by merchants and international markets far beyond our control. This has created a subsistence economy and a subservient life-style among our people. We are the victims of history."[1] The SMLD program intends to change this situation.

It is interesting to consider the project goals of the SMLD as stated in a 1976 report: (1) The land is a sacred trust. The church must help discover, on behalf of the people, the proper use of it for the benefit of the total community . . . help discover crop/animal production balances which best suit the available land potentials to provide food and income; (2) To be self-supporting by 1982 from the cooperative development of church lands . . . to where they will support the training and integration of the Samoan people into cooperative agricultural industries . . . beef and dairy are based on our most favorable resources: good climate, land, grass . . . contribute to the food needs of our nation; (3) Formal and informal education and training of people for agricultural leadership from top administrators and dairy technicians to 'barefoot' community workers and cooperative movement organizers . . . The training of school teachers, technological studies, etc., in the basics of the development process of our nation is a further dimension of this goal; (4) Using land development ministries as an integral part of the total ministry of the church. The key to 'land development' is identifying the appropriate way to use the land in a given area to maximize potentiality for the

---

[1]Lene Milo and Lalomilo Kamu, "Integrated Project Statement of the SMLD, Inc." Project submission to Foundation for the Peoples of the South Pacific, Inc., November 1976.

benefit of the people, both economically and socially but always within the limits of the sustainable capacity of the resource environment; (5) All workers and trainees to experience continuing education classes in areas of theology, animal and crop husbandry, group dynamics, and youth work.[2]

PROGRAMS OF THE SMLD

Crop Raising

Coconuts, cocoa, and bananas were being grown on the farm when the SMLD was formed. Copra, whole nuts, cocoa, bananas from Levaula, and taro, which is grown at Afiamalu, are used as food for farm staff and are marketed for cash returns to keep the farm running. These crops, especially the cocoa crop, needed a great deal of undergrowth clearing, pruning, and culling.

Bananas are not produced as a cash crop but are used to feed the staff and, along with pawpaw (papaya) and breadfruit, to supplement cattle feed. The banana industry throughout the South Pacific has had severe difficulties in the last decade, mainly as a result of the black-leaf streak disease. Also, New Zealanders began to prefer the cleaner-skinned Ecuador banana (although it reportedly does not taste as good). Another factor is that erratic Pacific shipping services have left farmers unable to export their crops at the moment of ripening.

Cocoa is harvested largely by women and family groups. South Pacific societies are communally organized, and this social structure reached its highest level in Polynesia, where the extended family still is very strong. In Samoa, "fa'a Samoa" (the Samoan Way) still rules to the exclusion of all else and has been absorbed even into church structures. A study from New Zealand indicates that these structures are strengthened when Samoan groups find themselves in a new milieu. Eighty acres of cocoa are harvested, with the women being responsible for breaking the pods, fermenting the beans for five nights, and drying the beans in the sun or in hot-air dryers. The SMLD sells both the cocoa and the copra in sacks to the government. These two crops are almost totally responsible for the farm income that supports all other activities.

The main use of coconuts is to produce copra (dried mature fruit), and this is used for various industrial and commercial enterprises, notably soap manufacturing. However, whole coconuts are sold in urban areas or are exported to Islanders living overseas in order to produce cream

---

[2]Milo and Kamu, "Integrated Project Statement."

Woman making buttermilk donuts for school snack program

for cooking from the squeezed flesh. The gathering of nuts is another activity and in some cases is being done by nuclear family groups living at the SMLD. In earlier times, the extended family worked together to help a family in difficulty. However, with the emergence of a cash economy, the extended family did not always have the financial resources for school fees, household items, and so on. One nuclear family has solved this problem by working together (both parents and children) after school or work, collecting coconuts for the SMLD. Passionfruit, another crop grown at the SMLD farm, has been only recently introduced. Two and one-fourth acres are now being cultivated. The flowers are hand-pollinated by families of workers or trainees living at Levaula to supplement family incomes.

Cattle Raising

With Australian church funds, a Brahmin bull was procured; today the beef herd at Afiamalu is flourishing. The first progeny of the Brahmin bull/mixed cow stock are now a few months old, and the herd numbers about sixty-five. Eventually, cattle will be supplied to villages of Western Samoa. An outreach worker will spend time in the villages helping to raise awareness in all areas of the development program. He also will decide if the land is suitable for a cattle project and if so, when the village can be ready for the arrival of the stock. After the cattle are established, the outreach worker will follow up at regular intervals and help the village to prepare for the next step in the development process.

Women slashing weeds with machetes to make paddocks

Dairy Program

A dairy herd of twenty cows is also maintained. The large, new milking shed, which eventually will have a solar water heater on the roof, is nearing completion. Three-fourths of the milk produced is used in the school snack program (see the section on nutrition and the snack program), while the remainder is sold as whole milk and butter.

The dairy program also has a training aspect. Some women are permanent workers in the milkroom, and one woman will soon go overseas for training in milk processing.

Nutrition and the Snack Program

The snack program is designed for basic training in nutrition and is intended to rectify an immediate imbalance in local diets. It is also aimed at providing better nutrition to the many children who attend school without having had breakfast. Traditionally, Samoans ate two main meals, one in mid-morning, another in the evening. It has been found that as children travel away from their homes or villages and enter higher education, they often do not eat during the entire day. And even if funds were available, there is nowhere to purchase food. The snack program, therefore, is not a supplement, but often a vital nutritional necessity, and further expansion is envisaged.

School snacks and lunches are delivered by the farm van and must be purchased. Snacks are bought by farmworkers on a weekly system of debiting purchases against wages, and some items are purchased from the milk room on the SMLD. Items sold are milk, milk pancakes (three pancakes cost five sene), milk icepops (frozen icicles made of milk and flavored with pure fruit juices), cheese, yogurt, butter, passionfruit ice blocks (five sene each), pure lemon drinks, and solar-dried fruit (pawpaw, coconuts, and bananas).

The solar drier for the snack program was built by a visiting appropriate technology advisor and consists of a wire frame under glass. With front wheels, it can be moved around like a wheelbarrow in and out of the sun. Air circulation is carefully planned to prevent mildew in the humid climate.

Upon returning to Australia, the appropriate technology advisor also built a texturizer, using the plans provided by Meals for Millions, and shipped it to Samoa. The texturizer will produce puffed biscuits from grains and vegetable protein, thus adding to the nutritional foods provided in the snack and school lunch programs.

Foreground, copra being sun dried; Background, copra burners and shed where the workers cut copra

As part of the nutrition and food program, a drum oven and smokeless stove were built. They now are used in training activities. This technology eventually will be duplicated in the various villages and with community groups as part of the outreach project. New recipes and methods of preparation and improvement of traditional foods using both the village open fire and the new technology are important components of the general nutrition project. The staff is concerned that, as in many developing countries, the people have a tendency to purchase canned and "trade store" goods, especially food products that not only have a deleterious effect on the national economy in terms of foreign exchange but also provide a diet less satisfactory than the traditional one. In fact, certain Western "junk" foods are being actively opposed by the national nutrition committees or their equivalents in all South Pacific countries.

Solar fruit drier adapted from Brace Research Institute plans

Problems of maintaining adequate supplies of ingredients for the snacks fraught this program. Yeast cultures for yogurt, raw fruits for drying, and sugar for juices fluctuate in their availability. Small amounts of sugar required for the snacks currently are difficult to obtain. However, this problem should improve because Fiji is facing a bumper harvest in 1979 and has little prospect of selling it all through traditional overseas channels.

Pasture Improvement

Pasture improvement is a continuing need, and labor-intensive community activity is being used for planting and weeding whenever possible. Samoa was actively volcanic in recent geological times, and large boulders stud the landscape. In two fields these were bulldozed out (with the aid of a small grant from the Australian Freedom from Hunger campaign) and sold for the construction of sea walls, house platforms, and other structures.

Pasture grasses were then hand-planted by the women of a nearby village, who, as a continuing communal fund-raising activity, weed the fields to the accompaniment of much singing, clowning, and general enjoyment. The continuing sale of rocks as more areas are upgraded has so far managed to pay for the annual farm audit. After next year's planting, thirty-five acres of soil and pasture will be improved. Pasture seeds are bought from stocks developed and held by the Department of Agriculture.

In connection with the cattle-raising project and dairy program, round ferro-cement tanks are being built on the farm and placed in each stock paddock and in the stockyards. One mile of piping now connects the tanks to the water source. Due to rapid tropical growth, field sizes are kept small so that vegetation can be controlled. Therefore, it is thought that many tanks are needed.

Training

Staff for the SMLD and other church programs have attended carefully selected overseas courses, and eleven persons now have returned. The nutritionist visited the United States for the Meals for Millions course, and both she and the farm manager had an observation visit to Papua New Guinea and Australia as well as training in special aspects of their work. Other staff members have attended the Navuso agricultural school in Fiji and Flock House in New Zealand. The outreach worker studied at the International School in Canada in Community Development, and another staff member trained in India.

A small mechanics workshop has opened as a training project. It currently is servicing church and farm vehicles in addition to performing a small number of commercial repairs. It maintains itself financially and may later provide income for the SMLD if commercial jobs are undertaken regularly. The instructor is a New Zealand churchworker who is being paid local salary rates. Tools were donated by one of the Commonwealth country's high commissioners in Samoa from a discretionary fund. It is hoped that a few young men will gain practical experience via this apprenticeship scheme. In addition to the mechanics workshops, a carpentry program is envisioned.

With the exception of the nutritionist, some of the instructors trained on site for the Au'uso fowl project (see the section on outreach programs), the one woman to be involved in the projected milk processing course, and the women who are trained in the milkroom, almost all training opportunities are offered to men. Less than 10 percent of overseas training (in time, about 1 percent) has been available to women.

Through contact with groups and individuals working in the same field, the SMLD has been able to use the expertise of short-term advisors, who often are volunteers receiving only room and board. The mechanics trainer came for two, possibly three, years. The new milking shed is being built by a retired Australian farmer who rented out his own farm to undertake this project and a previous one in Papua New Guinea. He is also a volunteer. The Australian Council of Churches' (ACC) Energy for My Neighbor program in 1978 provided a couple as appropriate technology advisors. Their presence resulted in the building of the texturizer, the solar drier, and a grinder, and in the general preparation of other systems and plans that could be implemented by the staff. ACC provided food, fares, and an allowance for equipment and tools and has since shipped the texturizer to Samoa.

Apart from such short-term help, all staff of the SMLD are local, and the one expatriate receives local salaries and wages. A New Zealander whose husband is working in Samoa voluntarily does all of the bookkeeping and paying of wages and will be training a local person to perform these tasks as soon as the systems are set up.

Outreach Programs

One of the stated aims of the SMLD is agricultural and community upgrading. A young minister is responsible for this program. As previously stated, the cattle program is being expanded to villages of Western Samoa through the outreach program. Also, a church high school for boys is developing its hill farm, with the SMLD providing transportation and planting materials for groups who travel to the farm to do development work. The SMLD also is furnishing agricultural training. This program already is providing food for the school, which has many boarding students.

The Au'uso, or Methodist Women's Fellowship, has a fowl project for which overseas funds were given. Village Au'uso groups are loaned laying and breeding hens after three or four women have trained either at the Government Agricultural College or the SMLD. The SMLD is responsible for the procurement and holding of the chickens until they are placed by the outreach worker. The president of Au'uso reported that research is being undertaken in nutritional uses of eggs and chicken meat, and some groups are selling eggs for income to obtain the feed. Au'uso is now looking at the possibility of working with the SMLD outreach worker to place two dairy cows with some village Au'uso groups for nutrition projects based on the SMLD model. These cows would come from the SMLD herd and would be supplemented by animals from the government herd.

Five families who will be trained every three to five years on the farm will act as nuclei for further cooperative agricultural enterprises utilizing church and village lands. It is hoped that the beef, dairy, and crop enterprises will be developed to a stage where sufficient profit will be made to finance a cooperative farming scheme. This program would involve training these key families initially, setting up short-term programs for pastors, and then training the families in the actual communities involved in the project. The cooperative farming scheme is considered by the SMLD to be a sound concept but is viewed as very experimental in its application to Samoan social conditions. Statistics show that Samoa is one of the Pacific countries with the least development in cooperatives, and no cooperative department exists in the government. Society itself, with the extended family structure, is essentially cooperative, but this mitigates against the business systems required. The outreach worker, together with the trained families, will be crucial to the development of the scheme.

FINANCE

Several financial sources already have been mentioned. Other sources include a grant from the World Council of Churches for the cattle program and funds from the government and churches to finance overseas training programs. Otherwise, the SMLD is financially self-supporting in its day-to-day activities, in the general running of the farm, and in the on-site training programs. Extra resources are sought for additional local or overseas training experiences, the setting up of new components within the program, research and development of appropriate technology systems, and the erection of new farm buildings and houses for trainees and their families. The government of Western Samoa has emphasized agricultural development in its five-year plan and therefore has included Levaula as a participant. This makes Levaula eligible to apply for training funds, seed discounts, and other benefits.

The Samoa Methodist Church does not require any income to be returned to it from Levaula and Afiamalu. However, the SMLD pays a token T1500 or T1200 annually as an acknowledgment of the use of church lands for their programs.

FUTURE DEVELOPMENTS

It is hoped that housing prototypes in villages and for the use of farm staff will be developed under the auspices of the Human Settlements-Cooperative Farmer scheme. Information and experience from a village women's project in Tonga may be transferred via the author to the SMLD. The Tongan women currently are making concrete panels for privies and concrete water tanks using coconut fiber for reinforcement. They also are fencing house yards against pigs and are developing vegetable gardens; one ferro-cement house has even been erected. It is hoped that this house will be used as a demonstration and training unit for the SMLD, and that if additional funding can be found, appropriate technology visits will occur between the two projects.

The role of the outreach worker is innovative, and many future developments depend on his expertise and on that of the returning families. An additional worker hopefully will be responsible in the future for the dissemination of domestic appropriate technology services and equipment.

It is hoped that use of solar energy will be augmented. In the next few years, solar panels will be installed on more buildings, and research will be started in the use of parabolic cookers.

Forestry work will be included in the training and will be implemented on the SMLD lands. Five hundred mahogany trees already have been planted.

Kava, a ceremonial and social drink, prepared from the crushed roots of the shrub Piper methysticm (Piperaceae), will be grown as a cash crop; it is slow-growing but lucrative.

During a recent visit to Samoa, the author, together with the SMLD nutritionist, the hospital nutritionist, and a woman doctor who also is the director of the family welfare center in the Health Department, met to discuss plans for an expanded school lunch program. The SMLD is seen as the place for experimentation with products and appropriate technology systems and equipment. The author is attempting to identify potential initial funding, while the mothers' clubs, rural and urban health sisters, and workers are investigating alternatives in the preparation of the foods. The SMLD sees this initial experimentation and setting up of small projects that later can be spread to other groups and areas as vital for its nutrition and appropriate technology programs.

Some of the work done at the SMLD is now becoming an example to the rest of Samoa, and to the South Pacific region in general.

REFERENCES

Australian Council for Overseas Aid. Pacific Dossier. Canberra, Australia: Progress Press, 1978.

Government of Samoa. Five-Year Plan. Apia, Western Samoa: Government Planning Office, 1976-1980.

Haas, Anthony, ed. New Zealand and the South Pacific: A Guide to Economic Development in the Cook Islands, Fiji, Niue, Tonga, and Western Samoa. Wellington, New Zealand: Asia Pacific Research Unit, Ltd., 1977.

Pitt, David, and Macpherson, Cluny. Emerging Pluralism: The Samoan Community in New Zealand. Auckland, New Zealand: Longman Paul Ltd., 1974.

Note: Photographs in this case study courtesy of Diane Goodwillie.

The author further recommends the following for a background on Samoa: novels by Robert Louis Stevenson; Sons for the Return Home (novel), Flying Fox in a Freedom Tree (novel), Pouliuli (novel), and "Inside Us the Dead;" (poems), all by Albert Wendt; My Samoan Chief, a bibliography by Fay Calkins; and The Coming of Age in Samoa by Margaret Mead.

QUESTIONS TO CONSIDER

1. This is an excellent example of an integrated development plan; does it have high replicability; that is, are the conditions favorable in other countries for development of this type of system, or are the Samoan characteristics in some essential way unique? What elements are present or missing in other countries which would affect the success of such an effort? Are there examples of similar endeavors in other locations?

2. The multiplicity of development and extension activities in this situation may explain its overall viability; do you agree?

3. What is the motivating force working toward the success of this project? What social aspects does it have, and what can be inferred about the attitudes and characteristics of the people?

4. A relatively sophisticated technology was introduced in the form of the texturizer for producing snack foods that was designed and produced in the United States; does this say anything about the role of the developed economies in these types of grass roots development efforts? What sort of mechanism should be established to aid the flow and application of appropriate technology from distant sources like this? What are the necessary information channels, and how should they be organized?

13
Tanzania:
Biogas Generator

Richard S. Roberts, Jr.

INTRODUCTION

A low-cost version of a generator of biogas (methane from animal waste) was developed and demonstrated in April 1978 by the Arusha Appropriate Technology Project (AATP) of Tanzania's Small Industry Development Organization (SIDO). Ten months later three local cooperatives and a number of individual entrepreneurs were producing these generators, and an estimated thirty to forty were in use in institutions and rural homes in the region.

The technology is well known, particularly in India with its "gobar gas" plants. Animal manure is mixed with water and placed in a sealed container; the fermentation process--particularly effective in hot climates--produces methane, which accumulates in the tank and is carried by hose or tube to cooking and/or lighting appliances. In the case of the AATP design, the container is a metal cylinder, low and wide, sealed on one end and fabricated by welding together flattened fifty-gallon oil drums. The cylinder can be of different dimensions, but the usual one is the height of an oil drum and measures about two and one-half meters in diameter. On the open underside of the tank, eight radii of angle iron connect a central vertical pipe to the edges of the tank. The angle iron radii, from which crossed bars protrude down at forty-five degree angles for stirring, help the tank hold its shape and hold the vertical pipe that serves as the hub around which the tank can be rotated to stir the slurry. The tank is placed open end down in a pit in the ground approximately three meters deep and twenty centimeters larger in diameter than the tank. Stakes in the side of the pit prevent the tank from dropping too far down, but the accumulated methane normally holds the tank with its top well above the slurry surface. Above the tank, a heavy branch or tree trunk crosses the diameter, with the branch supported by a stake on either side of the pit. A vertical pipe through the central hub of the tank is fastened to the middle of this branch to ensure that the tank remains centered and supported for stirring by rotation. Animal manure and water are mixed in a one-to-one ratio by volume and poured into the pit until the level is above the bottom of the tank, thus making the tank a sealed container. The outlet for the methane is a common piece of pipe situated on the top of the tank. A valve controls the release of the methane, and a rubber hose runs from the outlet to the hot plate or other appliance to be fueled (see Figure 13.1).

BACKGROUND

SIDO was established in 1973 as a national parastatal organization under the Ministry of Industries to promote and provide service to small industries. As such, it is responsible for planning, coordinating, promoting, and offering almost every form of service and technical assistance to small-scale industries on a national level. One of its activities, the AATP, was established in Arusha in northern Tanzania early in 1977 with financial and technical assistance from the Swedish International Development Administration. The aim of AATP is to assist villagers in developing self-reliant, self-development technical skills to meet their expressed needs. Through the use of village dialogue in planning, implementation, and evaluation, innovative technical solutions are developed with only those resources proven to be readily available to the villagers themselves. The work of AATP is carried out by a staff of twenty, including four U.S. technicians.

In one village visited by AATP staff in their village surveys early in 1977, cooking fuel was identified by the people as a major concern. The traditional fuel is firewood or charcoal made from firewood. However, collecting firewood is tiring and time-consuming, and buying charcoal is costly. Moreover, continued reliance on these resources by a growing population threatens the forests of the region.

Given the climate and the fact that most rural families have some livestock, the generation of methane from animal waste was an obvious potential source for cooking fuel in rural households. In fact, the technology had been brought from India and applied in a few instances in the Arusha region as well as in other parts of the country. There was known to be interest, but the available systems were far too costly for low-income rural families.

To meet the identified need in this case, a low-cost system that could be manufactured locally was wanted. AATP staff analyzed available resources and information during the second quarter of 1977.[1] They concluded that an unlined, or rock-lined pit could be employed (depending on local soil) and that galvanized sheets could be soldered together to make the tank. The sheets were readily available, as were the pipes and the valve. Many local craftsmen had the necessary skills and tools to solder the sheets together, even in small towns. Drawings of the proposed design were prepared in July 1977.

During the second half of 1977, AATP set this project aside as record rains and public health problems prevented pursuit of work with the village. The same circumstances also stimulated demand for other, high-priority uses of AATP's limited resources and staff time (e.g., developing low-cost latrine covers as part of the sanitation effort that resulted from the public health problems.) However, a model of the generator subsequently was built in January 1978, and by the end of April a full-scale prototype had been constructed and tested at AATP.

It was then field-tested with Lutheran missionaries in the region and at the home of the chairman of the village that had asked for the generator. It soon became evident that the soldered joints were not strong enough; leaks developed in transport and when the generator was used under village conditions.

Given this experience, AATP turned to a different metal and to welding to gain the requisite strength. It designed and built a new prototype by welding together flattened fifty-gallon oil drums in place of the galvanized sheets. The supply of oil drums obviously is not unlimited, but the information gathered locally by AATP indicates that they are readily available. Welding is easy to arrange in the area. The resulting tank proved to be much stronger and less costly than the tank made of soldered, galvanized sheet. This redesigned prototype was built and tested in the second quarter of 1978.

The prototype fabricated from flattened oil drums is in the yard of AATP, near the entrance and in full view of anyone passing near or entering the compound. AATP encourages visitors and estimates that, in addition to the more than one thousand recorded visitors in the year ending with June 1978, several hundred unrecorded villagers stopped to see what was happening. Demonstrations were and still are provided for such visitors. In the second half of 1978, visits and resulting publicity--much of it by word

---

[1] AATP has a small but comprehensive reference collection and regularly receives publications from some two dozen organizations around the world doing work related to appropriate technology.

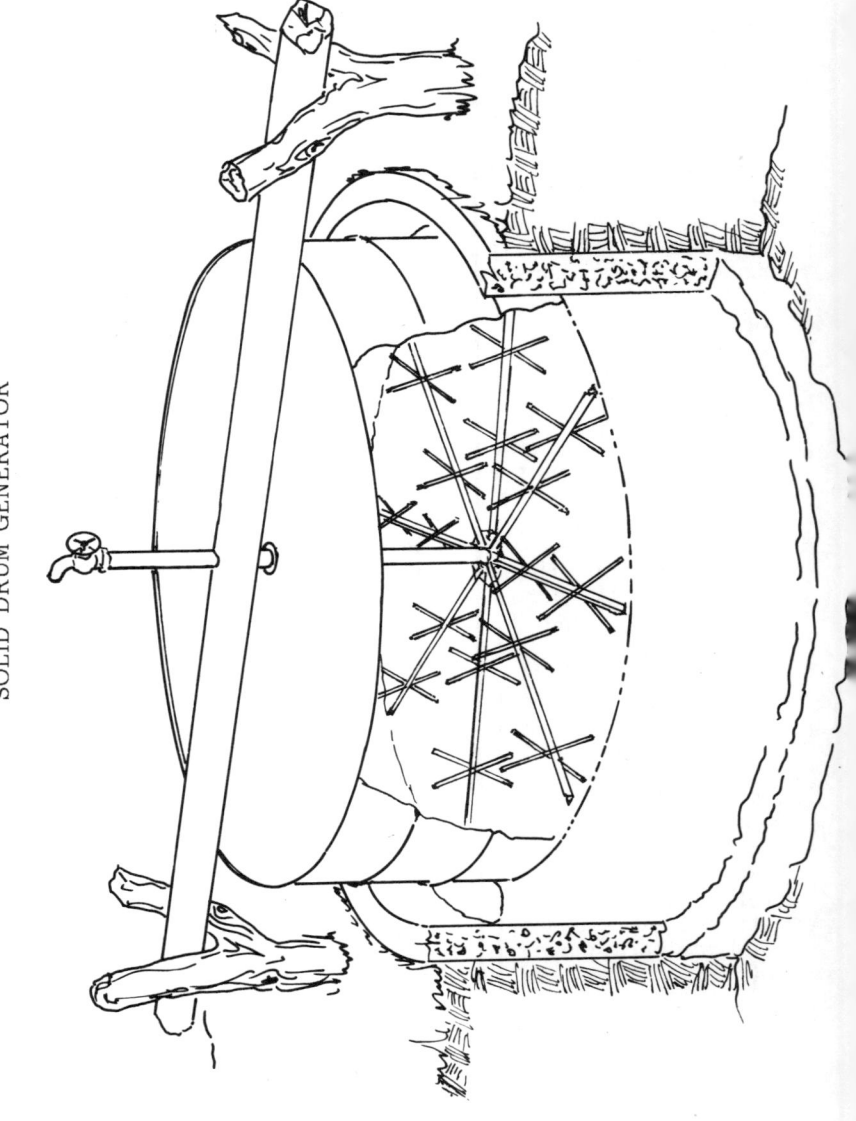

FIGURE 13.1
SOLID DRUM GENERATOR

of mouth--resulted in a number of requests for methane generators. These were referred to Ujuzi Leo, a small cooperative located next to AATP in Arusha, and to others as well.

Ujuzi Leo is a ten-member cooperative that since 1976 has been planning to manufacture windmills for pumping water. First introduced in the mid-1970s, these windmills have since been redesigned by AATP, which is assisting the cooperative. Faced with a variety of delays in obtaining financing, Ujuzi Leo began to make and sell candles to keep itself going. In 1978, confronted by the prospect of considerable delay in obtaining promised (foreign) financial assistance, Ujuzi Leo sought other products that it could make to keep itself in business. The cooperative had obtained funds with which it had built a modest structure next to AATP and purchased some essential equipment for metalworking, including welding equipment. Given this capacity, and in consultation with AATP, Ujuzi Leo decided to build methane generators (as well as an AATP-modified version of the Cinva Ram brick press). Co-op members learned how to fabricate the methane generator by building one under AATP guidance in April 1978.

In the meantime, two other cooperatives and a number of individual entrepreneurs expressed interest in making methane generators. In each case, AATP showed them how to do it, provided measurements and drawings, and assisted wherever necessary. (Very little help actually was needed; for example, co-op members learned how to test for leaks and where to apply pitch to ensure a good seal at joints.) AATP stressed the importance of backing the product with a readiness to repair faulty work. (AATP would not refer people to manufacturers about whose work it heard complaints.) As a result, in early 1979 three cooperatives and several individual entrepreneurs were making methane generators to order in the Arusha region; an estimated thirty to forty generators of AATP design are in use. The users are not the "poorest of the poor;" they are less poor village farmers, cooperative societies, missionaries, school headmasters, and the like. As will be noted later, cost was still found to be a barrier to the adoption of the system by the poorest segments of the rural population.

THE TECHNOLOGY

The manufacture of this methane generator requires no special skills other than an ability to do oxyacetylene cutting and welding--a skill readily found in the larger towns of the region. A welding unit, a few common hand tools, and, preferably, a good vise are the equipment requirements. These represent an investment of approximately $4,000 at early 1979 exchange rates and Arusha prices. Of course, with this equipment many things other than methane generators can be made, and such equipment normally is found in even small metalworking shops in the Arusha region.[2]

The raw material requirements are easily met in the Arusha area (a design prerequisite from AATP's point of view). A typical methane generator two and one-half meters in diameter requires six fifty-gallon oil drums, which are flattened after their tops and bottoms are removed. It also requires ten meters of two to three centimeter angle iron or one and one-half to two centimeter-diameter pipe for the "radii" on the open end of the tank. One meter of four-centimeter-diameter pipe is needed as a central hub, and a three-meter piece of small-diameter pipe is used as a guide over which the central hub rides. Other needed parts include a valve (common plumbing supply), a few pieces of hardware, and some heavy tree branches or small tree trunks. Two men can make a methane generator in about four days (one day preparing sheet metal from drums; one day forming; one day welding; and one day tarring and testing).

The methane generators are made only on the basis of firm orders, not for stock. This, and the fact that the down payment generally required with a firm order amounts to approximately the cost of the raw materials, means that making methane generators places no working capital (or cash flow) burden on the manufacturer. The cost of fabricating a two and one-half meter-diameter methane generator in early 1979 was $140 to $170, and the selling price was $180 to $220.

Fueled with manure from two cows (most village families in the region have more than two cows per person, and the cows are kept in pens at night, making manure collection relatively easy), this methane generator will provide gas for the normal cooking needs of a village family of eight persons. It should do so indefinitely, the only maintenance requirement being semiannual cleanings and a recoating with fresh tar after perhaps five years.

The gas can be used to fuel a commercial range or hot plate or a very simple AATP-developed cooker that can be placed in the traditional wood or charcoal brazier. This AATP cooker (borrowed from a Malaysian design) is a tin (can) fifteen to twenty centimeters high and eight to ten

---

[2]The tank weighs about 100 kilograms. Transport to installation site is arranged with local owners of 1/2-ton vehicles such as Land Rover pickups and other light trucks. Some manufacturers have their own vehicles.

centimeters in diameter (other sizes are quite possible) with multiple holes punched in its top. The bottom is sealed, and a rock inside acts as a weight to provide stability. The gas hose is attached to a tube soldered in the side of the can.

BENEFITS

According to AATP estimates, the family of eight would consume approximately forty kilograms of charcoal per month, which would cost some five dollars; thus, the methane generator would in effect pay for itself over a period of about forty months. Many villagers do not use charcoal but rather firewood that they gather themselves; in this case, the saving would be in time and effort spent, mostly by women and children, in gathering the firewood an estimated two hours or more daily per family.

The benefits resulting from the introduction of this technology accrue to those who make the generators, to those who use them, and to society at large.

(1) Those who make them benefit by having income and employment; they also receive a profit as cooperative members or as individual entrepreneurs (SIDO-assisted cooperatives such as Ujuzi Leo are allowed a markup of 30 percent over all direct and indirect costs of making the product).

(2) The users benefit in that their cooking fuel costs are reduced to zero once the cost of the methane generator has been paid; or, for those with no cash cost for fuel gathered by themselves, the time and effort that go into collecting are immediately freed for other activities, and one chore is removed from the daily routine. AATP has not attempted to quantify the significance of such saved time; AATP staff feel that the real test of technical feasibility is more extensive use of the technology in the region and that only then can results be precisely assessed. They also point out that through their work with the methane generator, as with other technologies, they are trying to develop a methodology for technological diffusion. Thus, their prime concern is with the process of introduction and acceptance and of village-level diffusion thereafter, rather than with the individual technologies per se.

Another benefit the system offers its users is the possibility of using the slurry in the pit (and tank) as nitrogen fertilizer. The slurry has more nitrogen per unit of weight than manure, the use of which is not now common but is being encouraged. No data are available at present on either the extent to which purchasers of methane generators are using the slurry as fertilizer or the "real" value of what AATP considers a side benefit of the system.

Finally, methane can be used other than for cooking. Another obvious use is lighting. Pressure lamps made for kerosene can be modified by soldering a gas input tube into their reservoir to admit the methane. The lamps--like the oil drums--are imported, but they are readily available around Arusha at a cost of about fifty dollars; the modification involves an expense of perhaps two dollars.

(3) The benefit of methane generators to society depends on the extent to which they come into common use. If this occurs, a reduction in the destruction of forests and other ground cover should follow, since the extensive use of firewood as a household fuel by a growing population should diminish. In some countries this destruction has become a major ecological problem, causing serious soil erosion and the loss of arable land. This is occurring rapidly in many parts of Tanzania, and an increasing use of methane generators can play a role in slowing it.

The policies of the government of Tanzania are and have long been to encourage self-reliance and cooperative forms of organization. They have similarly stressed rural development and improving the life of the rural poor. The present case is an example of these policies being effectively reflected in practice.

Only ten months have passed since the first prototype of this methane generator was tested--and therefore seen--at AATP. The record is encouraging; people are buying the generators and the numbers of generators are increasing. However, SIDO and AATP recognize that it is too early to conclude that this technological innovation has become an accepted part of life in the Arusha region.

FIGURE 13.2
SEVEN-DRUM BIOGAS GENERATOR

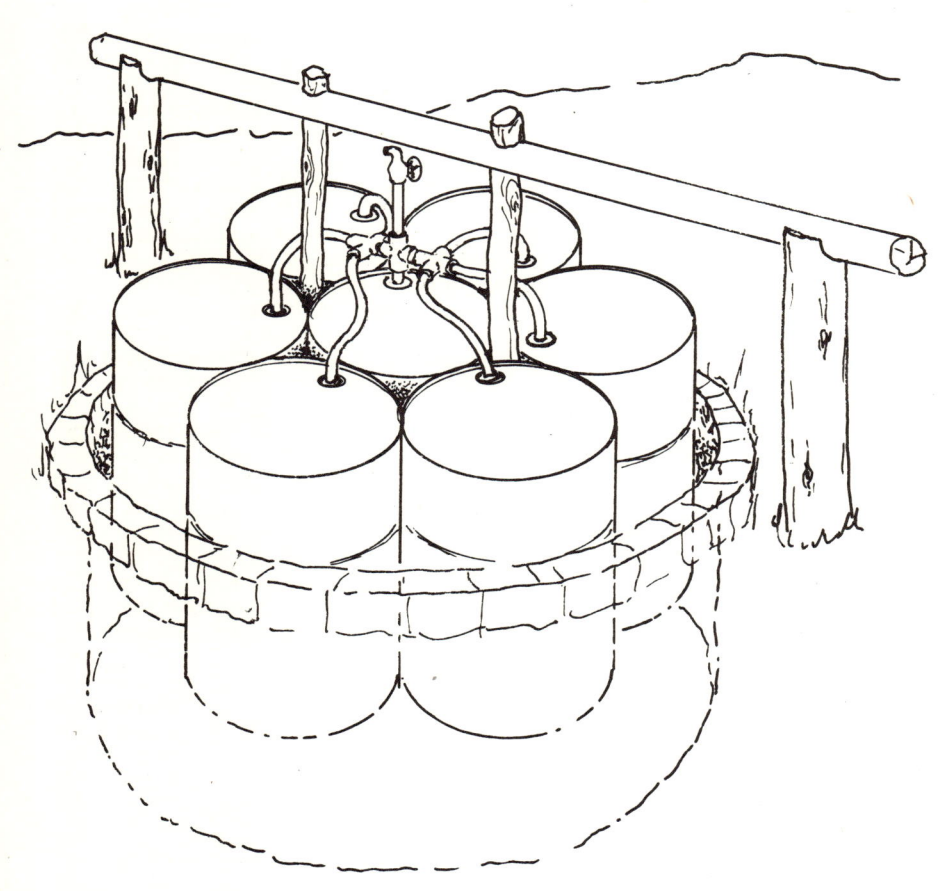

They also recognize that many rural families still cannot afford the system, particularly in terms of cash outlay, and that it cannot be fabricated in most villages. Because of this, AATP developed a lower cost design that requires no welding and only minimal metal-cutting facilities. Development began nine months later than in the case previously described, but by mid-1978 both systems were being tested at AATP.

The low-cost design consists of seven fifty-gallon oil drums bound together with baling wire or something similar (see Figure 13.2). One drum is in the center, surrounded by the other six. The bottoms are removed from all the drums, and a nipple is attached to the threaded fitting already in the top of each drum. Hoses from the outlet nipples meet in the center of the assembly, where they are attached to a common pipe that becomes the outlet for the system as a whole. A simple valve (as with the other design) opens or closes this system outlet, controlling the flow of methane via an attached hose to the appliance to which it is attached. This cluster of drums is placed in a pit in much the same way as the solid-drum generator. The latter system seals in more of the slurry surface and thus captures more gas, but despite the loss, the seven-drum design provides cooking fuel for a family of five under ambient temperature conditions in the plains.[3] The seven-drum system is offered by Ujuzi Leo and the other cooperatives at approximately one-third less than the solid-drum design. However, it can be built at the village level, and families can do most or all of the construction work, thus reducing the monetary cost still more.

A prototype of the seven-drum system is on display and is being demonstrated at AATP. It also has been used at a village school for several months. This system was installed at AATP at the same time as the other design, but in the

---

[3]Field tests at a higher altitude than at AATP demonstrated that lower temperatures at the higher elevations reduced system efficiency, making it clear that prevailing ambient temperature, taking into account drops from day to night, is very important in determining the local appropriateness of the biogas technology.

ten months that have passed, it has not caught on with the populace.[4]

The seven-drum unit was developed to see if its potential for village-level production and the low cost of acquisition will be accepted by the rural "poorest of the poor." AATP staff observe that, although it is obviously premature to predict such acceptance, by comparing their experiences with the two designs they hope to learn useful lessons regarding innovation. As they see it, this technology, thus developed, is designed to help them learn--and understand-- at what level appropriate technology development can and should start, . . . one of several unanswered questions to which an answer is needed to make a reality of large-scale appropriate technology efforts.

REFERENCES

Arusha Appropriate Technology Project. **Annual Report**. 1977-1978, p. 70.

PERSONS INTERVIEWED

Staff members of AATP and of Ujuzi Leo Cooperative.

QUESTIONS TO CONSIDER

1. This case is an example of an effective appropriate technology project initiated by a government-sponsored institution: what general conclusions can be reached concerning the utility of these types of organizations? Should they be generally propagated in other nations?

2. Some say that the range of simple, low-cost solutions to problems in poor regions are really not very great and that there is a finite and limited need for this kind of activity. Do you agree? If so, does this argue for perhaps limiting efforts in these subject areas and

---

[4]Since fieldwork was done for this case study, AATP has learned and informed the authors of several instances of village installations of seven-barrel cluster biogas generators. Apparently they have been built in the villages by people who have seen the demonstration unit at AATP and have picked up the drawings and description of it (in Swahili). AATP has now begun a survey to assess the extent to which their ideas are spreading in this manner.

devoting the energy instead to more sophisticated technology applications? Has the importance of low-cost, "poor man's technology" been oversold?

3. What might be done in the Tanzanian instance to further encourage the use of the waste slurry from the gas generators as fertilizer?

4. The AATP states that its interest, broadly, is in the processes of technology diffusion. What further activity in this regard could be suggested? Are there other organizations that have pursued this interest and could provide some guidance?

5. What more might be done to propagate the methane generator technology? Is there need for providing a financial assistance system that would permit the poorest families to acquire units?

14
Brazil:
Explosive Metalworking Program

James M. Miller, Jim D. Mote,
and Henry E. Otto

ORIGINS OF THE PROBLEM

The Instituto de Pesquisas Tecnológica (IPT), originally a state-supported research institute, has been in operation for over fifty years. The director of IPT, Dr. Alberto Pereira DeCastro, began formulating plans in 1969 to develop methods by which IPT could better assist Brazilian industrial growth.

Since World War II, the economy of Brazil has been changing from an agricultural to an industrial base. The state of São Paulo has been the focal point of development. During the postwar era, Brazil's Gross National Product (GNP) increased as much as 16 percent in some years. Such dramatic industrialization gave rise to a multitude of problems. Many small- and intermediate-sized concerns lacked resources, human or technological, to adequately solve these difficulties. Brazil, therefore, was importing goods and services that required large capital expenditures and foreign currency.

Another Brazilian problem was the growth of multinational concerns that brought in their own technology to produce the goods they marketed. Most of this technology, captive to the particular multinational firm, is not disseminated to Brazilian-owned industries. Utilization of this technology to provide goods for the Brazilian market requires that royalties be paid, which, in turn, results in a capital outflow from Brazil. In addition, a large number of small concerns that are suppliers of parts and components to the multinationals have problems meeting the quality-control requirements of the large organizations. These small- to medium-sized firms need a better technological foundation to participate in economic growth.

One method of developing this base was discussed by Dr. Castro with personnel at the United States Agency for International Development (USAID)/Brazil and with the state of São Paulo. The government recognized that science and technology were fundamental in the overall development of Brazil. Under the program, the state of São Paulo proposed to give its research institutes the necessary financial support to improve their research and development (R&D) capabilities. They realized that the demand for new processes would increase, which would require an expanded capability for diversifying production and an efficient scientific and technological information service.

A loan agreement between the state of São Paulo and USAID was established on 30 May 1973 for $15 million to further the goals of the overall science and technology program. This loan agreement provided the mechanism to use personnel from the United States and to provide contract support for research institutes. One of the participating institutions was IPT, which was to initiate four demonstration projects, one of which was identified as "Explosive Forming, Cladding, Bonding, and Welding." The state placed these demonstration programs and others under the jurisdiction of a Program of Science and Technology for Development (PROCET) to support engineering development, product design, market analyses, and applied research.

Personnel from the Denver Research Institute (DRI) visited IPT in January 1974 to discuss a formal linkage with IPT and to prepare a proposal under the PROCET objectives. A direct link between the metallurgy divisions of the two research institutes was proposed. The demonstration projects selected were explosive metalworking, packaging, the surface quality of deep-drawing steels, and weathering steels. A proposal was submitted by DRI through IPT to the state of São Paulo to establish a contract with the Instituto de Pesquisas Tecnológica.

The program was approved, and work started on the Explosive Metalworking project in July 1974. The scope was: to develop processes to re-form metal plates or sheets into controlled shapes; to laminate dissimilar metals, such as titanium or stainless steel, onto steel, or to bond lead to steel; to weld oil and gas pipelines; and to produce dished heads of aluminum and steel for fuel tanks and chemical processing tanks.

This demonstration project indicated that the goals were viable and that IPT-industry interaction could be broadened. At the conclusion of the first two years, in 1976, the project was evaluated, and IPT, DRI, and PROCET decided to continue the effort, which was funded by a follow-on contract still operative under the USAID loan agreement.

DESCRIPTION OF THE TECHNOLOGY

Explosive metalworking, in its broader sense, may be defined as the use of the energy released by a detonating explosive to change the shape of a metal part (explosion forming); to join similar or dissimilar metals (explosion cladding or welding); to cause changes in the metallurgical and engineering properties of a material (explosion hardening); to separate metals (explosion cutting, shearing, and punching); and to compact metal or ceramic powders (explosion powder compaction). Explosives provide a relatively cheap and readily available source of power. Hence, for the types of metalworking mentioned above, the use of explosives as the source of power is attractive because of the versatility, low capital investment, and unlimited growth potential.

As applied to current commercial practice, the widest areas of use of explosive metalworking are explosion forming, explosion cladding, and explosion hardening.

Explosion Forming

Explosion forming, as usually practiced, is a standoff procedure (see Figure 14.1). It is a dome-forming operation where the metal blank (workpiece) is positioned over a die cavity and clamped into place by a hold-down ring. The air is removed from the cavity behind the blank through the vacuum link via an auxiliary vacuum pump to allow the unimpeded flow of the metal blank into the die cavity after the explosive is detonated. The entire assembly rests on the bottom of a tank filled with water, which serves as an energy transfer medium.

An explosive charge is placed at a predetermined distance from the workpiece and detonated. The pressure produced is transmitted through the water and becomes a ram, forcing the workpiece into the die cavity. In addition to forming domes or leads for tanks, variations of the process can be used for tube bulging, leading or corrugating panels, forming or sizing of conical preforms, and other functions.

Explosion Cladding

In explosion welding, or cladding, two metal surfaces, driven together by explosive forces, weld together as a result of their high-velocity collison. Neither of the metals melt during the collision. However, a jetting action occurs between the colliding plates that cleans the metal surfaces of contaminants just prior to their being forced together under great pressure.

In parallel plate cladding, the metals are arranged so that the cladder metal is positioned exactly over the backer metal at a predetermined standoff distance (see Figure 14.2). An explosive of the required detonation velocity is placed on top of the cladder metal. The quality of explosive, explosive detonation velocity, and standoff distance are selected

277

FIGURE 14.1
ESSENTIAL ELEMENTS OF AN EXPLOSION
FORMING OPERATION

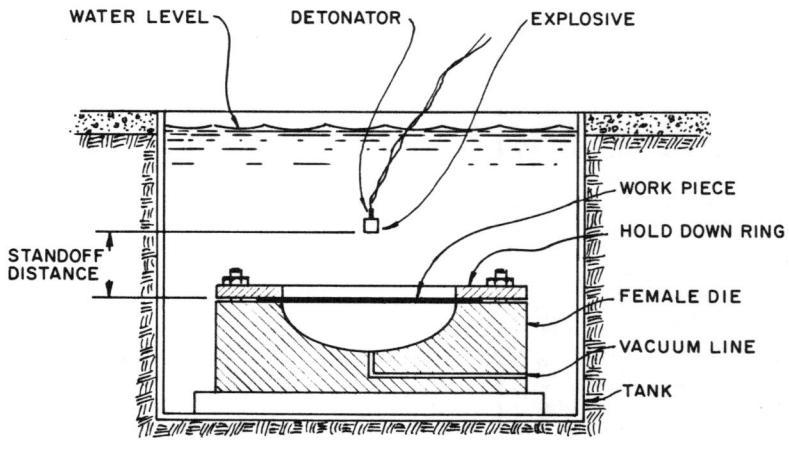

FIGURE 14.2
SCHEMATIC FOR PARALLEL PLATE CLADDING

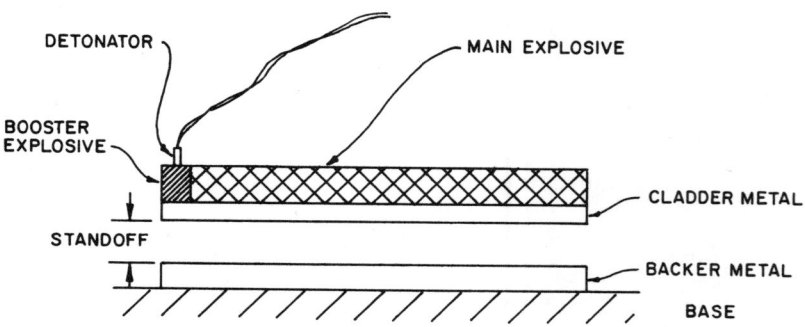

so that the two plates will collide at a characteristic velocity and angle. The main explosive charge is detonated by a booster charge, and the two plates collide (see Figure 14.3). A very high pressure is generated near the collision point, and the metal surfaces can flow as a spray of metal from the apex of the angled collision. The surfaces are stripped of contaminants, which are expelled in the jet, thus removing bond-inhibiting surface films. The resulting contaminant-free surfaces are pressed together by the high pressure so that atomic attractive forces are established and a metallurgical bond is formed.

The greatest tonnage of explosion-clad products is in the form of flat plates that are used for corrosion protection of chemical processing equipment, electrical transition joints, and structural transition joints.

Explosive Hardening

Explosive hardening is accomplished by placing an explosive with a fast detonation velocity in contact with a metal. The hardening is effected by high shock pressures that are generated in the metal, which causes internal microstructure change resulting in much stronger metal with little alteration in material dimensions.

FIGURE 14.3
SCHEMATIC OF COLLIDING PLATES SHOWING JETTING
ACTION DURING EXPLOSION WELDING PROCESS

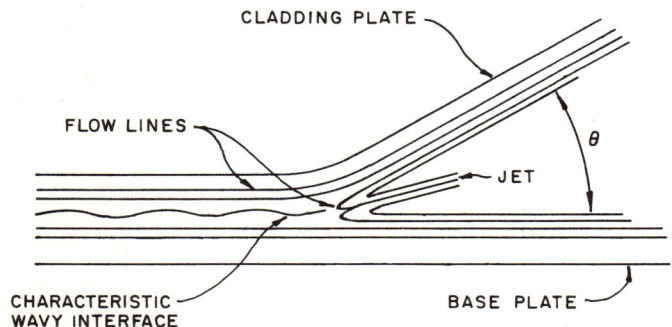

The process has been employed successfully in hardening railroad frogs made from Hadfield steel. It has also found use in hardening hammers for rock-crushing equipment. Explosive hardening has the advantage of being easily used in remote locations on massive machinery.

TECHNICAL SOLUTIONS

In the beginning, IPT had no capability in the field of explosive metalworking but had an on-going metallurgy division with forty years of experience. The implementation of the program was divided into: (1) original market survey; (2) training of IPT personnel; (3) facility development; (4) production of clads; and (5) related production.

Market Survey

The initial market survey was conducted during July and August 1974 in the United States to determine products and users of the technology. This preliminary survey provided an understanding of the basic marketing and economic factors involved in the explosive-metalworking technologies and offered a model for a similar survey in Brazil.

A survey followed in Brazil from August to October 1974. In-depth interviews were held with thirty-four industrial organizations, the majority of which were considered potential users or specifiers of the technology. Width of application was stressed rather than depth, to give a broad view of the market situation, which, in turn, made it easier for IPT to contact the firms after the IPT staff had acquired the necessary training.

The initial survey indicated that the easiest market to penetrate was for clad products. This market was highly variable, but general trends and forecasts were made in market size projected on the annual increase in Brazil's GNP in 1974. Another factor was the large expansion underway in the Brazilian steel industry, which was projected to reach 30 million tons per year by 1980. The closely allied heavy-mechanical industry, a user of clads, had been growing at a rate of about 16 percent a year between 1969 and 1972. It was concluded that the demand for clad products would follow the trend for primary metal users and heavy equipment manufacturers.

Other factors were considered. Brazil was planning to start the production of stainless steel sheet and plate in 1976, which would assist in developing the clad market. However, the projected stainless steel mill did not start operations until September 1977. Also, expansion of the Brazilian petrochemical industry would increase the demand for clads.

In Brazil the most logical explosive-forming area was for storage and pressure-vessel heads. At the time of the initial survey, many firms were using gore segments, in which the gores were welded together to form large heads. Of the 1,000 heads per year being made, it was thought that 750 could be produced by explosive forming. Other heads that were manufactured by spinning and thin-stainless heads could be explosively formed, too. The initial survey, however, indicated that the market would be hard to penetrate due to the existing technology.

Subsequent surveys were conducted by IPT personnel. On one survey, the market was found to be unfavorable. The manufacture of heads in Brazil was based upon a wide number of international specifications that varied enough to give a large product mix. The product mix in turn meant that a large number of dies would be required rather than just a few of standard size. The result was that it was not economically feasible to explosively form heads in the one to two meter diameter sizes. Subsequently, a 6,000 ton press, capable of making most heads, was installed in Brazil. This press precluded the development of explosive-forming facilities by IPT.

A market survey was also conducted for explosive hardening of railroad cross-overs (jacares). The jacares for heavy-duty use are made of manganese (12 percent) steel and are explosively hardened by many railroads and suppliers throughout the world. This survey indicated that about five hundred jacares a year would constitute the Brazilian market when it is developed. One drawback in introducing the technology, though, is the lack of data comparing explosive hardening with other hardening methods, such as hammer or roll hardening. The potential market value was estimated at $50,000 to $100,000 per year (1974) based on a cost of $100 per jacare. However, costs for hardening per jacare have risen since the survey to better than double the original figures (1979).

Training

Seven IPT personnel were trained in explosive metalworking at DRI in two different groups. The first group of four Brazilians was trained in all aspects of explosive metalworking during August to December 1974. Three IPT engineers were assigned to the Explosive Metalworking Project. They received on-the-job training at Campo Experimental do Lorena in explosive welding. These Brazilian engineers subsequently were sent to DRI for training in explosive forming and theoretical aspects of explosive welding. However, a turnover in trained personnel occurred, indicating one of the problems in insuring continuity of the project. Progress was consequently delayed. Moreover, none of the engineers who left IPT were able to use their training in their new jobs.

Another phase of the technology transfer was the presentation of course work in explosive metalworking at the University of São Paulo (USP). Seven graduate students completed a course in explosive metalworking, but alternate schools for the presentation of course work in explosive metalworking were not instituted.

Facility Development

A facility for explosive metalworking was developed by IPT near Lorena, São Paulo, in the Vale de Paraiba. The site, called Campo Experimental do Lorena, is midway between the industrial centers of São Paulo and Rio de Janeiro and is accessible to the state of Minas Gerais.

The four most important explosive manufacturers in Brazil are in the vicinity of Lorena. The largest steel mill in Brazil, CSN, is located at Volta Redonda, less than one hundred kilometers from Lorena. The second largest steel mill, Usiminas, is located near Belo Horizonte, as is Acesita, the mill that produces stainless steel plate. Lorena is also midway between the ports of Santos and Rio de Janeiro, which would be the ports for exporting explosive clads if the requirement arises.

Plans have been formulated for the construction of a new port eighty-five kilometers from Lorena, which will serve the southern part of the state of Minas Gerais as well as the Vale do Paraiba. A railroad spur is located three kilometers from the site. Thus, Campo Experimental do Lorena meets all of the requirements of being remote but has good access to supplies and market.

The site at Lorena, consisting of 360,000 square meters of land, was donated to IPT by the municipal government of Lorena. A local engineering firm was employed to develop the site. The roads, drainage, and welding site were developed first, followed by the buildings for setups and explosive storage. The planned explosive-forming facility was not built due to market conditions. However, a large materials-handling building was erected. The facilities at Campo Experimental do Lorena include laboratories for complete metallographic examination and nondestructive testing of clads.

Production of Clads

The main product produced by IPT at Campo Experimental do Lorena has been tube sheets for use in heat exchangers. IPT takes the manufacturers' materials and performs the explosion bonding. The clad composite is then returned to the manufacturer for fabrication (see Figures 14.4 and accompanying photographs). Materials clad on the carbon-steel substrate have included naval brass, stainless steel, and cupronickle.

The process has allowed for a reduction in the importation of clad tube sheets into Brazil. Also, some manufacturers using weld overlay and brazing, which are not too satisfactory, have supplemented these processes with explosion bonding and obtained a cost reduction.

The first tube sheets were explosion bonded by IPT in June 1975. Since then, the number of tube sheets produced and the quality of bonding has steadily increased as has the number of customers. Early tube sheets were not tested nondestructively, and the only quality control was whether or not tapping on the surface gave evidence of voids. Ultrasonic inspection equipment is now used to determine if the tube sheet is totally bonded. Reject rates are on the order of less than 3 percent of the total tube sheets welded.

Other Production

Explosion-clad sheets of 304 stainless steel to carbon steel were made for evaluation by one fabricator. These clads were formed with no problem into the desired configuration.

This type of clad was evaluated for applications by Petrobras. Conventional tests as specified in the American Society of Testing Materials (ASTM) 262 were conducted on the clads. These tests indicated that clads of nineteen millimeters and greater met all specifications.

Tube to tube sheet expansion was carried out on two heat exchangers. Over four thousand tubes were expanded successfully. Hydraulic tests were conducted after expansion to insure the quality of the expansion process. Eight leaks were repaired mechanically. Wax was used as the transmission media, which is less costly than the molded polyethylene plugs used in other countries.

Bimetallic thermostat materials were explosion bonded. Billets were first melted and rolled at IPT, then explosion bonded. After welding, the composite was rolled to size. These materials provided another market for IPT that formerly required the material to be imported by Brazilian industry.

The hardening of jacares for a manufacturer of railroad hardware has been done successfully. Several jacares have been hardened and are currently being evaluated on Brazilian railroads. It is anticipated that this market will expand after further evaluation.

FIGURE 14.4
PROJECTED INDUSTRIAL DEMAND FOR CLAD PLATE
IN BRAZIL

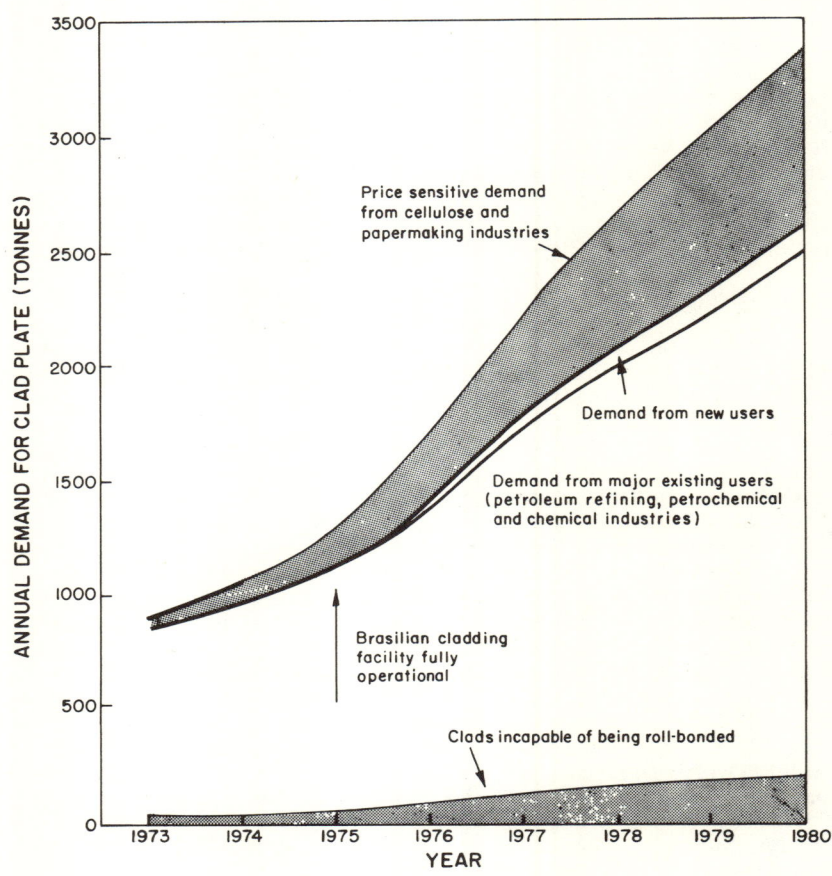

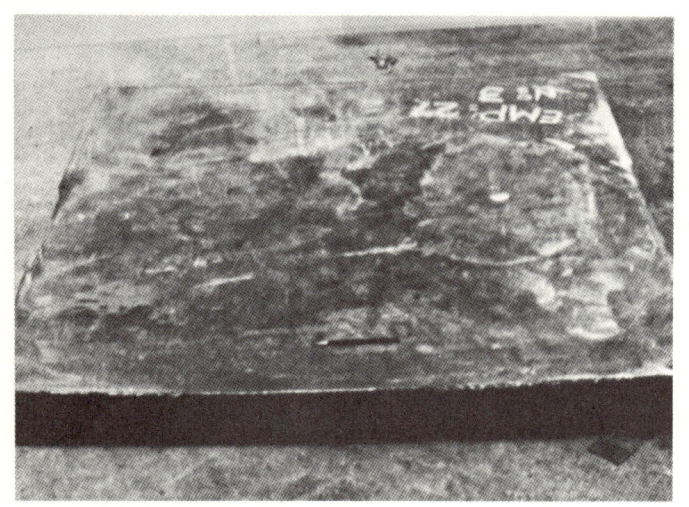

As clad tube sheet composite

Tube sheet composite after machining

Tubesheet During Fabrication Into Tube Bundles

Stainless Clad Plate After Explosion Bonding

TECHNICAL COMPETENCE OF THE STAFF IN TERMS OF OBJECTIVES

IPT products currently rely on explosive metalworking, cast-products technology, and foundry techniques. An increase in the volume of these products has required IPT to demonstrate their capabilities to industry.

In the Explosive Metalworking Project, which introduced an entirely new product to Brazil, time and perserverance were required to show that good-quality "clads" could be made. The project staff needed to gain necessary experience, but present evaluations indicate that the IPT personnel recognize the problems of their industry. A sampling of industrial sponsors indicates a high level of confidence in this IPT group. The staff is now stable and has a capable marketing ability.

The Explosive Metalworking Project was one example of introducing totally new products and services into Brazil. IPT and DRI agree that the success in this area was due in part to the training program.

INTERNAL/EXTERNAL COMMUNICATIONS

Because of the distance between the Lorena field station and IPT (approximately two hundred kilometers), problems existed at the former of which headquarters was unaware. Strengthened communications and closer personal relationships have resolved many of these difficulties.

Communications with other divisions have not been so successful and need improvement. The Package Engineering Laboratory, which requires assistance from the wood, plastics, and ceramics division, has experienced difficulties on interdivisional projects.

Sponsor relationships have been good, as reported by the contacts made with industrial sponsors. The individual project managers are responsible for these relationships. Difficulties do occur on most research and development projects, but the sponsors have been kept aware of problems as they arise.

Promotional efforts concerning IPT's capabilities have been made through the presentation of papers, seminars, and classes in some disciplines. Coordinated promotional efforts have helped toward this effort, but if a particularly industrious project manager leaves, it may jeopardize the program. Industry awareness of IPT's capabilities could insure success of the new technologies.

ECONOMIC ASPECTS

Starting with 1976, the project income has been:

1976	234,243 Cruzeiros	($ 22,098)*
1977	1,518,781 Cruzeiros	($106,357)**
1978	971,027 Cruzeiros	($ 57,644)
		(to 31 August 1978)

A reported backlog of orders for September 1974 totals 600,000 Cruzeiros.

In terms of customers in this project, there were:

1975	7
1976	9
1977	10
1978	13

---

* approximate conversion 1976 and 1977, June rate.
**approximate conversion 1978, April rate.

IMPACTS OF THE TECHNOLOGY

Explosive metalworking is an example of the transfer of a new technology to Brazil. The goods produced have been accepted by Brazilian industry. However, industry must continually be convinced that a quality product is being produced, which means that manufacturing techniques and continued experience must continue to be developed by IPT.

Objectives of this project (and others) were:

(1) Increase the interaction between IPT and industry to assist in the industrial growth that is required to increase Brazilian industries' export capacity;

(2) Develop a greater competence on the part of IPT to undertake programs to solve industrial problems;

(3) Develop a managerial capability in IPT to assume leadership in the above considerations;

(4) Use specific metallurgy and materials projects to stimulate, increase, and train IPT staff in mechanisms for institute-industry interaction.

Originally, four demonstration projects were undertaken by IPT and DRI. Two survived the entire period of the five-year loan agreement.

The IPT-DRI program has demonstrated that technology can be transferred by the interaction of two research institutes, and that the technology can create new services and products for recipient industries. If the transferring process is to be successful, however, a high degree of interest is required of the recipients. They must actively promote the technology with the industries concerned in addition to acquiring a working knowledge of the new technology.

REFERENCES

Denver Research Institute, <u>Proposal, Program of Assistance in Institutional Development to Instituto de Pesquisas Tecnológica (IPT)</u>. Denver: DRI, 1973.

Denver Research Institute, <u>Phase Report - A Preliminary Investigation into the Market for Explosive Metalworking Technologies in Brazil</u>. Denver: DRI, 1974.

Denver Research Institute, <u>Final Report: Program of Assistance in Institutional Development, IPT</u>. Denver: DRI, 1978.

Instituto de Pesquisas Tecnológica (IPT), <u>Programs Metalurgin</u>. São Paulo: IPT, 1978.

QUESTIONS TO CONSIDER

1. Does the reader believe that this case presents an example of the transfer of appropriate technology? What are the factors in this situation that conform to the concept of appropriate technology? What were the effects on employment, income distribution, and industrial productivity?

2. Considering the intensive training activity that took place, are there any conclusions as to the necessity or effectiveness of this process? What about the turnover rate of those who were trained? Was it normal? Could anything have been done to reduce it?

3. Was the market survey work conducted as a preliminary to the project adequate? Is there any indication that alternative technologies were considered and cost/benefit analyses conducted?

4. Is an R&D institution such as IPT the most effective place to introduce and seek commercialization of sophisticated technology? What should be the role of government-supported institutions with regard to proprietary technology from foreign sources?

5. What do you think of the appropriateness of explosive metalworking technology in other developing country situations?

6. What was the importance of the relationship between the two institutions? Should this be a pattern for other situations? Who should bear the cost of assistance programs of this type?

15
Ghana:
Small-Scale Sugar Processing

Gary D. Kilmer and David L. Sussman

INTRODUCTION

Technoserve, Inc., is a nonprofit development assistance agency headquartered in Norwalk, Connecticut. Program offices are located in Ghana, Kenya, Nicaragua, and El Salvador. The corporate purpose of the organization is to increase the economic and social well-being of low-income people in developing countries through the self-help enterprise development process. In pursuit of this objective, Technoserve provides host country project sponsors with a complete range of technical and managerial services in the conceptualization, planning, implementation, and management of self-help enterprise projects.

This case study concerns the development, modification, and replication of a labor-intensive, small-scale sugarcane processing technology in southern Ghana. Since 1973, the economic environment affecting the development of such projects in Ghana has been a difficult one. Many basic goods such as petroleum, industrial/agricultural equipment, spare parts, and food products must be imported. Serious balance of payments deficits and foreign exchange shortages make the importation of these essential goods very difficult, and they are normally in short supply, if available at all. Annual inflation has been high, generally in excess of 100 percent, and economic growth and employment opportunities have not kept pace with population growth.

The technology development process that is described in the following pages was carried out by Technoserve at the request of, and in close cooperation with, groups of Ghanaian farmers and entrepreneurs who own and operate two small-scale sugarcane processing plants. Two projects were undertaken, one beginning in 1973, the other in 1976. The first was aimed at developing a technology to produce raw crystal sugar, the second at producing sugar syrup. The constraints noted above were considered carefully in the development of the technology now utilized by these projects.

PROJECT BACKGROUND

Ghana currently is forced to import more than 90 percent of the sugar consumed annually by its population. Total consumption of sugar in Ghana through the mid-1970s was near 100,000 tons per year.[1] This figure represents the sum of locally produced sugar and imported sugar rather than real demand, which may be considerably higher. A state-owned company operates two large-scale sugar mills with a total capacity of 40,000 tons per year; however, in recent years they have operated at less than 30 percent capacity, and their current productivity is significantly below that level.[2]

Technoserve became involved in its first sugarcane processing enterprise in December 1973 in response to a request from a Ghanaian agronomist who for several years had worked with the government-owned sugar mills. This man already had initiated a simple sugarcane crushing operation in an area where a number of small-scale farmers had planted cane that they intended to sell to one of the government-owned mills. However, since the mill was far from their fields, high transport costs prevented the farmers from selling their crops to this mill.

The agronomist, who was the original sponsor, and Technoserve viewed this project as a means of providing a market for this sugarcane, encouraging increased production of a profitable cash crop by small-scale farmers, creating jobs in an area with no industrial activity, and satisfying a small portion of the total demand for sugar. The project was undertaken in cooperation with the traditional authorities of the area, a group of individual cane growers, and two major commercial banking institutions. Technoserve provided professional staff (Ghanaian and expatriate) residing in Ghana and, as required, home office and short-term technical personnel. This technical and managerial expertise was not otherwise available to the project sponsors.

DEVELOPMENT OF THE TECHNOLOGY

In the first project, initial efforts of Technoserve were directed at strengthening the operation of the cane-crushing facility established by the Ghanaian agronomist and surveying the technological alternatives for small-scale crystal

---

[1]Government of Ghana, External Trade Statistics, 1968-1975.

[2]Interviews, executives of Ghana Sugar Estates, Ltd.; and public records, 1975.

sugar production. The original project concept was based on the open pan "khandsari" process that was used for many years in India and modified by technology employed elsewhere.

A comprehensive project study was completed in April 1975, and a plant was constructed that same year. This pilot test project was based on the use of equipment and techniques from a number of different countries (see Figure 15.1 for a schematic drawing of the production process).

Sugarcane is cut by hand and carried to the plant site, where the juice is extracted by crushing. The bagasse is then spread in the sun to dry, after which it is used to fuel the furnaces. The juice is pumped from the crushing pits to an elevated tank, where the pH is adjusted by the addition of milk of lime. From the liming tanks, the juice flows to a preheat tank, where the juice is brought to a near-boiling temperature and floating impurities are removed.

From the preheat tank, the juice flows into settling tanks, where nonsugar solids ("muds") are allowed to settle. The clarified juice is then passed through filter bags and into the evaporating pans. The "muds" also are filtered, and the clear juice is sent to the evaporating pans. In the evaporating pans, the juice is boiled until it reaches a semisyrupy consistency. Impurities are skimmed from the boiling juice at this point. The semisyrup is then transferred to a smaller finishing pan, where the boiling process is completed.

The heavy syrup (raab) is then moved to the crystallizer, where seed crystals are added and crystallization is promoted by slowing stirring the raab as it cools down. From the crystallizer the raab is moved to the centrifuge, where the molasses is purged from the sugar crystals. The resulting molasses is then sent back through the finishing, crystallizing, and purging processes while the crystal sugar is dried and packaged.

Appropriate diesel-powered three-roll crushers manufactured in India were obtained from a dealership in Ghana. The cooking pans, which could be manufactured in Ghana, were adapted from designs of similar pans used in the maple and sorghum syrup cottage industries in the United States. The crystallizers, based on Indian designs, also were manufactured in Ghana. The original centrifuges used for purging the molasses from the crystal sugar were laundry centrifuges manufactured in Denmark, although a small-scale industrial centrifuge later was imported from Germany. The remainder of the equipment was manufactured in Ghana.

The various pieces of machinery were located and modified by Technoserve advisors with the assistance of technical experts throughout the world. Two short-term consultants were employed for several weeks to assist with

FIGURE 15.1
RAW CRYSTAL SUGAR PROCESS

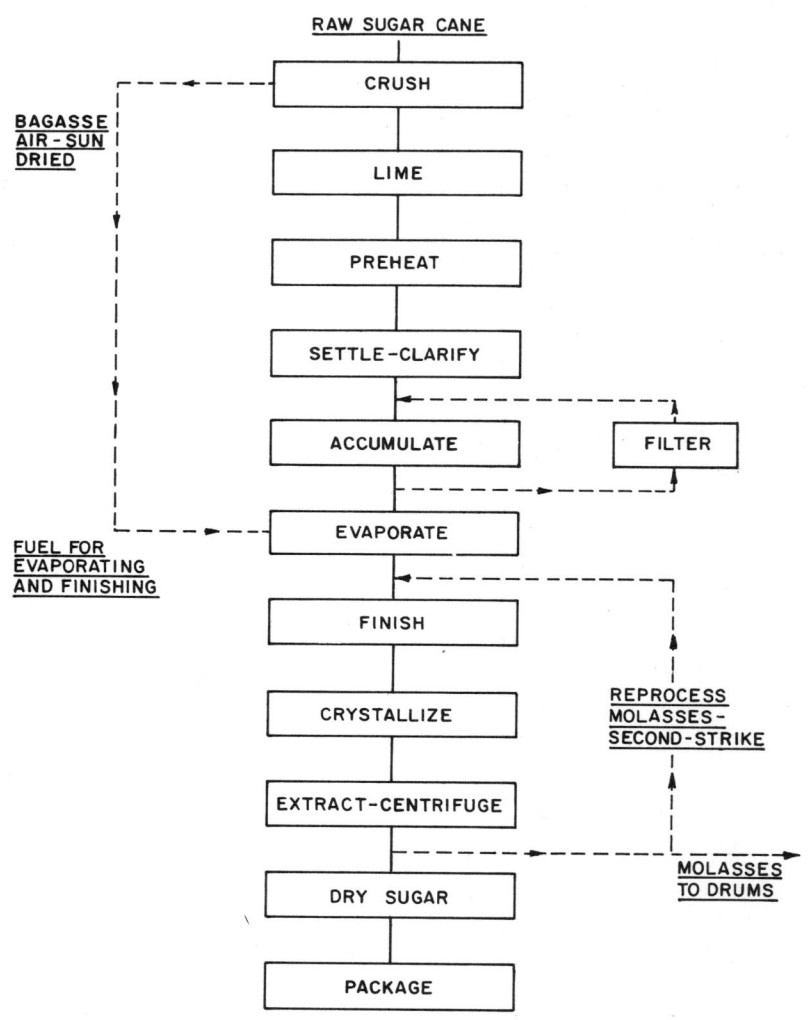

Harvesting of sugarcane by hand

Cane being crushed at plant site

equipment design and installation and with the start-up of the pilot test. One was a food technologist with many years of experience in commercial food processing. The other was the retired manager of the United States Department of Agriculture sorghum syrup research and development center.

Four young Ghanaians who had received "khandsari" sugar production process training in India also were hired. This team, along with the local project sponsors and Technoserve advisors, resolved many of the technical problems that arose. The production of marketable raw crystal sugar during the test period was accomplished.

The application of the technology selected was facilitated by the availability of electrical power and pipe-borne water at the project site. It also was possible to place the production equipment in existing buildings, further reducing the capital cost. When full-scale production was initiated in 1976, the plant had an estimated capacity of 3,000 tons of sugarcane per season (20 tons per day for 150 days).

Early in 1976, Technoserve was contacted by another group of farmers and entrepreneurs who requested assistance in implementing a similar project in another part of the country. It appeared as if the technology utilized in the first project could be employed in this undertaking, and a comprehensive project study was initiated. Like the circumstances surrounding the first project, a significant amount of sugarcane already was being produced by small-scale farmers in an area distant from the government-owned sugar mill. However, the soil and climate of the new area were more suited to sugarcane production.

During the second project study, certain factors suggested that the original project concept and technology required modification. The reasons included the following: electricity and water were not available in the area; more sugarcane was available at the beginning and a greater number of farmers were to be involved in the project; capital costs for implementing the same technology on a larger scale would be too high; the complexity and cost of installing and operating an electric generator as a power source would be too great and could not be relied upon, given shortages of spare parts and fuel in the country; management and transport constraints would adversely affect the increased logistical problems associated with getting the cane from the field and into production on a timely basis; and a market for low-polarity sugar syrup was identified.

Based on the experience of the first plant, it was important to consider another set of factors in the design of the second: (1) the economic viability of the first plant was marginal due to government-controlled prices for crystal sugar and relatively low raw sugar recovery rates; (2) considerable training and experimentation would be necessary to increase the recovery rate; for example, scheduling the cane deliveries from the farms to the plant would require tight control since crystal yields decline with the lapse of time between cane harvesting and processing, strict supervision of production personnel would be necessary, and improved varieties of cane would have to be introduced; (3) the supply of cane from small-scale farmers and the sugar content of the cane would vary significantly from year to year depending on rainfall, the availability of labor, cane diseases and pests, and so on; (4) the design of the original cooking pans and furnaces would present problems of heat distribution; and (5) inefficiencies in the original furnaces would necessitate the use of expensive wood fuel to supplement the sugarcane bagasse.

With these factors in mind, the second project was planned so as to minimize the capital cost and operational complexity of the production process. This was accomplished by redesigning the production technology to produce low-polarity sugar syrup (see Figure 15.2 for a schematic diagram of this production process).

FIGURE 15.2
SCHEMATIC DIAGRAM OF LOW-POLARITY
SUGAR SYRUP PROCESS

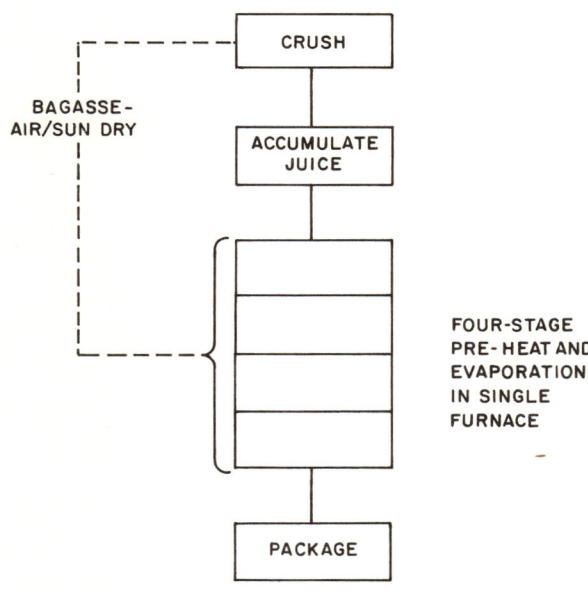

Low-polarity sugar syrup has a variety of applications in Ghana as a substitute for imported refined sugar and other imported sweeteners in the production of bakery products, ice cream, and other confectioneries. A food technologist worked for a brief period with potential users and trained them in substituting sugar syrup in their operations. Sugar syrup also is used in the manufacture of various pharmaceutical products and alcohol.

The second project was put into production within eleven months after the original request for assistance was received by Technoserve. The capacity of the second plant is 3,400 tons of cane per year (17 tons per day for 200 days). Many of the design improvements of the second plant have now been applied in the first plant, and the economic viability of both enterprises has been improved considerably.

In the second plant, cane is delivered to the crusher; from here, the extracted juice is pumped to a holding tank, where a flocculating agent may be added. The juice then is pulled by gravity to a preheat/settling tank built around the stack for heat efficiency. Impurities settle out and the clarified juice is passed to the skimming pan. Gravity pulls this juice to the skimming/concentration pan, where it is concentrated. The syrup is then filtered and stored for packaging. Floating impurities are removed as the juice flows through the evaporating pans (see Figure 15.3).

The bagasse (fibrous cane residue) generally is sun dried and stored for fuel, although it can be burned with reduced efficiency immediately after the juice is extracted. The furnace is fueled completely in this way during the dry season and requires little, if any, external source of combustibles during the short rains.

Beyond the savings in capital costs, this modified process offers several operational advantages: (1) the necessity of tightly scheduled harvesting, delivery, and processing operations is reduced; (2) the need to carefully control the pH level of the sugarcane juice is eliminated; (3) the costly and difficult sugar crystallization process is eliminated; (4) the centrifuging (purging), drying, and sugar packaging steps in the process are eliminated; (5) the redesigned furnaces and cooking pans eliminate the need for fuel other than bagasse and enable the cooking to be done in a linked, four-stage, gravity-flow process; (6) it is possible to construct small-scale plants in relatively isolated rural areas in close proximity to the sugarcane fields; and (7) the finished product can be stored for long periods without high risk of deterioration.

The technology that resulted from this experience is very simple and labor intensive. It can be replicated in Ghana and other countries, but such replication requires a comprehensive assessment of many social, environmental,

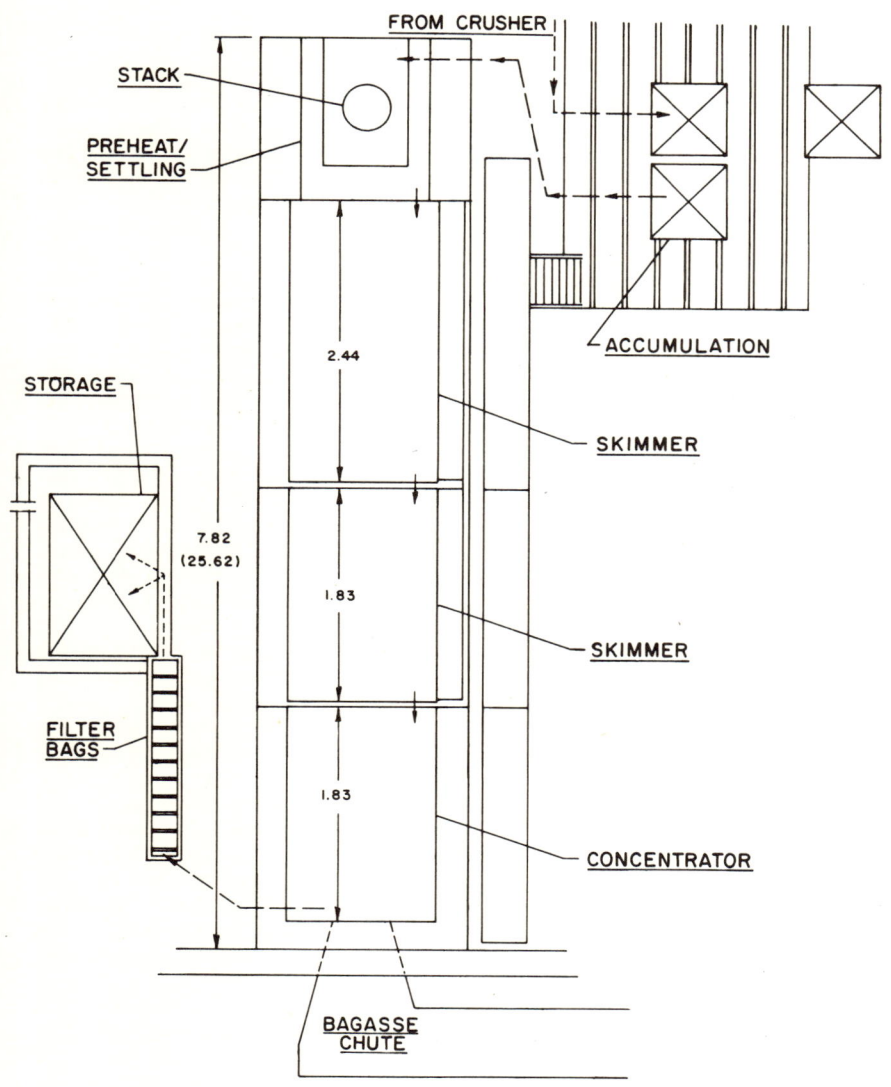

FIGURE 15.3
FURNACE GROUND PLAN

economic, infrastructure, and market factors that vary from country to country. The "appropriateness" of such enterprise-related technologies in diverse socioeconomic environments must be analyzed carefully to assure that such technologies can be utilized in an economically viable and socially beneficial manner.

ECONOMIC AND ORGANIZATIONAL ASPECTS

The total cost of developing the technology and implementing these two projects was shared by Technoserve and the project sponsor groups. Sponsor costs included partial reimbursement of the expenses incurred by Technoserve in assisting the projects. Technoserve provided no direct grant or capital inputs to the enterprises. It did assist in the preparation of loan requests submitted to local commercial banks and in the location of necessary foreign exchange. Technoserve costs not met by project reimbursements are met by private and public sector grant sources that support Technoserve operations worldwide. Technoserve has spent some $300,000 on the development of the two enterprises.

The total capital invested in the second enterprise (at February 1979 Ghanaian cedi-U.S. dollar conversion rates) was approximately $87,000. At 1979 prices, the capital cost per 1,000 tons of sugarcane processed per year is approximately $25,600. It is estimated that this cost can be lowered to $15,300 per 1,000 tons of sugarcane processed if the scale of the plant is increased to 17,000 tons of cane per year.

The present technology requires approximately twenty-five plant workers per 1,000 tons of sugarcane processed. Per unit labor requirements decrease only slightly as the capacity of the plant is increased. The capital cost per job created is approximately $1,000 at 3,400 tons of cane per year, but this decreases to about $700 once the capacity reaches 17,000 tons per year (see Table 15.1 for a comparison of this technology with more capital-intensive technologies).

The capital costs of the second project were met primarily from local sources, including share capital investment and loans from a local commercial bank. The total foreign cost of the project was $56,000, which included the cost of sugarcane crushers and diesel engines, tractors, trailers, farm implements, pumps, and laboratory equipment. In other countries some of this equipment might be available locally.

In this relatively high-risk, rural-based enterprise, the debt/equity ratio of the capital financing was 2.5. The share capital of the enterprise was provided by the sponsor group, which includes Ghanaian entrepreneurs and smallholder sugarcane farmers. Both enterprises are organized

TABLE 15.1
CAPITAL INVESTMENT AND LABOR REQUIREMENTS

Plant Capacity Tons Cane Per Year	Capital Investment[1]		Number Employees[1]		Capital/ Labor Ratio
	Open-Pan Technology (syrup)	Capital Intensive Technology (refined sugar)	Open-Pan Technology (syrup)	Capital Intensive Technology (refined sugar)	
3,400	$25.6[2]	--	25	--	$ 1,000
17,000	15.3[3]	--	22	--	695
50,000	--	$138[4]	--	5	27,600
150,000	--	60	--	2.5	24,000
250,000	--	50	--	2.2	22,727

(1) Per 1,000 tons cane per year
(2) Actual investment at February 1979 Ghanaian cedi-U.S. dollar conversion rates
(3) Estimated at 1979 prices
(4) Development Center for Economic Cooperation and Development, Paris, 1969 (adjusted for inflation)

as limited liability companies with no single shareholder controlling a majority of the shares. Small-holder farmers participate in the ownership of both enterprises.

Due to their proprietary nature, the actual operating statements of the enterprises have not been included. Additionally, it is important to note that such information is of marginal value since input and market prices vary considerably across countries. Similarly, government intervention through control prices and marketing mechanisms, as well as international trade agreements, significantly affect the cost/price structure of the sugar industry throughout the world. In Ghana this is particularly true. Significant changes have been made in the control prices of inputs (cane and labor) during the relatively short lives of the enterprises. The cedi has been devalued, and financial authorities agree that it is significantly overvalued to this day. The country also has experienced inflation in excess of 100 percent per annum during the last several years.

However, in order to demonstrate the financial viability of this business concept a summary analysis of current financial performance is included. The following data have been derived from operations during the fourth quarter of 1978 (October, November, and December). This represents the most current financial data available at the time of writing and also the first quarter of the enterprise's 1979 fiscal year.

TABLE 15.2
INCOME DERIVED

		(¢000's)	($00's)
Sales		¢103.3	($37.6)
Raw Materials	19.1 ($6.9)		
Labor	16.4 ($6.0)		
Indirect and Overhead	6.5 ($2.4)	42.0	($15.3)
Gross Margin		¢ 61.3	($22.3)
General and Administrative		-18.1	($ 6.6)
Income		¢43.2	($15.7)

Using this period as a basis for analysis, the gross margin approximated 59.3 percent of sales (61.3 ÷ 103.3 = 59.3). Since annual fixed costs at current prices are estimated to be ¢72,000, annual sales of ¢122,000 would be required in order to break even. This sales volume equates to 3.54 months of normal production. Thus, it is readily apparent that the concept is highly profitable within the current economic environment, and there is considerable margin for change.

ECONOMIC IMPACT

The two plants currently in operation directly employ about forty people each. At full production capacity they will employ eighty-five each. The job categories include harvesting crews, manual labor in the plants, and semiskilled and skilled workers in production, maintenance, and management. These figures do not include the indirect jobs created on the small-scale cane farms in the outgrower areas. Since the plants create a market for cane, more rural jobs become available.

On-the-job training is provided for plant workers. Agricultural extension and input supply services from the enterprises and Technoserve provide additional training to sugarcane outgrowers and facilitate the expansion of the acreage planted to sugarcane. In a land surplus economy such as exists in Ghana, sugarcane cultivation has not displaced food crops. One plant paid its workers a total of ¢34,000 ($29,500) in the 1977-1978 season and paid local farmers an additional ¢22,000 ($19,000) for their sugarcane. These figures will increase substantially in the 1978-1979 season. The plant operation represents a major source of income for the local economy.

The impact of these projects on Ghana's foreign exchange reserves also is positive. Foreign exchange capital costs are moderate, and foreign exchange requirements to meet operating costs are insignificant. Furthermore, the sugar syrup produced is a direct substitute for sugar-based imported projects.

These projects have facilitated the flow of agricultural loans to small-scale farmers through local financial institutions. The availability of crop loans is further enhanced by the supplies and technical assistance needed by the farmers. Technoserve and the project sponsors have facilitated the formation of farmer associations to increase cane production and benefits to farmers.

SOCIAL IMPACT

A comprehensive evaluation of the impact of these projects on the economic and social fabric of the rural communities that they affect has not been attempted. Several comments can be offered, however.

(1) Increased employment opportunities have been created in these rural areas, reducing the migration of labor to the urban areas.

(2) Increased incomes for low-income workers and farmers has resulted from the creation of jobs and markets for sugarcane.

(3) A large proportion of the labor force is composed of women who have been trained to increase their job skills.

(4) Technoserve has developed a sugarcane extension project through which it is working directly with small-scale farmers to increase their productivity.

Other benefits also result from the implementation of projects:

(1) Increased rural economic activity often leads to increased governmental services and infrastructural development in such areas.

(2) The experience of direct participation and successful cooperation in innovative, productive activities of this type may increase the confidence and creativity of rural communities in undertaking additional economic and community development activities.

(3) The deployment of appropriate technology of the type described may increase the opportunities of rural communities to effect positive change in their lives on their terms and to participate more fully in the national development process.

RELATIONSHIP TO NATIONAL GOALS

This technology is consistent with the development priorities of the government of Ghana. Government authorities have supported strongly the development of this technology since 1975. In fact, Technoserve was requested by the Ministry of Economic Planning to help facilitate the accelerated development of the small-scale sugarcane industry in Ghana.[3]

REFERENCES

Government of Ghana, External Trade Statistics, 1968-1975.

PERSONS INTERVIEWED

Various executives of Ghana Sugar Estates, Ltd.

QUESTIONS TO CONSIDER

1. What may be concluded from the case concerning the need or the desirability to directly fit the technology to the local conditions? Is this an important aspect of technology application, or can standardized systems be used in a wide variety of locations?

2. What ingredients in this situation made it possible to introduce this appropriate technology? Is it the general experience that such winning combinations can be effected everywhere and that, therefore, the introduction of technological innovations can be a widespread activity? If so, what should be done about initiating programs on a large scale?

---

[3]Correspondence with Dr. Robert Gardiner, Commissioner for Economic Planning, Government of Ghana, 1975.

3. The importation of an appreciable amount of foreign technology was required in this situation; what would need to be known to determine the possibility of more local production of sugar plant equipment?

4. Finance availability was vital to the success of this project; is it believed that such financial resources are generally available in the developing countries, or is finance a severe problem that requires considerable expansion of facilities?

5. Do you feel that the cost per workplace created in these two situations was appropriate, or was it excessive? What criteria do you use in making this determination?

6. The importance of government intercession in price regulation in Ghana made important differences in the viability of the project. Is government typically such an important factor in such situations?

16
Thailand: The Introduction of Mint Agriculture and Processing

Ronald P. Black and Sachee Piyapongse

INTRODUCTION

In 1970 and 1971 several parties became interested in *Mentha arvensis*, var. *piperascens*, commonly known as Japanese mint, as an agricultural product that held potential opportunities for agriculture and industry in Thailand. A major activist and catalyst for the development of the nation's mint agriculture and processing industry was the Applied Scientific Research Corporation of Thailand (ASRCT), the country's major multidisciplinary applied research organization.

The mint agriculture and processing industry of Thailand was introduced from 1970 to 1974, and its potential for the country was projected. From 1974 to 1979, Thailand attempted to commercialize the technology. By 1974, ASRCT had an active research project concerning mint agriculture, mint oil extraction, menthol production, mint economics, and a nationwide promotion and extension program. A number of farmers had begun to grow mint; oil distillation plants had been set up; and a firm for the production of menthol had been established. ASRCT had projected a 200 ton annual rate of production of mint oil by the end of 1974 and a 1,000 ton-per-year rate by 1978. The production of 1,000 tons of mint oil per year would occupy approximately 116,000 farmers, 1,400 workers at distillation plants, and 130 at menthol production firms. This level of oil production would contribute U.S. $15.4 million to the national economy annually as a direct result of mint agriculture, oil extraction, and menthol production.

However, in 1978 Thailand actually produced only slightly more than thirteen tons of mint oil; many mint oil distillation plants had gone out of business; and the original menthol production firm had halted operation. How had the development of such a promising "appropriate technology" turned out so dismally? Or had it?

DEVELOPMENT OF MENTHOL AND MINT OIL PRODUCTION UP TO 1974

Of the several varieties of mint grown in the world, the most valuable is peppermint. Peppermint (Mentha piperita) and spearmint (Mentha spicata and Mentha cardiaca) are grown in temperate zones and do not grow well, if at all, in tropical climates. The essential oils of peppermint and spearmint, which are produced by distillation, are used as a flavoring in foods and candies. Although it is possible to extract crystallized menthol (the major aromatic chemical constituent), from peppermint, it is not economically desirable. A higher value is placed on peppermint oil as a flavoring; also, peppermint contains a lower proportion of menthol, 50 to 55 percent by weight, compared with alternative mint varieties.

The primary source of commercial menthol is Japanese mint (Mentha arvensis, var. piperascens), which is grown mainly in temperate or subtropical countries. This is the original source of menthol and was produced in Japan for menthol export prior to World War II. When the war caused Japanese exports to cease, Brazil began supplying menthol to the Western Alliance and replaced Japan as the leading world supplier. In 1974 Japan produced much less mint oil than it had before, relying mainly on mint oil imports and on synthetic menthol. Although several countries, e.g., Argentina, Australia, the People's Republic of China (PRC), Indonesia, New Guinea, the Union of South Africa, the United States, and Thailand, produced some Japanese mint for their own use, in 1974 the bulk of the world supply came from only two exporting countries, Brazil and Taiwan. More recently, the PRC has begun exporting menthol while Taiwan's role has decreased.

The essential oil of Japanese mint, which constitutes approximately 3 percent of the dry weight of the plant, can be recovered by a steam distillation process. The crude mint oil is further separated by a freezing process into menthol crystals and "dementholated" mint oil, in which about 55 percent menthol by weight remains. By more intensive and expensive processing techniques, greater amounts of menthol can be recovered from the mint oil.

The proportion of menthol in crude mint oil varies by the variety of Mentha arvensis grown and by the location where it is grown. Brazilian mint oil contains about 80 percent menthol by weight. Taiwanese varieties in 1974 contained between 60 and 80 percent. More recently, however, a Thai variety, So Wo-1, has been used in Thailand and Taiwan and is yielding 80 to 90 percent menthol.

The menthol crystals, with an average 1974 world price of about $42 per kilogram, are used in the manufacturing of cigarettes, cough drops, toiletries, balms, and medicinal preparations. Dementholized mint oil, with an average 1974 world price of about eighteen U.S. dollars per kilogram, varying by menthol content, is used for flavoring food, candy, chewing gum, and so on. Dementholized Japanese mint oil is inferior to peppermint oil as a flavoring, although the two types of mint oil are chemically similar. In many parts of the world, Japanese mint oil has been used to adulterate peppermint oil.

In the years just prior to 1974, Brazil produced about 6,000 metric tons per year of mint oil, while Taiwan produced about 1,000 tons per year. In the 1973-1974 season, however, floods in Brazil reduced production to only 2,000 metric tons, while in Taiwan's fertilizer-intensive agricultural economy, the shortage of petroleum and the resulting high cost of fertilizers caused the 1973-1974 mint oil production to drop to only 300 tons. In 1974, world demand for menthol and dementholized mint oil had been growing by 10 to 15 percent a year, and a world shortage of mint oil and menthol developed.

INTRODUCTION OF JAPANESE MINT TECHNOLOGY

ASRCT was formed in 1964 and by 1965 had initiated an essential oils program with the purpose of identifying and testing available natural materials as potential sources of commercial perfume or flavor materials. One of the materials on the ASRCT list for study was mint.

In 1970 the Nissho-Iwai Company of Tokyo sent representatives to Thailand to discuss the possibility of a joint venture between two Thai firms, Hong Huat Company (HHC) and T. Unichem, and Nissho-Iwai for the development of a menthol industry in Thailand. The Thai firms, owned by the Tangtrongsakdi family, were both involved with chemicals. It was the family's opinion that between the two companies the technical problems associated with processing the mint and the marketing of the mint products could be managed; however, the family felt it would need assistance in promoting the growth of mint. Therefore, the family approached the owners of a Thai fertilizer firm, the Metro Company, of which one member of the Tangtrongsakdi family was a major shareholder. It was proposed that Metro join the discussions with the Nissho-Iwai representatives.

In early March 1971 a Metro employee informed a senior staff member of ASRCT's Tropical Agricultural Products Institute about the interest of the Thai and Japanese firms in introducing mint to Thailand. Later in the month a representative of the Japanese firm brought mint cuttings to Thailand; he remained in the country for approximately two weeks to organize the venture and to survey possible areas for growing the mint.

Later that month the Metro representative again approached his friend at ASRCT to inquire about the possibility of ASRCT working on the project. He was informed that ASRCT would be interested but that the services of ASRCT would cost about 20,000 baht.[1] This was the last that ASRCT heard of the joint venture's interest in assistance for some time.

Shortly after the representatives of Nissho-Iwai returned to Japan, the Thai investors and Nissho-Iwai reached an impasse in their negotiations and discussions were terminated.

The Thai investors, however, did not give up the idea of producing menthol. They decided to form a corporation called Thai Chemicals Company, Ltd. (TCCL), and in June 1971 they purchased a former pharmaceutical plant in Bangkok. The facility had been closed for about one year and could easily be modified to produce menthol.

The Ming Sheng Chemical Company Ltd. of Taipei, Taiwan, whose Ice Berg brand of menthol had been distributed in Thailand by HHC, was approached with an offer to join TCCL in a venture to produce menthol and mint oil products. Ming Sheng was Taiwan's largest producer of menthol. It not only used Taiwan-grown mint oil but also purchased dementholized mint oil from Brazil and extracted the remaining menthol. The company's major markets were South Asia, Canada, and the United States.

After considering the HHC offer, Ming Sheng decided to accept. TCCL would own 51 percent of the venture and Ming Sheng 49 percent. The new manager of TCCL, Pirote Tangtrongsakdi, was to visit Taiwan for several months to study menthol processing, and Ming Sheng was to send one of its men, who also was an investor in the new venture, back to Thailand to be production manager. The TCCL manager spent October and November 1971 at Ming Sheng then returned with his new partner and several engineers to begin the project. Crude mint oil was to be imported from Taiwan.

Also during the same October-November period, a trip was provided by Metro for several major users of their fertilizer to Japan, Taiwan, and Hong Kong. One of these users was Somchai Lohajoti, the general manager and principal shareholder in the family-owned Nanleaf Tobacco Company Ltd. (NTCL). In 1971 NTCL had approximately 5,000 families working at its tobacco-curing facilities or growing tobacco for the company. Somchai was an elected representative to the Thai National Assembly for the province of Nan. Nan, located in North Thailand on the Laotian

---

[1] One U.S. dollar equals approximately twenty baht.

border, was a rural province far "up-country" from Bangkok, the center of Thai business and government. Even in 1974 it required one day of traveling over stretches of unpaved roads to reach the province; nor had telephone service penetrated the area.

Despite the remote location, Somchai's company employed the latest agricultural techniques: he had his own experimental farm that grew tobacco cuttings from around the world. In 1971 Somchai was looking for a crop to rotate with tobacco.

During his Metro-sponsored trip in late 1971, Somchai was introduced to the possibility of growing mint. This immediately caught Somchai's attention because mint was an important raw material used by the tobacco industry. As a result, the NTCL general manager arranged to visit a mint farm in Taiwan.

In the meantime, several ad hoc attempts had been made to grow Japanese mint, brought to Thailand by the Nissho-Iwai representative, in various locations in that country. These experiments generally were unsuccessful.

Also during this period, the coordinator of the ASRCT essential oils program had become more actively interested in the possibilities of Japanese mint for Thailand. He had discovered that a Bangkok candy company, Adams (Thailand) Ltd., used menthol in its candy. The possibility of local sales suggested a readily accessible market. This appeared particularly attractive because the import duty on menthol was 30 percent of the price. The essential oils program coordinator subsequently visited Adams and obtained the specifications for the menthol used by the firm.

Other interest was found in a Thai source for menthol when an ASRCT group leader met with the production manager of Colgate Palmolive (Thailand) Ltd. at an Australian trade fair. This growing interest encouraged the ASRCT essential oils program coordinator, and in December 1971 he decided to attempt to crystallize menthol from Japanese mint.

The first step was to obtain a source. Another ASRCT staff member--the one who had earlier met with the Metro representative--was aware that the Division of Agricultural Chemistry of the Ministry of Agriculture and Cooperatives had obtained some of the Nissho-Iwai mint cuttings and had been growing, curing, and drying mint. A batch of dry mint leaves was obtained from this source. The essential oils program coordinator distilled oil from the leaves and crystallized menthol from the oil. The crystals were given to the acting plant manager of Colgate Palmolive (Thailand) Ltd. for tests. This manager formerly had spent time at ASRCT as a UNIDO associate expert, so he was familiar with the corporation's capabilities. ASRCT was informed that the quality would need to be improved but that Colgate was interested. Following this, ASRCT entered a systematic research program aimed at investigating Japanese mint.

It was just at this time that the Metro representative arranged an informal meeting between the general manager of NTCL and his friend at ASRCT.

The ASRCT staff member gave Somchai six cuttings of the mint from the Division of Agricultural Chemistry and agreed that ASRCT would work with NTCL. In May 1972 Somchai made two trips to Taiwan to observe mint farms and distillation facilities. In October the ASRCT essential oils program coordinator prepared what he considered to be a good menthol crystal from mint that ASRCT had grown at an experimental farm in Pakchong. Somchai took this to the Thai Tobacco Monopoly, where he was encouraged.

Somchai's next step was to distill samples of crude mint oil from mint he had grown in Nan. In November samples were sent to firms in Taiwan and Japan that produced and exported menthol. They found the quality high.

In the meantime, Somchai established formal contact with ASRCT, which led to discussions of a possible technical assistance contract between NTCL and ASRCT. ASRCT's first offer was to provide assistance over a period of one year for a sum of 15,000 baht. The NTCL general manager thought this amount too high, however, and negotiations were about to break down when the research director of ASRCT's Technological Research Institute, which ran the essential oils program, offered an alternative proposition. He suggested that ASRCT would provide the requested technical assistance if NTCL would pay ASRCT 1.5 percent of its sales of crude mint oils for five years, plus out-of-pocket expenses incurred by ASRCT in providing the assistance. The NTCL general manager immediately accepted this offer, and ASRCT began a program of technical assistance that involved agricultural aspects, such as growing mint in Nan, and distilling mint to extract the crude oil. In November 1972 ASRCT established an experimental mint farm in the northern part of Nan on land donated by NTCL.

The ASRCT essential oils program coordinator, however, still wanted to involve TCCL in the project. In February 1973 he met with the three brothers of the Tangtrongsakdi family, who also were the major shareholders of TCCL. Although the meeting concerned other business, the subject of menthol arose. It was suggested that ASRCT assist TCCL by promoting mint farming and providing technical assistance to the farmers and crude oil distillers who would thereby provide TCCL with an indigenous source of crude mint oil. After considerable negotiation, a contract was signed in November 1973. ASRCT was to promote the farming of mint and provide technical assistance to mint farmers and distillers of the crude mint oil. In return, ASRCT would receive

2 percent of the cost of all mint oil purchased by TCCL, other than that purchased from NTCL, for a period of seven years commencing 1 January 1974 (see Table 16.1 for figures on target crude oil production and projected ASRCT revenue for crude oil at two different price levels).

It was also in 1973 that NTCL made its first sale of mint oil. This was to TCCL.

TABLE 16.1
PROJECTED CRUDE OIL PRODUCTION AND
ASRCT REVENUE AT DIFFERENT PRICE LEVELS

Year	Tons of Crude Oil	Thousands of U.S. Dollars at 1973 Prices*	Thousands of U.S. Dollars at 1974 Prices*
1974	200	36	80
1975	400	72	160
1976	600	108	240
1977	800	144	320
1978	1,000	180	400
1979	1,200	216	480
1980	1,500	270	600

*These calculations are based on ASRCT receiving 2 percent of the price paid for TCCL's purchases of crude mint oil. For any purchased from NTCL, however, ASRCT would receive only 1.5 percent for the period of their contract with NTCL. Therefore, these projected income levels were slightly inflated.

Source: ASRCT Essential Oils Program

THAI MINT AGRICULTURE AND INDUSTRY

At the time of ASRCT's optimistic 1974 projections, the international situation undoubtedly looked promising to a newcomer to the field. The price was right, experiments had demonstrated that high-quality mint could be produced in Thailand, and national policy was directed toward diversifying agriculture. The following three subsections report the results of Thailand's agricultural and industrial efforts in the growth of mint and in the production of oil and menthol.

Growth of Japanese Mint in Thailand

As previously noted, by 1974 ASRCT had two major contracts for the production of mint oil. Both have had a different but profound impact on mint agriculture in Thailand. Under the contract with NTCL, ASRCT operated an experimental farm on NTCL land and therefore developed a basic knowledge of mint agriculture in Thailand. The contract also helped NTCL to become the most successful producer of mint oil in Thailand--reportedly producing 80 percent of Thailand's mint oil in 1977.

The second contract, between ASRCT and TCCL, may have influenced clients in a less happy direction. Following the signing of the TCCL contract, ASRCT launched a major nationwide television, radio, and newspaper campaign to induce farmers to grow mint. The campaign met with immediate success, with requests for assistance coming from 480 potential mint farmers or oil extraction plant owners. ASRCT selected approximately fifty applicants; however, of this original group, all but two have given up attempts to grow mint or extract oil.

In retrospect, one ASRCT staff member noted that the corporation had made a mistake by promoting the farming prior to having done sufficient basic agricultural experimentation. Further, he noted, the project did not have enough staff to provide technical assistance to such a large group of farmers located in many different parts of the country. Others at ASRCT noted that the problem resulted from inexperienced farmers, poor farm management, poor soil and adverse environmental conditions, stiff competition from other crops, and fluctuation in the price of menthol in the world market.

In a 1976 ASRCT survey, five persons or companies were found to own distillation facilities in nine Thai provinces. The facility owners and farmers covered by the survey projected that 11,800 rai would be planted in mint (see Table 16.2). Based on the amount of mint oil reportedly produced in 1977 (see Table 16.4), the amount of mint required to produce a kilogram of mint oil (see Table 16.5), and the amount of mint produced per rai on an efficient farm (see Table 16.3), it is believed that the actual area cultivated in mint was considerably less than the projections.

ASRCT estimated that the total labor force for mint farming during the 1976-1977 season was 6,500 workers. The labor force normally consists of a farmer's family. Women and children do approximately 60 percent of the work, which includes applying fertilizer, weeding, and planting. Heavier labor such as plowing, harvesting, and transporting the mint is performed by men. Nan farmers harvest the mint leaves three times per year. This is a common practice in Thailand even though alternative procedures are being examined.

Primitive still composed of an oil drum and pan,
Prae Province

NTCL still--most common type in Thailand

TABLE 16.2
LOCATION AND AREA OF FARMS SERVING DISTILLATION
PLANTS DURING THE 1976-1977 SEASON

Companies of Persons Owning Distillation Plants	Plant and Farm Location	Area Projected To Be Under Cultivation* (rai)**
Nanleaf Tobacco Co. Ltd.	Nan	3,000
Sai Khangkit	Chiangmai	500
Manop	Tak	200
Menthol Thai Import Export Co., Ltd.	Chanburi	4,000
	Nakornpathom	2,000
	Chacherngsao	100
	Supanburi	100
	Krabee	400
Producer in Amphoe Chiang-Kum	Payao	1,500
TOTAL AREA PLANTED		11,800

*From an ASRCT survey in 1976.

**One rai is equivalent to 1,600 square meters.

Source: ASRCT Survey 1976, report published in 1978 by ASRCT.

  An ASRCT study of production costs and returns of twenty-seven farmers in Nan province revealed that Nan farmers obtained an average annual yield of 4.3 tons of mint per rai (see Table 16.3). The production costs per rai were 3,018 baht while the income was 4,348 baht, yielding a difference of 1,330 baht, or $66.50, per rai. Of course, since the labor was done by the farmer and his family, this difference included the cost of labor and so was not "net" profit. Nan farmers sell their mint to NTCL and thus have benefited from the experimentation and extension activities of ASRCT. Also, NTCL has set a price of one baht per kilogram of "standard" mint leaves that is not altered with the fluctuating prices of menthol. This is not a universal practice in Thailand.
  Most of the farmers own their land for growing mint; only a minority has to rent. Most farmers obtain operating funds, usually in the form of fertilizer, stolon, and cash, from the oil distillers. Repayment of the loan occurs when the farmer sells his crop to the distiller. At this time the loan is deducted from the total sale price of the mint. The

TABLE 16.3
COST AND REVENUE OF MINT GROWING IN NAN PROVINCE 1976

Harvest	Land Rent (baht/rai)	Plowing by Tractor (baht/rai)	Labor Cost Weeding (baht/rai)	Labor Cost Harvesting (baht/rai)	Other (baht/rai)	Material Costs Fertilizer (baht/rai)	Material Costs Stolon (baht/rai)	Transportation (baht/rai)	Others (baht/rai)	Cost of Production (baht)	Production per rai (kg)	Revenue per rai (baht)	Net Revenue per rai (baht)
I	110	60	414	146	317	149	33	122	80	1,431	1,539	1,539	108
II	-	-	429	196	34	166	-	151	40	1,016	2,016	2,016	1,000
III	-	-	300	93	24	79	-	67	8	571	793	793	222
TOTAL	110	60	1,143	435	375	394	33	340	128	3,018	4,348	4,348	1,330

Source: ASRCT Survey of Mint Plantations in Nan Province in 1976.

TABLE 16.4
PRODUCTION AND IMPORTATION OF MINT OIL IN THAILAND

Year	Local Mint[1] Oil Production (Kg.)	Imported[2] Mint Oil (Kg.)
1972	---	65,283
1973	1,930	61,854
1974	10,199	85,882
1975	15,704	22,439
1976	12,869	67,275
1977	13,246	132,218

Source: (1) Industrial Statistics 1977. Ministry of Industry.
(2) Department of Customs.

TABLE 16.5
PRODUCTION COST OF MINT OIL

Bt/distillation

Raw material	Mint, 500 Kg @ 1.30	650
Labor	One Foreman x 40 bt/day (2 distillations)	20
	Two Unskilled @ 25 bt/day each (2 distillations)	25
Utilities	Firewood	30
	Diesel oil (for water pump)	5
Transportation		70
Administration salary		25
Other expense		15
Depreciation		44
Production cost for 3 kg. of mint oil (1 distillation)		844
Average production cost per kg.		281
Sale price in 1978		310
Gross profit before income tax and interest		29

Note: This data is based on a mint oil factory with three stills. The capacity of each still is three kg. of oil per distillation; each still is operated for two distillations a day, 120 days per year.

Source: ASRCT Survey of Production Cost of Mint Oil in Tak Province in 1978.

TABLE 16.6
FLUCTUATIONS OF PRICES OFFERED BY THE THAI CHEMICAL COMPANY
FOR LOCALLY PRODUCED OIL FROM 1974
(Baht/kg.)

	1974	1975	1976	1977	1978
January	400	350	280	310	320
February	400	365	290	310	320
March	406-412	320	300	320	320
April	406-412	320	300	320	320
May	406-412	320	280	320	320
June	406-412	260	280	320	320
July	406-412	280	280	320	320
August	406-412	280	280	320	320
September	406-412	280	280	320	320
October	406-412	310-320	280	320	320
November	406-412	310-320	280	320	320
December	400	300	280	320	320

Source: The Thai Chemicals Co., Ltd.

cost of creating a job place on a mint farm is calculated to be seventy U.S. dollars[2] (see Table 16.3).

Mint Oil Production in Thailand

While Thailand did not experience the 100-fold increase in mint oil production that some at ASRCT had hoped for in 1974, the country did witness an approximate 7 fold increase in oil output from 1973 to 1977 (see Table 16.4). One factor that affected the oil output negatively was a decrease in price from a high of 412 baht per kilogram in November 1974 to a low of 260 baht per kilogram in June 1975 (see Table 16.6 for price fluctuations from 1974 through 1977). International markets commonly demonstrated even greater variation;

---

[2]From Table 16.3, the operating costs for the first crop, including rental of land and equipment, is seen to be 1,431 baht or approximately seventy U.S. dollars. Assuming income generated from selling the first crop could finance the inputs for the second crop, and so on, then seventy dollars is all that would be needed to begin farming a rai of land. While approximately five different people would normally be involved in farming the land, ASRCT experts estimate that one rai of mint will require the equivalent of one full-time person. Thus, seventy dollars would provide one full-time job place.

for example, in early 1973 the price for menthol was 266 baht per kilogram, but in September 1974 the cost had risen to 987 baht per kilogram.

In 1976, sixty-seven mint oil plants were operating in Thailand, with 200 stills. Of these, eleven plants with 86 stills were owned by NTCL.[3] The other fifty-six plants with 114 stills were located in other parts of the country.

Studies of the investment and production costs of mint oil distillation were conducted by ASRCT in 1976 and 1978 (see Tables 16.5 and 16.7). Based on these figures, the cost per job place in a mint oil distillation plant was calculated to be approximately eight hundred U.S. dollars. The plant operates about four months per year, normally during March, April, June, July, and August. Oil extraction plants are operated mostly by men.

Menthol Production in Thailand

In 1977 Thailand consumed 120 tons of menthol (see Table 16.8). If the country were self-sufficient, mint oil distillation plants would have had to produce 200 tons of oil. This would have consumed approximately 40,000 tons of mint, which would have occupied 20,000 farm laborers. In fact, only thirteen tons of mint oil were reportedly produced during 1977. During the year, 148 tons of mint oil were imported, and 28 tons were exported. The perhaps curious phenomena of exports and imports of mint oil taking place at the same time apparently results from the actions of at least two involved groups. First, the menthol producers were attempting to obtain mint oil for the production of menthol that they would sell locally or internationally, depending on where they could obtain the best price. Also buying mint oil were Thai traders speculating in the world commodity market for mint oil. The traders would, of course, sell where they could obtain the best financial return, and this was often abroad.

The products of menthol companies are menthol and dementholized oil. Since 1974 the combined value of the local production of these has been modestly stable (see Table 16.9). Imports and exports of dementholized oil seem to have been more erratic, but due to the Thai Customs Department's acquisition of several types of oil under one code, it is difficult to be sure exactly what is happening with respect to the movement of dementholized oil.

In early 1979 two companies in Thailand had the technology to produce menthol. One of these, TCCL, formed early in the decade, had temporarily ceased operation due to

---

[3]Based on an ASRCT survey in 1976.

TABLE 16.7
INVESTMENT REQUIREMENT FOR FACTORY WITH EIGHT STILLS

Requirement		Cost (baht)
Land (10 rai) @ 5,000 per rai		50,000
Building		25,000
Machinery		
- Distiller (8 units) @ 35,000 per unit	280,000	
- Set of water pumps	16,000	
- Total		296,000
Equipment		
- Pulley	1,800	
- Scale	1,200	
- Container	600	
- Total		3,600
Total Fixed Capital		374,600
Average investment cost per still		46,825

Note: This information is obtained from a factory with eight stills having a capacity of 48 kg. of mint oil per day, when operating two distillations per day and 120 working days per year.

Source: ASRCT survey on the Economic Condition of two Mint Oil Factories in Nan Province, 1976.

a lack of mint oil for processing. The other, the Menthol Thai Import Export Company Ltd., has been in existence since 1975. Only thirteen tons of the companies' mint oil came from Thai sources; the remainder was imported (see Table 16.8).

Pirote, the manager of TCCL, noted that, as a result of an ASRCT-sponsored seminar in 1977, the government placed a 20 percent surcharge on imported menthol in addition to an existing 30 percent import duty. According to Pirote, this only led to a rush of smuggling menthol that had been "dumped" on the Hong Kong market at below what he believed could be true production costs. Pirote said that the government could help the most by encouraging the growth of indigenous menthol. When self-sufficiency is reached, an effective ban could be placed on the importation of mint oil and menthol. According to Pirote, until this

TABLE 16.8
PRODUCTION AND CONSUMPTION OF MINT OIL AND MENTHOL IN THAILAND 1972-1977

Year	Mint oil				Menthol			
	(1) Local production (kg.)	(2) Imports (kg.)	(3) Exports (kg.)	(4) Quantity locally processed into menthol (kg.)	(5) Local production (kg.)	(6) Imports (kg.)	(7) Exports (kg.)	(8) Local consumption (kg.)
1972	-	65,283	-	65,283	39,170	94,521	-	133,691
1973	1,930	61,854	-	63,784	38,270	151,546	-	189,816
1974	10,199	85,883	5,040	91,041	54,625	33,532	1,362	86,795
1975	15,704	22,439	-	38,143	22,885	9,549	-	32,434
1976	12,869	67,275	4,788	75,356	45,213	41,981	3,930	83,264
1977	13,246	148,350	27,939	133,657	70,741	66,474	17,839	119,376

Source: (1) Industrial Statistics 1977, Ministry of Industry.
(2), (3), (6) and (7) Department of Customs.
(4) Data is obtained by adding (1) and (2) and subtracting (3).
(5) Production of menthol is calculated from (4).
(8) Data is obtained by adding (5) and (6) and subtracting (7).

TABLE 16.9

PRODUCTION, IMPORT AND EXPORT OF MINT AND MENTHOL IN THAILAND 1973-1977

Year	(1) Mint oil production (kg.)	(2) Production of menthol from locally produced mint oil			(3) Production of dementholized oil from locally produced mint oil			(4) Imports of mint, dementholized and peppermint oils **			(5) Export of peppermint oil ***			
		Quantity (kg.)	Average imported price (bt/kg.)	Value (1,000 bt)	Quantity (kg.)	Average World price * (bt/kg.)	Value (1,000 bt)	Quantity (kg.)	Average imported price (bt/kg.)	Value (1,000 bt)	Quantity (kg.)	Average exported price (bt/kg.)	Value (1,000 bt)
1973	1,930	1,158	182.66	211.5	579	139.1	80.5	61,854	121.33	7,504.8	-	-	-
1974	10,199	6,119	372.11	2,276.9	3,059	358.5	1,096.7	85,882	218.05	18,726.3	5,040	508.15	2,564.1
1975	15,704	9,422	414.5	3,905.7	4,711	188.2	866.6	22,439	290.12	6,509.3	-	-	-
1976	12,869	7,721	333.4	2,574.2	3,860	151.2	583.6	67,275	246.12	16,557.9	4,788	343.63	1,655.3
1977	13,246	7,947	365.2	2,902.5	3,973	194.0	770.8	148,350	234.27	34,753.5	27,939	218.50	6,104.9

Note:
* The average world market price is used for calculating the value of dementholized oil because the average imported price is not available (import prices comprise cost, freight and insurance).
** Imports of mint oil, peppermint oil and dementholized oil are all included under the same code of the Thai Department of Customs.
*** Export of peppermint oil is comprised of peppermint oil or dementholized oil (imported), locally produced dementholized oil, and a small quantity of mint oil.

Source:
(1) Industrial Statistics 1977, Ministry of Industry.
(2) and (3) The production of menthol and dementholized oil was calculated as follows: weight of menthol equals 60 percent of the weight of mint oil; weight of dementholized oil equals 30 percent of the weight of mint oil; ten percent of the weight of mint oil is lost during processing.
(4) and (5) Department of Customs.

condition in mint oil production is obtained, the government cannot do much except attempt to reduce the level of smuggling.

Based on the capital required to set up TCCL and initiate operations, the TCCL manager estimated the cost of creating a job place in the menthol production industry to be approximately $15,000.[4]

AGRICULTURAL, ECONOMIC, AND SOCIAL FACTORS

Agricultural Factors

From its experiments and observations, ASRCT has found that the major agricultural problem in growing mint is weeds. ASRCT has demonstrated that a lack of weeding can reduce the yield of mint oil and fresh herbage by 80 to 90 percent. Weeds affect the smell of the oil, and the work necessary to harvest the mint increases. However, by weeding the plants once every two weeks, optimum yields[5] can be achieved. This requires the full-time labor of one person per rai of planted lands during the mint farming period.

A common experience observed by ASRCT is that a Thai farmer will experiment with planting one rai of his land in mint. It is given adequate attention, and good income is obtained. The next year, ten rai are planted, but the farmer is unable to provide adequate attention to the crop. His yield falls drastically and he gives up mint as a crop. Despite ASRCT's explanation of the problem, the farmer is no longer interested in mint.

The second most important factor affecting the crop is the need for irrigation. While mint can grow during both the wet and dry seasons, irrigation becomes critical during the dry season. The manager of TCCL also pointed out that the wet season causes problems. The farmers are inhibited in their weeding activities at this time, plus harvesting and drying the mint for processing become difficult.

ASRCT suggests that the third major factor affecting the success of mint farming is the application of fertilizer. In its experimental program, ASRCT has discovered which fertilizers and quantities provide the best return on investment. However, the ASRCT staff involved in this program numbers only four, and it is impossible for them to act as extension agents for the country. Consequently, information

---

[4] This estimate was in 1971 currency.

[5] This practice, however, may not give the maximum income if hired labor is used.

on fertilizer usage is not readily available to the average Thai farmer. Also, ASRCT points out that many farmers are reluctant to invest in fertilizer even when they are informed of ASRCT's results. This may be due to the lack of capital and inability to obtain it. An exception to these circumstances is Nan, where NTCL has the benefit of ASRCT's experimental program and where farmers may obtain the necessary information. Also, NTCL provides credit to farmers for purchasing recommended formulas and quantities of fertilizer.

Finally, ASRCT has discovered that the best soil for growing mint is sandy loam or loamy sand. Although this type of soil is found in most parts of Thailand, it is not the only kind. Farmers, who are unaware of soil effects on mint production, sometimes plant their mint in less than the optimum type of soil.

Economic Factors

In many parts of Thailand more attractive agricultural or industrial economic opportunities than growing mint may exist. For example, in 1976 farmers could get such a high price for growing chili that many mint farmers switched from mint to chili. As a result, many mint oil extraction plants went out of business. Now there may be no one to whom these farmers could sell mint if they wished to switch back.

Another example is that a few years ago the Thai government established a policy of encouraging new industrial ventures to be located outside of Bangkok. As a result, an industrial estate was located at Min Buri near Bangkok, an area that contained many mint farmers. The farmers wished to take advantage of the opportunities offered by the new factories. Mint, however, is a labor-intensive crop requiring constant attention. Other crops such as rice and corn can be planted and to a large extent left alone until harvest, allowing the farmer to work in a factory. Consequently, farmers in the Min Buri area switched from mint to crops that would allow them to work in the new factories.

A different type of economic problem that may affect the mint farmers has recently been reported. According to oil distillers in Thailand, mint oil is now being smuggled into the country.

Social Factors

The farmer and mint oil extractors are very dependent upon each other, and this relationship often has not been realized. As a result, farmers have shifted crops to maximize short-term profits and oil extraction plants have offered low prices for mint when the world price for menthol is low and mint is abundant. The former causes the oil extraction plants to go out of business; the latter discourages farmers and they switch to new crops.

In Nan, however, NTCL recognized this interdependency and established a policy of paying the farmers a set, and presumably adequate, price for a kilogram of "standard" mint. This price is maintained by NTCL even when world mint oil prices would suggest a financial loss, and the farmers bear with the price when the opposite is true. This recognition of mutual interdependence may well account for 80 percent of all Thai mint oil coming from NTCL.

Another social factor relates to historical Thai agricultural practices. Thailand has always been relatively wealthy by world standards in terms of the land available per farmer. This has led to less intensive farming practices than are followed in many other parts of Asia. However, mint must be intensively farmed to be successful.

Finally, many Thais have indicated that they place different values on such things as freedom, the nature of work, and money than perhaps other cultures do. Freedom is rated high, as the name of the country suggests-- Thailand meaning "land of the free." Drudgery in work may be shunned more than in some countries, possibly as a result of the abundance of food on trees and plants and in the streams of Thailand. For the same reason, money may not seem so important.

All of these factors, if true, might argue against the Thai going into mint farming. It certainly limits freedom, it is somewhat onerous work, and sufficient income is often available in Thailand from other sources.

CONCLUDING SOCIOECONOMIC CONSIDERATIONS

Mint farming in Thailand has provided work for 6,500 workers including 3,900 women. Oil extraction plants have employed nineteen.[6] Menthol-producing firms have employed a number of others. Mint agriculture and industry now contribute $.6 million annually to the Gross National Product. They have helped to diversify both agriculture and industry in Thailand. ASRCT says that it has learned much about how its research and services can affect the country. Despite the fact that early goals have not been achieved, some claim that mint growth and processing in Thailand still has been successful.

---

[6]The calculation is based on the present production of thirteen tons of mint oil per year.

REFERENCES

Black, Ronald P., and J. Gordon Milliken. <u>The Applied Scientific Research Corporation of Thailand Commercializes Its Research and Services: A Case Study of ASRCT's Essential Oils Program.</u> (Denver: Denver Research Institute, 1974).

PERSONS INTERVIEWED

Nitasna Pichitakul, director, Project Development Department, ASRCT

Narong Chomchalow, deputy governor 1, ASRCT

Lumduan Maprasert, acting director, Economic Research Department

Songkhit Vitsuthipitakul, project leader, Mint Production Project, Agricultural Research Department, ASRCT

Pirote Tangtrongsakdi, manager, Thai Chemical Company, Ltd.

QUESTIONS TO CONSIDER

1. World commodity market fluctuations grossly affect Thai mint production; is there any course of action to reduce this problem? Does the stabilization action of Nanleaf Tobacco Company Ltd. (NTCL) in setting a fixed price for mint purchases from farmers seem to hold the answer? What about international cartels or consortia?

2. Does there appear to be any means of bringing down costs of mint oil/menthol production? Is there any apparent need for technological innovation?

3. What about the effect of government import duties; is any policy change indicated?

4. Credit availability seemed to play a significant role in Nan province; should this be extended elsewhere? How?

5. The paradox of simultaneous import and export of the same commodity suggests something improper with the market mechanisms; how might these be adjusted?

6. Do you concur that, shortfalls in production estimates notwithstanding, the "mint experience" in Thailand has had a net beneficial effect?

17
Central America: Fungal Fermentation of Coffee Waste

Suellen Sebald Edwards

INTRODUCTION

The idea of producing a protein source for animal feed from waste is an intriguing one. To the coffee processor who has to figure out what to do with 78 percent of each coffee cherry that comes into the processing plant, the alternative of producing something useful out of waste is exciting. In Central America research has been going on for nine years to achieve exactly that--the production of protein from coffee wastes. During those nine years, various phases of work and directions have been pursued to solve one of Central America's biggest headaches--what to do with mounds of coffee pulp and how to keep rivers and lakes near coffee beneficios (processors) clean and free of pollutants. Progress has been made, and the next few years seem very promising in terms of achieving these goals.

BACKGROUND

The program for converting agro-industrial wastes into animal protein supplements began as a National Aeronautics and Space Administration (NASA) supported program, 1969 to 1973, directed toward demonstrating the technology transfer process through direct participation in the conversion and application of aerospace technology for commercial use. The objectives included undertaking additional research and development on technology available through the national space agency so that the technology could be adapted or applied to nongovernmental use, and attempting to stimulate the introduction of the technology into commercial usage by working with industrial organizations.

The Instituto Centroamericano de Investigación y Tecnología Industrial (ICAITI) and the Denver Research Institute (DRI) coupled their expertise to initiate a program of joint research, initially funded by NASA. Two objectives were to develop useful by-products from waste materials of the coffee processing and alcohol distilling industries in Central America and to prevent pollution of water and soil by these wastes. ICAITI and DRI found that out of twenty-one cultures tested the fungal strain Aspergillus oryzae was the most suitable organism for converting coffee wastes or spent wash to mycelium, a potential food. The use of this fungus has two advantages over the use of bacteria or yeasts, also used to process wastes: (1) it appears that the wastes can be treated successfully using continuous fermentations carried out under nonaseptic conditions; and (2) the product, fungus mycelium, is filamentous in nature and can be recovered easily by coarse-vacuum-screen filtration. These two advantages mean that facilities can be built for a much lower capital cost than a bacterial or yeast fermentation plant.

Since the initiation of the project, several other fungal strains have been developed: Penicillium orustosum, Gliocadium deliquescenus, Trichoderma harzianum. All of these strains are being used in the current research project.

The project also resulted in the formation of an international network of research institutes active in the conversion of agricultural wastes. The consortium included: Federal Institute of Industrial Research (FIIR), Nigeria; (ICAITI), Guatemala; Instituto de Investigaciones Tecnológicas (IIT), Colombia; Instituto Nacional de Tecnologia, (INT), Brazil; Instituto de Tecnologia de Alimentos (ITAL), Brazil; and DRI, United States. Several newsletters were published by DRI containing research results from the participating institutes and were distributed through the network.

CHOICE OF THE SUBSTRATE

Coffee-processing waste water was chosen specifically as the substrate for growing biomass. An understanding of how coffee is processed in a typical beneficio will illustrate some of the unique problems that Central America has to solve.

The processing of coffee fruits, which involves the use of large amounts of water that become high in organic matter content, yields 4.5 tons of wastes (which can be useful or nonuseful) for each ton of dehulled coffee. In some beneficios, such as in El Salvador where fresh water is scarce, plants are centralized and the process water is recycled several times, thus producing an even greater pollution concentration in the waste water. These waste waters are often put into evaporation ponds where aerobic digestion and bad odors develop.

Coffee processing typically involves the following steps (see Figure 17.1):

(1) Coffee berries are dumped into a receiving tank of water.

(2) Coffee berries that float are channeled to separate processors, for they are of inferior quality. The berries that sink are transported with water to the pulpers, where the pulp is separated by mechanical action.

(3) The de-pulped berry (grain and mucilage) is sent to a fermentation tank. (The partially pulped or unpulped berries are recycled back to the mechanical pulper.)

(4) The grain is kept in the fermentation tank without water for several days, allowing for the natural biodegradation by coffee enzymes and microflora and thus removing the mucilage.

(5) After fermentation, the coffee grain is washed in water channels, where the grains that float are separated from the grains that sink.

(6) The washed grains are put on large cement or tile drying patios, where the moisture is evaporated by solar energy (during this time the grains are "raked" to ensure even drying).

(7) The grains, when they reach a critical level, are transferred to dryers, where moisture is reduced to about 10 percent.

(8) The dry grains are then dehulled mechanically.

(9) The result, green coffee, is bagged and shipped to consumers

Two factors led to the decision to use coffee-processing waste water as the substrate: the increase in contamination caused by those "wastes," and the growing need for food. Thus, development of a process that would combine treatment of effluents from coffee processing with controlled growth of biomass was chosen as the best option. The ultimate objective was to develop a simple, nonaseptic, continuous fermentation system that could be installed at beneficios. It was believed originally that the biomass production would cover the cost of water treatment.

FIGURE 17.1
PROCESSING OF COFFEE BERRIES

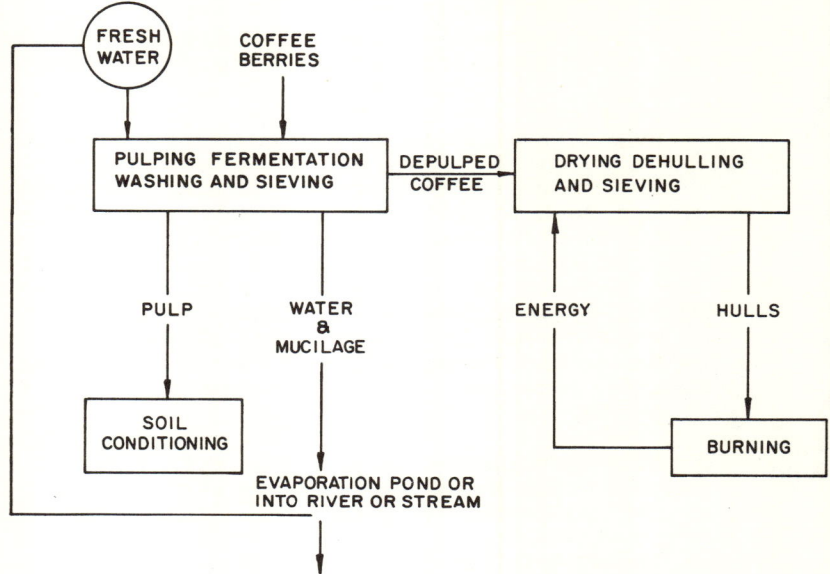

Source   "Protein from Waste: Growing Fungi on Coffee-Waste", *Chemtech* (October, 1976): 637.

The project began with the selection of adequate strains from shake flask experiments. Microorganisms were screened and selected according to their chemical oxygen demand (COD) consumption, nutrient requirements, and growth. Promising results were obtained using various designs of bench-scale fermentors under nonaseptic conditions in both batch and continuous processes. Rat-feeding studies were undertaken at ICAITI to compare the growth rate of rats fed the protein supplement with rats fed commercial protein. Preliminary results showed that the product was nontoxic and that the rats grew at approximately the same rate as those not fed the biomass.

The development of biomass from unsophisticated procedures in a developing country is viewed as a significant achievement by the project leaders. Usually this experimentation takes place with costly, sophisticated equipment; it was rewarding to the researchers to have success with bench-scale fermentors under nonaseptic conditions in both batch and continuous culture.

PILOT PLANT -- 1974-1975

Next came the design, construction, and operation of a pilot plant fermentation unit in El Salvador, beneficio Curazao. ICAITI sent personnel to work at and to supervise the building of the plant; the beneficio contributed by providing labor, building concrete tanks, and purchasing an air compressor. The beneficio was interested in working with ICAITI and in utilizing the process for producing biomass. The fact that the process would significantly reduce the COD content of the waste water was only a side bonus and not the main concern of the beneficio.

The pilot plant consisted of a 2,000-gallon equalizing tank, two 5,000-liter fermentors, and two 450-gallon seed fermentors (see Figure 17.2). An equalizer was needed because the nature of the coffee pulp depends on the time of year and the altitude at which the coffee cherries are picked: those picked at lower altitudes have less water, as do those picked near the end of the coffee season, when the atmosphere is very dry. Thus, the substrate varied greatly with the amount of water in the cherry pulp. Controlling the substrate would have involved changes in the beneficio process, for the volume of water and of coffee processed fluctuated widely.

The waste water entered the pilot plant via a sedimentation tank, where large, suspended particles were removed, and continued through the equalizer (for twenty-four hours). The fill tanks received the substrate, with nutrients added according to the carbon/nitrogen/phosphorus ratio. The inoculum, a 450-gallon culture of T. harzianum, was produced nonaseptically for twenty-four hours on a 2 percent blackstrap molasses supplemented with phosphoric acid and ammonium salts. The pH was adjusted to 3.5 with sulfuric acid.

FIGURE 17.2
PILOT PLANT FOR WASTE WATER TREATMENT

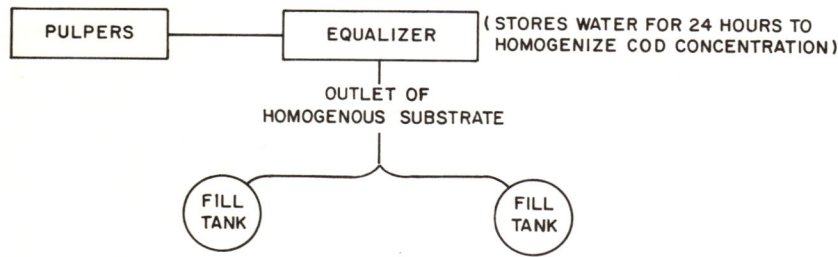

The pilot plant operated for one coffee season (six months) under nonaseptic conditions (which was possible due to the high acidity of the waste), resulting in good fungal growth. However, the amount of protein was low; therefore, it was difficult to recover the biomass because it was so dilute.

Several factors contributed to the decision by the beneficio to terminate the pilot plant after only one season. One reason involved the economics of recovering the biomass. When calculations were made for the scale-up to production level, it was shown that the biomass recovery and drying were too expensive compared to the price at that time of soya and fishmeal feed supplements; they were selling at the same price that it would cost to market the coffee waste biomass. Second, additional research was necessary to solve some of the recovery and drying problems before a marketable product could be produced. The beneficio was unwilling to contribute financially for further research. The El Salvadoreans were disappointed in the whole experiment. ICAITI may have partially contributed to the disillusionment by not explaining well or making clear from the start the nature of a research project, i.e., that when the pilot plant operations were over, there would be no guarantee of a biomass-producing-plant blueprint. Third, the process water being used by the beneficio was contaminated. Whether this affected the animal feed was not determined; additional tests would have been necessary. The process reduced the COD content in a significant way, so the process was cleaning the water, but this was not of interest to the beneficio (no government regulations existed at that time regarding water pollution).

Efforts to encourage other Central American funding agencies to continue the project were not successful, mainly because of the unique characteristics of El Salvadorean coffee beneficios. El Salvador, unlike other Central American countries, has large, centralized coffee beneficios. In addition, they utilize scarce water resources by recycling the water through the processing plant, resulting in a large concentration of wastes in the discharge water. Because the results of the pilot plant project were not transferable to other beneficios with other production methods, the alternative sources for funding were not interested in continuing the project.

At this point, the project leaders went "back to the drawing boards" and reviewed the direction of the project. Discussions were held to decide whether the project should choose a different substrate. The answer was "yes."

SWITCH TO ANOTHER SUBSTRATE TO PRODUCE BIOMASS

Because of the high viscosity of the mucilage (mainly composed of pectin, carbohydrates, and water) and the fact that aeration is difficult except by adding water, it was decided not to use mucilage as a substrate. (Recovering the pectin is the only use for the mucilage, and this research already is being done throughout Central America.) Also, ICAITI spent eight months working on laboratory fermentations of the mucilage and of combinations of mucilage and sugarcane molasses. The preliminary economic analysis showed that the substrate was not practical; it was recommended that ICAITI fund something else.

In July 1976, ICAITI began working with the coffee pulp. Coffee pulp comprises a major portion of the cherry and is thrown away immediately after the berries are depulped at the beginning of the production process. Several methods have been tried to find a use for the pulp. Traditionally, the pulp has been used as a fertilizer or compost. It also can be ensiled and used for animal feed bulk. (Research by a Central American nutrition foundation, INCAP, has found that 20 to 25 percent pulp is the maximum amount that can be added to animal feed.) Ensiling involves storing the pulp in silos, where it is compressed manually to remove all oxygen, and keeping the pulp in the silo for one year, after which it can be used for animal feed. This solution creates storage costs and runs the risk of producing toxic animal feed, for unless all the air is squeezed out, pockets of aerobic digestion will occur, resulting in fungal growth or alcohol formation. If the coffee pulp is left in large piles, aerobic digestion begins immediately, fouling the air, plus the pile squeezes the pulp on the bottom, causing the pulp juice to permeate the soil and eventually contaminate the water surrounding the beneficio.

Because aerobic fermentation begins as soon as the pulp is dumped into the mound, ICAITI's researchers decided to press the pulp and use the pulp juice as the substrate. The juice is homogeneous, consisting of carbohydrates and 5 percent sugar--seemingly an ideal source. Also, because all Central American coffee beneficios have the same process for removing the pulp, the results would have universal application, unlike the original substrate. Since the second quarter of 1977, ICAITI has been testing the pulp juice with the same fungi used with the waste water. The tests have been encouraging: the fungi grew well and consumed the carbohydrates (80 percent).

In 1978 ICAITI researchers started a new fermentation process, stopping the fermentation before all the carbohydrates had been consumed. This resulted in carbohydrates and biomass, which were concentrated to a syrup.

Coffee cherry depulpers and fermentation tanks
at a coffee beneficio in Guatemala

Coffee waste pile at a large beneficio in Guatemala,
to become compost or animal feed supplement

The reasons for concentration were threefold: (1) to separate the fungi required a filtration energy; (2) to preserve the fungi required drying and thus greater amounts of energy; and (3) the concentration procedure was a one-time cost; it also resulted in a low-water-activity product, thus retarding the growth of yeast. Thus, the syrup could be stablilized without refrigeration, and the coffee juice molasses could be mixed with feed to enrich the protein content.

Tests at ICAITI with a 500-liter tank will provide enough data for market studies to be completed by 1980. Also, the switch to pulp juice has benefits in terms of the size of the market. The Central American beneficios all have the same process for removing the pulp, and all have the problem of what to do with the pulp. The low viscosity of the substrate makes it excellent for fermentation.

SECOND PILOT PLANT -- 1978-1979

A pilot plant is being installed at beneficio CAFOSA in El Pino, Guatemala, to test the coffee pulp juice substrate program (see Figure 17.3). The technology of the U.S. citrus industry, a continuous screw process, purchased with OAS money (a Vinzent machine, costing $17,000) is to be installed by August, in time for the coffee season (September-March). The output is one and one-half tons

Large open-air drying patio for coffee beans

per hour. The pilot plant, with the continuous press, will accommodate large beneficios; the alternative of a hydraulic press for smaller beneficios, yielding smaller outputs, is being studied at ICAITI.

The question of economy of process is still a problem. The following facts have to be considered:

(1) The coffee season is from September to March; therefore, two alternatives may be considered when a decision is to be made on how to complete an economic feasibility study (one year of fermentation data is needed).

   (a) Alternative 1: When the season ends, use another substrate, e.g., sugarcane molasses.

   (b) Alternative 2: Have pressing capacity higher than fermentation capacity, concentrating coffee pulp juice and storing it for year-round use.

(2) ICAITI is gambling that the biomass will be economically feasible with current prices of livestock feed.

   (a) ICAITI has studied many processes for producing single-cell protein (SCP), and a large percentage of the cost was for the raw material.

   (b) However, ICAITI is utilizing a substrate that is a waste from coffee production. The product is looked upon as a savings.

   (c) Other processes utilize sophisticated, aseptic processors. ICAITI will utilize inexpensive, unsophisticated materials and equipment, the major expense being the aeration equipment and energy requirement. The process is nonaseptic.

(3) ICAITI has not tested any yields from a continuous press, nor any yields from the substrate--it has only initial test results. Thus, the project is operating with hope.

IMPACT

The impact of the coffee pulp fermentation project is hard to assess at this point, for long-term results are

FIGURE 17.3
PILOT PLANT FOR COFFEE PULP JUICE PROCESSING

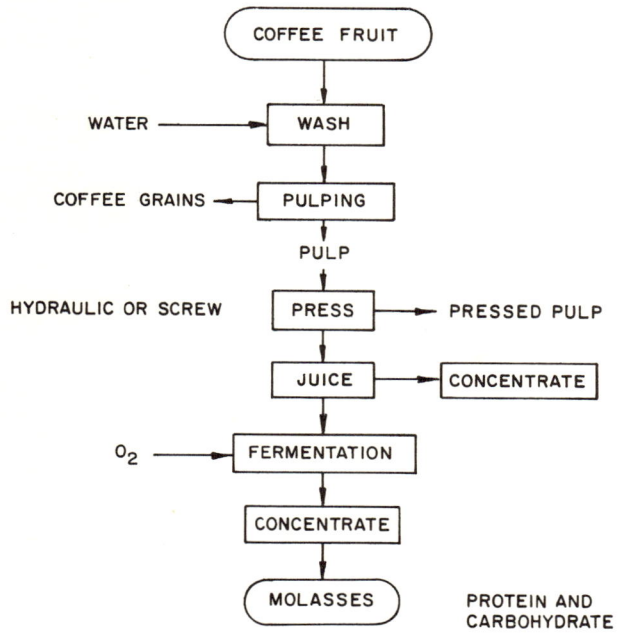

nonexistent. However, the impact will be felt in several areas.

To the campesino, the biomass will add value to the coffee crop; hopefully, this value will be reflected in the price that the beneficio will have to pay for the crop. The livestock producer will have a greater availability of protein sources that hopefully will be less expensive than those currently available. This project also will have an impact on energy consumption for protein production. Because the protein will be produced from waste, the impact versus output will create a balance.

Impact on employment generation will be felt, assuming that the pilot plant design is successful. Beneficios can have the system installed (with two or three crop seasons of technical assistance from ICAITI) and can utilize five people for its operation. This reflects a 10 percent increase of total labor for an automated beneficio. Finally, there is and will continue to be an impact on the contribution to scientific knowledge. Many scientific papers, distributed worldwide, have resulted from the project and have created much interest in the process. They also have contributed to a body of literature that was very small.

ECONOMIC ASPECTS

It is difficult to determine exactly how much money has supported and is supporting this project, because ICAITI contributes what is not provided by OAS funds; this amount varies from month to month. Support from outside ICAITI for the past five years has varied in amount and source. The OAS contributed $45,000 in 1974; the current budget includes $35,000 from OAS. The Regional Office Central American Program (ROCAP) of the Agency for International Development (AID) contributed to the project until 1975; this support totaled approximately $7,000. The money contributed by OAS was spent on equipment, materials, and some salaries.

SPIN-OFFS TO THE ORIGINAL PROGRAM

The program has created wide interest among coffee growers, processors, associations, and various individuals. The interest from these groups is not exclusively for biomass production but also for waste water treatment and methane generation. Professors and biologists also are asking to be trained in the technologies at ICAITI. This interest has been generated by word of mouth, through ICAITI's participation in coffee associations, and through the publication of scientific articles.

Examples of this interest are as follows:

(1) After ICAITI made a presentation of current research projects at the National Coffee Growers Association of Guatemala (ANACAFE), growers from El Tumbador, Guatemala, requested a presentation of technical innovations for by-products of coffee waste.

(2) Brazil, Venezuela, Costa Rica, and El Salvador have asked to have people trained in ICAITI's microbiology laboratories.

(3) Through the Coffee Research Center in Costa Rica (CICAFE), information on ICAITI's research results and projects will be disseminated, creating awareness by a very large group of enterprises.

(4) A Costa Rican beneficio is asking the Costa Rican Development Bank (CODESA) to invest in a $2 million plant to convert the coffee pulp juice into a useful product. ICAITI has been approached to provide research and technical assistance.

(5) Several beneficios in Guatemala have approached ICAITI to provide them with technical assistance for cleaning up their waste water. Even though waste water regulations are not enforced in Guatemala, some communities are bringing suit against the beneficios, and they are turning to ICAITI.

REFERENCES

Aguirre, F. et al. "Protein from Waste: Growing Fungi on Coffee Waste." Chemtech (October 1976): 636-642.

Church, Brooks, and James P. Blackledge. "Extending the Universities' Role in the Exploitation of Aerospace Technology." Final Report to NASA, Grant No. NGL 06-004-096. Denver: Denver Research Institute, April 1974.

Dengo, J. Gabriel. Production of Fungal Protein in Lesser Developed Countries. Denver: Denver Research Institute, August 1973.

Espinosa, R., B. Moldonado, J.F. Manchu, and C. Rolz. "Aerobic Nonaseptic Growth of Verticillium on Coffee Waste Waters and Cane Blackstrap Molasses at a Pilot Plant Scale." Biotechnology and Bioengineering Symposium 7 (1977): 35-44.

"Fungal Treatment of Agricultural Wastes for Water Re-Use and Conversion to an Animal Feed." Denver: Denver Research Institute Invitational Conference, 16-17 August 1973.

PERSONS INTERVIEWED

Francisco Menchu, project leader, ICAITI

Cheryl Schneider, biochemist, ICAITI

Roberto de Leon, biochemist, ICAITI

Francis Aguirre, division head, ICAITI

Ricardo Garcia, project investigator, ICAITI

Oscar Maldouado, electrical engineer, ICAITI

Rodolfo Espinosa, Destiladora de Alcoholes y Rones

Three beneficio managers

QUESTIONS TO CONSIDER

1. This application of "space age" technology is interesting; would it be useful to make systematic searches of advanced technology inventions looking for appropriate applications for developing countries?

2. Failures in research are, of course, common; was the coffee producer justified in withdrawing support for the research after the initial disappointing results? Why does it seem so difficult to get funding for development programs in developing countries?

3. Pollution abatement was an important side effect of this research. Can developing countries afford the cost of pollution programs?

4. The role of the research institute in this instance was central to the program; should there be more institutions of this type working in the developing world? Or is there currently a sufficient number? What is your impression of the efficiency and effectiveness of those with which you are familiar? What might be done to improve the availability and usefulness of R&D in the developing nations?

5. Are there other situations in which agricultural waste might be treated by the fungal conversion process?

18
The Philippines: Fish Preservation Techniques

Melinda L. Cain

INTRODUCTION

The traditional fish preservation techniques of smoking, drying, and fermentating are used widely by lower income families in both rural and urban areas of the Philippines. Most of the methods are not scientifically based, but rather are passed down through generations. Women, in particular, play an important role as fish processors, supplying a much-needed source of income to the family.

Since World War II, many Southeast Asian governments have tried to increase fish production for local consumption. However, with the exception of Japan and the Philippines, the increases have barely kept pace with population growth. Therefore, although current traditional techniques of fish processing and preservation are adequate, new methods to standardize production and to ensure better product quality would definitely benefit the consumer. These methods, however, must be affordable by the traditional processors, who process about one-half of the annual catch of the Philippines, in order to affect production. Sophisticated methods are likely to be used only by the commercial sector.

SETTING

Millions of people in developing countries suffer from inadequate diets, and particularly from a lack of protein-rich foods. To people in Southeast Asia, fish provides a readily available and important protein source as well as a major source of income. Fish contains 15 to 25 percent protein, although actual percentage composition depends on the form of the consumed fish. However, current rates of consumption are lower than would be imagined due to the relatively high cost of fresh fish, the limited supply of fish products, inadequate marketing systems, and the perishability of fish. In developing countries, the short storage life of fish is a major technical problem, especially in the hot climate.

Bacteria causes rapid spoilage of fresh fish in temperature ranges of ten to twenty degrees Centigrade. Deterioration is retarded as the temperature is reduced. For example, rates of complete spoilage from small vessels (that are not equipped with ice as protection) vary from five to twelve hours after the catch.[1]

Dr. Arroyo provided a technical description of the spoilage process:

> Fish spoilage is caused mainly by fish enzymes and bacteria. Later during the state of rigor, autolytic enzymes may act on the fish, causing softening and a general breakdown of the nitrogenous components, particularly in the area adjacent to the visceral cavity if the fish is uneviscerated. This is because when the fish dies, the walls that hold the digestive tract enzymes may be digested, the enzymes working their way to neighboring flesh. In the case of dressed fish, where the viscera has been removed, the autolytic change is less important provided the gutting procedure was done properly. After rigor has passed, bacterial decomposition proceeds at a rapid rate until spoilage is complete.[2]

The Demand for Fish

A major problem in estimating the demand for fish is the lack of comprehensive and uniform reporting of fish production and consumption statistics. This is particularly true of the small and scattered fishing areas in Southeast Asia. Furthermore, due to the 2,000 or more edible fish found in the Philippines, there also is a lack of knowledge on the science of fish as food and on the relationship of the many varieties to consumption patterns. However, the annual per capita consumption of fish in the Philippines is estimated to be twenty-seven kilograms[3] (see Table 18.1 for consumption patterns of seafood). As with most food commodities,

---

[1] M. G. Hunter, "Fish Curing and Drying." UNIDO, Regional Consultation on Promotional and Technical Aspects of Processing and Packaging Foods for Export. ID/WG. 172/16, p. 1.

[2] Fish Processing Handbook, Food Nutrition Research Center, NIST, Manila (1966), p. 15.

[3] Ibid., p. 3.

TABLE 18.1
ANNUAL PER CAPITA USE (KILOS) OF SEAFOOD BY REGION, 8 SURVEYS, MAY-JUNE 1974-MARCH 1976, PHILIPPINES

Region	Number of Families	Fresh and frozen fish					Dried and smoked fish	Crustaceans and mollusks	Canned fish	Grand total
		First Class*	Second Class	Third Class	Unknown Class	Total				
1. Ilocos	611	5.8	2.0	2.4	0.9	11.1	5.2	2.9	0.2	19.4
2. Cagayan Val.	344	6.4	0.9	2.0	0.6	9.9	5.2	3.4	2.5	21.0
3. C. Luzon	885	8.0	2.4	2.6	1.2	14.2	3.8	3.3	1.3	22.6
4. A. Gr. Manila	792	9.8	2.4	6.0	1.0	19.2	2.9	5.8	1.4	29.3
4. B. C. Luzon	968	6.4	3.4	8.0	1.8	19.6	2.4	3.1	1.5	26.6
5. Bicol	640	3.4	2.3	7.1	2.2	15.0	2.8	4.0	1.0	22.8
6. W. Visayas	800	7.1	4.4	3.9	3.8	19.2	4.1	6.1	1.3	30.7
7. C. Visayas	640	3.9	3.7	6.8	5.7	20.1	4.8	3.5	1.1	29.5
8. E. Visayas	560	5.0	4.8	6.3	6.1	22.2	4.3	4.8	1.4	32.7
9. W. Mindanao	246	4.0	3.3	6.0	5.1	18.4	5.5	1.1	0.5	25.5
10. N. Mindanao	712	4.7	3.8	5.3	3.6	17.4	6.8	2.3	1.4	27.9
11. E. Mindanao	547	3.4	2.5	11.0	1.6	18.5	6.3	5.6	2.8	33.2
12. C. Mindanao	255	5.8	3.2	6.1	2.9	18.0	6.6	3.0	1.1	28.7
Philippines	8,000	6.0	3.1	5.7	2.7	17.5	4.4	3.8	1.3	27.0

Note: 1. The annual per capita rate of use of all sea food averaged 27.0 kilos with Eastern Mindanao having the highest rate, 33.2 kilos, and Ilocos the lowest, 19.4 kilos.

2. The per capita rate of use of all fresh and frozen fish was 17.5 kilos. It ranged from a low of 9.9 kilos in Cagayan Valley to a high of 22.2 kilos in Eastern Visayas.

3. Dried and smoked fish had an average rate of use 4.4 kilos per capita and ranged from a low of 2.4 kilos in Southern Luzon to a high of 6.8 kilos in Northern Mindanao.

4. The use of crustaceans and mollusks averaged 3.8 kilos per capita with Western Visayas being highest, 6.1 kilos, and Western Mindanao lowest, 1.1 kilos.

*First-class fish includes: milkfish, mackeral, mudfish, tuna, and others; Second-class fish includes: tilapia, slipmouth, sardines, and nemiptend; Third-class fish includes: round scad, big eyed scad, bonito, anchovy, and others.

Source: Dr. Arroyo, College of Fisheries, University of the Philippines.

factors influencing this demand include perishability, distribution, marketing, price, competition with other foods, taste, and income. Processing methods, in turn, promote certain patterns of consumption, depending upon available materials.

Fish products come in many forms: fresh, frozen, canned, cured, or smoked fish; fish meal; fish oils; and fish sauce called patis or bagoong, used to flavor rice. A very common fish is bangus, or milkfish. Approximately 106,000 metric tons of this fish were produced in 1975, more than 90 percent of which was pond-raised.[4] Herring, tuna, and round scad are popular salt-water fish.

Generally, more fresh fish are consumed when the supply is readily available to large population centers. However, in the tropics fish must be treated if they are not consumed the same day that they are caught. Therefore, dried or smoked fish are common in the Phillipine diet. Most of the frozen and canned fish are for export. The oldest and least expensive form of preserving fish is curing, which includes salting, drying, smoking, pickling, and fermenting.

TRADITIONAL TECHNIQUES OF FISH PRESERVATION

Drying and Smoking

About 64 percent of an annual catch of 1.3 to 1.5 million metric tons is eaten fresh; about 33 percent is processed by the traditional methods of salting, drying, or smoking. These techniques may be done separately but may also be combined (e.g., drying includes salting as the first step; smoking includes both salting and drying before the actual smoking is done). Regional differences in processing the dried or smoked fish are few. Therefore, the simple steps described here are generally applicable to most regions in the Philippines, although slight variations in equipment or process may occur due to local custom or taste preferences.

Salting, drying, and smoking, as already mentioned, often are done together. The use of salt or brine extracts moisture by osmosis, lowering the water content of the food below the point where bacteria can live and grow. A saturated solution of at least 25 percent salt is used. Not only is the amount of salt important, but the purity of the salt also makes a great difference in the quality of preservation and taste.

Drying also preserves by lowering the water content to about 20 percent. The usual practice is to first soak the fish in the brine solution to extract surface moisture and

---
[4]Ibid.

preserve the fish during the drying process. The fish is then placed in the sun to extract the remaining moisture. Drying trays made from bamboo are most often used. However, plastic to cover the fish has been introduced recently to increase the drying power of the sun and to protect the fish from dirt and insects.

Smoking has very little preservation action, but it does add a distinctive flavor and attractive color. Smoking can be done as an additional step after salting and drying. Some type of airtight container into which smoke can be introduced is needed. A large tin barrel (about four feet tall with a diameter of two feet) or large concrete smoking bins may be used. In either case, a fire is built in the bottom of the container with wood chips or sawdust. After soaking in brine and drying in the sun for eight to twenty-four hours, the fish are placed on a rack and smoked for varying periods of time, depending on the species and size of fish.

The bangus, or milkfish, is very popular in the Philippines. Its preparation is as follows. The fish are cut through the belly, and the gills and entrails are removed. The eviscerated fish are washed thoroughly. A slight cut along the backbone facilitates the intake of salt. The prepared fish are soaked in 25 percent salt solution for thirty minutes, are drained and rinsed to remove excess brine, and are then cooked in boiling brine (20 percent salt) until the eyes become opaque white. The cooked fish are dried and are finally hot-smoked for two hours or until they are golden brown.

Rural and Urban Preservation Processes

About 200 smokehouses are operated in a rural area near Salinas in Cavite. These smokehouses process large and small herring, bangus, and round scad. Many processing businesses are operated next to the house of the business owner.

The fish are bought directly from the fishermen in the Salinas fish market and are transported by tricycle to the processing houses, where they are washed and cleaned but not scaled. The fish are then sun-dried for about one-half hour before they are cooked in a vat of boiling brine for five minutes. (Twenty kilograms of salt are added to thirty gallons of water to make the solution.) The fish are then rinsed and spread on perforated wooden or bamboo trays and are set over smoking ovens for about five minutes. They are then packed into basket containers lined with banana leaves and are covered with paper.

In the urban area of Navotas the fish are also bought early in the morning from the local fishermen. There is

TABLE 18.2
VARIATIONS IN THE PRESERVATION PROCESS

Bangus	Milkfish	Chanos chanos (Forskal)	Split fish from anus to gill opening, remove gills, internal organs and black membranes. Wash and drain. Cut flesh through body cavity.	Rub 1 tbsp. salt in body cavity of fish and let stand for 10 to 15 min. for salt to penetrate. Immerse fish in boiling 10 percent brine solution for 5 min. or until flesh near tail portion is soft. Arrange fish in smoking tray and cool until firm.	Smoke fish at 32.2° to 38°C until golden brown in color.
Galong-gong	Round Scad	Decapterus macrosoma (Bleeker)	Wash fish thoroughly.	Soak fish in saturated brine solution for 30 minutes. Immerse fish in boiling 10 percent brine solution until partially cooked. Arrange fish in trays and drain for 10 min.	Smoke fish for 50 min. to 2 hours at 32.2° to 38°C or until golden brown in color.

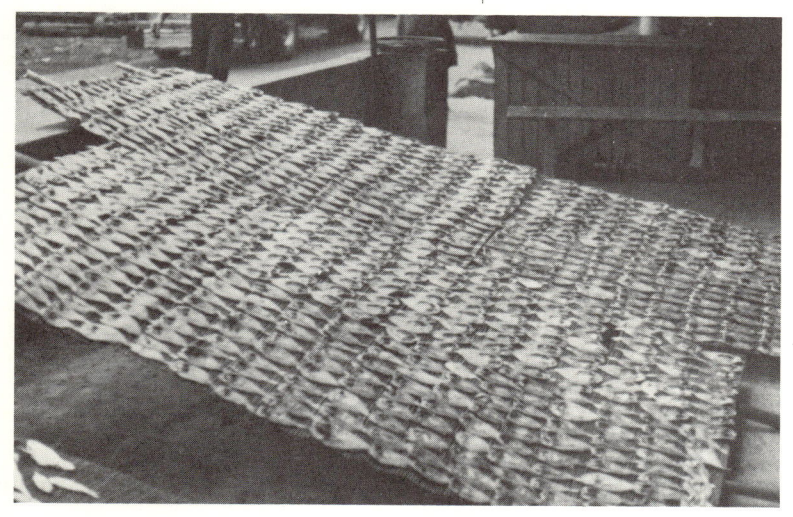

Fish drying in the sun

Smoking fish in a rural smokehouse in Cavite

even less room in an urban setting, and therefore the preservation process also is conducted in conjunction with the living quarters of the owner. Anchovies, smoked herring, and tuna are products processed here, and the smaller fish are not usually eviscerated before being soaked and boiled in brine. The salt used is an unrefined or solar salt that is purchased in an area of metro Manila called Paranque. The amount of salt, calculated from experience, is about one sack per five bañeras (each bañera is composed of about forty kilograms of fish.) After the fish are soaked and boiled, they are then dried and/or smoked.

OUTPUT AND MARKET

The output ranges from 90 to 200 kilograms per day when an adequate supply of fresh fish is available. If a processor has a direct outlet in a market (e.g., in Quezon City), her products are always sold, usually by midday.

Examples of costs and retail selling prices (based on an exchange rate of 7.3 pesos equals U.S. $1.00) are: bangus and round scad are bought at U.S. $1.66 per kilogram and are sold at U.S. $2.77 per kilogram. The herring is bought at U.S. $1.95 per kilogram and is sold at U.S. $2.77 per kilogram.

The processor also may serve as an intermediary and prefer to sell wholesale rather than retail. People may come to the smokehouse to buy fish wholesale and then sell it door-to-door or in local markets. The daily gross income from a wholesale processor would be about U.S. $42.

Smoked and dried fish bring about U.S. $0.04 per fish in the market. Since dried fish involves more labor in cleaning and splitting, many prefer to process smoked fish.

EMPLOYMENT

Traditional fish processing is often a family business, although outside workers are also employed. Most businesses are very small, using from one to five employees. The owner or manager of the processing operation is most likely to be a woman.

One rural smokehouse employed four women and one man. The man was responsible for keeping the fire ready to boil the brine solution and for cooking the fish. His salary was U.S. $2.75 per day. The women cleaned, dried, smoked, and arranged the fish for market and were each paid U.S. $2.05 per day. As most of the women had large families (from four to eight children), children often accompanied them and helped with the work.

At this income level, no luxury such as not working existed. Therefore, women must work for the family to have

Woman processor packing fish for market

an adequate income. In most cases, the woman controlled the family budget. The man gave her his salary, and the wife made the appropriate family purchases. On the other hand, for female-headed households, the processing business could be a woman's sole source of income. The urban wholesale operation was as such, and the owner (a woman) had an average monthly gross income of about U.S. $125.

COSTS

Because many of the processing operations were handed down through generations, little initial capital investment was required for the current operators. Operating costs include the purchase of the fresh fish, wages to laborers, and fuel. There are no real maintenance or repair costs to consider. Estimates of daily costs are as follows: labor, $10 to $13; fresh fish, $333 to $383; and fuel, $5.50. The only normal upkeep is the annual replacement of the wooden or bamboo trays used to hold the fish while smoking. Although the fat from the fish coats the wood so that it does not burn, after a year the trays usually need to be replaced. Most processors can usually recover their daily costs because the smoked and dried fish bring a higher price in the market than fresh fish. The amount of profit depends upon the available supply of fish and good weather.

Fish processing is a seasonal activity. When the catch is low, either due to high winds, stormy weather, or fish migration, there is less fish to process and the price per kilogram rises. Therefore, a major problem is the need for a dependable supply of fish. In many cases, the output of the processors could increase if more fish were available. At such times, smokehouses may process as little as one-half of their normal output. When the catch is too low, the smokehouse employees are asked to go home and thus lose their income for that time.

Late February to March, July and September are reported as bad weather months and therefore are a bad time for the fishing business. April and May are reportedly the best months. On the other hand, a small, home-run operation may already be operating for a sixteen-hour day and would not be willing to increase production even if the fresh fish supply were available.

Other problems that face processors are a readily available supply of water to wash fish and a better means of transporting the fresh fish from the market. Also, the sun is an important factor--so important that without an artificial drying capacity in case of bad weather, the business has to shut down. The amount of sunlight also affects the quality of the fish. If dried in direct sun, the scales on the fish remain and help to hold the fish body together, giving the product a much more appealing appearance. If there is little or no sun, the skin peels off, and the fish begins to fall apart.

FISH FERMENTATION

Other products commonly produced in the Philippines are patis and bagoong, made by a fermentation process. Fish such as anchovies, slipmouths, and other small fish are usually used to make bagoong. The following is a general procedure for producing these foods.

(1) The fresh fish are thoroughly washed in clean, fresh water, and all seaweed, sticks, and other extraneous materials are removed. The use of harbor water or stream water should be avoided because it may add microorganisms and fly larvae that will lower the quality of the bagoong.

(2) The salt is mixed thoroughly with the fish at the rate of one part of salt to three parts of

fish, or two parts of salt to seven parts of fish by weight.

(3) The fish and salt mixture is placed in earthenware pots, wooden or steel barrels, or vats to undergo fermentation. Care should be taken to keep the containers tightly covered to exclude flies which may reduce the quality of the bagoong.

(4) The containers with the salted fish are allowed to stand for periods from two weeks to a year to develop the characteristic aroma and flavor brought about by the breakdown of the fish proteins. The fermentation rate during the early stages of digestion will be twice as fast if pure salt is used instead of impure solar salt. Pure salt of about 99.5 percent NaCl content can be prepared from freshly harvested impure salt. Impure salt contains large amounts of magnesium and calcium salts, and these impurities are found mostly on the surface of the large, solid salt crystals. Simple washing with brine, then with a little fresh water and draining, will remove most of the impurities, thereby producing salt that is relatively pure.

(5) After the characteristic slightly fishy, cheese-like odor has developed and the fish flesh has disintegrated considerably, the proteinaceous liquid similar to soy sauce can either be skimmed off the top of the fish and brine mixture or drained off through a spigot opening near the bottom of the container. This liquid, which is the patis, may be strained or filtered until it it quite light in color. Good patis should contain 9 to 10 percent protein.

If the major product from the bagoong is to be patis, a number of extractions are usually made by adding concentrated brine solution (one part salt to three parts of water by weight) to the fish residue, letting it age for about two more weeks, and again drawing off the liquid. The protein content of the patis decreases after each addition of brine so that the fourth extraction will generally contain less than 1 percent protein.

The general practice, after drawing off a fair amount of patis, is to grind up the bagoong in a meat grinder and pack the material into cans or bottles for sale. If the bagoong is not fluid enough, additional concentrated brine is added.[5]

CONCLUDING REMARKS

Strategies in the five-year Philippine Development Plan, 1978-82, to enhance fishery resources include improvement of the fish marketing and distribution systems to minimize gaps between production and consumption on local and regional levels while retaining the quality of fresh fish and fishery products for consumers.

Targeted salaries for workers in agriculture and fishery are:

1977	2,559 pesos/worker ($355)
1978	2,635 pesos/worker ($365)
1979	2,719 pesos/worker ($377)

From the sites visited, and assuming that employees would work six days per week for an eight-month period, they would receive the targeted salary for 1979.

The Bureau of Fisheries and Aquatic Research has regional offices that are active in trying to improve the techniques used by the municipal fishermen and to explore ways to enhance their productivity. They have developed an articifical dryer that costs about U.S. $410 to $550, and given the salaries and incomes described here, this price is still much more than the municipal processers could afford. NIST, also, has produced on an artificial multi-source dryer that can be run by gas, sun, charcoal or wood. The cost however is about U.S. $1,370, again, far beyond the income of the small processor. Other research involves the improvement in the salt quality used by the processors. Just recently, researchers at NIST have invented a process by which fish is fermented into patis and bagoong in only four to thirteen days instead of 150 to 365 days. The process involves the utilization of the natural enzymes of the fish and/or fortification with microbial, animal, or plant propeolytic enzymes. The patis contains 12 to 14 percent protein and the bagoong around 16 percent.[6]

---

[5]Ibid., pp. 91-92.

[6]Mrs. Luz Arcega, consultant, Microbiological Research Department, NIST.

A current project underway at the College of Fisheries, the University of Philippines, Quezon City, is attempting to develop standardization of fish products--classifying them according to species, size, moisture, and salt content. This is very important because most of the processors do not follow strict procedures. Techniques handed down by custom are the rule, rather than those based on technical or scientific standards.

REFERENCES

Borgstrom, George. Fish as Food. 3 vols. (New York: Academic Press, 1965).

Burgess, G.H.O. et al. Fish Handling and Processing. (New York: Chemical Publishing Company, Inc., 1967).

Darrow, Ken, and Rick Dam. Appropriate Technology Sourcebook. (Stanford, California: Volunteers in Asia, 1976).

Everington, D.W. "Methods of Freezing Fish." UNIDO ID/WG 172/14, 1974.

Food and Nutrition Research Center. Fish Processing Handbook. (Manila: NIST, 1966).

Hunter, M.G. "Fish Curing and Drying." UNIDO, Regional Consultation on Promotional and Technical Aspects of Processing and Packaging Foods for Export. ID/WG 172/16, 1974.

Kasemsarn, Bung-orn. "Fish Preservation and Processing in Thailand." Technical Data Files, Applied Scientific Research Corporation of Thailand, Bangkok, Thailand, 1972.

Librero, Aida R. and Elizabeth Nicholas. "Socioeconomic Aspects of Milkfish Farmers in the Philippines." (Program Lar + Res As SEAFDEC-PCARR Res Program), Los Baños, Philco Ag. Resources Research.

Philippine Co. For Agriculture and Resources Research. The Philippines Recommends for Baugos 1978. Los Baños, Laguma, Philippines.

Philippine Development Plan, 1978-1982.

Rosob, R.L. and F.A. Larlcin. Malnutrition: Its Causation and Control. (New York: Gordon and Beach, 1970).

Van Veeu, A.G.. "Fermented and Dried Seafood Products in Southeast Asia," in *Fish as Food*, George Borgstrom, ed., 1965.

PERSONS INTERVIEWED

Dr. Estrella Alabastro
Department of Food Science and Nutrition
College of Home Economics
University of the Philippines
Quezon City

Dr. Patricia T. Arroyo
Department Chairman
Fish Processing Technology
College of Fisheries
University of the Philippines
Diliman, Quezon City
Philippines

Mrs. Olympia N. Gonzales
Science Research Supervisor
Food Research Department
Industrial Research Center
National Institute of Science and Technology (NIST)
Manila

Ms. Ileana Cruz
Commissioners Office
NIST
P.O. Box 774
Manila

QUESTIONS TO CONSIDER

1. Given the traditional nature of the fish-drying industry, what, if anything, might be done to increase its productivity based on the information in the case? Is there an apparent area where technological improvement may be possible? If so, who would bear the cost?

2. Is there opportunity to form some type of cooperative in this situation? What would be its function? Is there any means of stabilizing market prices in this type of situation?

3. Would it be feasible to introduce some secondary occupation for the workers in slack seasons?

4. Is it possible that alternative sources of protein could be developed to augment the use of fish?

5. Are traditional technologies impeding the productivity of the fish-drying industry? If so, how can this be ameliorated at an affordable cost to villagers?

19
Singapore: The Development of a Design Consulting Engineering Firm

Ronald P. Black and Chan Beng See

As Chan Chee Wah sat in the motor launch taking him across the water back to the ferry terminal from the small refinery island, he was deep in thought. He had just left a meeting with the officials of an oil company. His firm had been asked to carry out the preliminary study of a new berthing wharf for the refinery to cope with the increasing need of wharfage space. The chances appeared good that his firm may be asked to continue with the detailed design of the wharf if the study proved the project feasible and economical. His firm was now well established and was even gaining a reputation throughout the ASEAN region. Reflecting back, however, Chan remembered it had not always been easy--indeed, for him and his staff of twenty to reach their present position it had been a long and hard struggle.

The firm had its beginning in 1968 with the setting up of a Singapore branch of an international group of consulting engineers (the Group) with offices in the United Kingdom and Australia. The Singapore office began its design consulting engineering operations with a British aid project to design improvement for a ship repair yard. The office was managed by a partner of the Group. He was also very active in promoting the special areas of expertise of the Group in the fields of urban roads, bridges, interchanges, harbour and marine structures, urban development, and planning. He travelled widely throughout the Asian and Pacific regions and identified Hong Kong (HK) as another center of construction growth. In 1970 the Group was successful in securing a large highway design project from the HK government, and he moved to HK to manage the office there. The Group invited Chan Chee Wah (who was a university colleague of his) to join the Singapore office as manager with the status of associate partner.

At the time, civil engineering projects for private consultants, such as providing designs for bridges and port facilities, were few and far between. The Singapore office had to rely on designing industrial buildings and structural engineering works associated with such types of construction. Chan had the advantage that he was registered not only with the Singapore Professional Engineers Board but also with the Architects Board. This permitted him to practice in both disciplines. The office was therefore able to undertake modest industrial projects without having to act as a subconsultant to an architectural consultant. This also was advantageous to clients as they could entrust an entire project to a single consultant. Relying mainly on industrial building projects, the office grew slowly from a staff of four in 1970 to ten in 1975.

During this period, the Group intensified its activities in Southeast Asia to secure more civil and structural engineering projects. It formed a number of alliances with local engineers in Jakarta, Kuala Lumpur, Kota Kinabalu, Kuching, Bangkok, and Manila. This was considered necessary because the process of chasing after a potential project in each country is very costly, time-consuming, and requires close contact with the right officials. In this setup, the local engineer became the local contact, and the office served as a regional office for the surrounding areas.

Through such alliances and with the promotional assistance of the Singapore office, the Group obtained a number of projects in the ASEAN region, some of the more interesting ones being:

Malaysia	Temerloh Bridge, Pahang Jalan Larkei Overpass, Johore Bahru Container Movement Study, Peninsular Malaysia Beach Erosion Study, Penang Foundation Problem, Kuala Lumpur Four Bridges, Segamat/Kuantan Highway
Indonesia	Engine Testing Laboratory, Djipuliar Deep-Sea Beacon, Makassar Straits Port Rehabilitation, Tanjong Priok Container Berths, Tanjong Priok Ferry Services Study, Java
Thailand	New Town Development, Nava Nakon Study of Twenty-two Coastal Ports, Thailand
Philippines	Port Feasibility Study, Cotabato Ro-Ro Loading Ramp, Manilla Port Facilities Study, Negroes

The staff of the Singapore office was involved technically in some of these projects, resulting in close ties being formed between it and the other ASEAN offices.

The Singapore office continued to be active in industrial building design works and in structural engineering for multistory buildings, but civil engineering works such as jetties, wharves, and quay-walls began to take on added importance to the company's project mix.

One of the bigger Singapore civil engineering projects in the pipeline in 1972 was that of the design and construction of the largest dry dock in Southeast Asia (dwt. 477,000 tons) contemplated by Sembawang Shipyard Ltd. (SS). This shipyard complex was a former UK Admiralty naval base, which had been converted to a Singapore shipyard on the withdrawal of British troops.

According to Chan, the competition for the consultancy work for this dry dock project was very keen. The Group and the office submitted to the SS board a proposal based on the unconventional concept of using an on-site joint venture engineering team, including SS engineers, to design and supervise the project. The resulting team was to consist of the Group's engineers from London and Melbourne, local engineers and draftsmen from the Singapore office, and the engineering staff of SS. The team's engineering capability would be augmented with regular visits of overseas partners and backed up by specialists from other associated offices. The SS board, comprising several government representatives, was receptive to this scheme since it would ensure that a significant proportion of design and drafting expertise in dry dock technology would be transferred to Singapore.

Another favorable feature was the proposal to use a new dock gate design based on the cantilever principle. According to Chan, this was not only an economically attractive design for the dock but would enable SS to gain experience in the fabrication and installation of this new type of gate with assistance from the Group. This proved to be the deciding factor, and the Group won the contract in the face of strong competition from several international consultants.

When the new dock-gate design was completed, the consultants proposed that a scale model be built and laboratory tested to check its performance characteristics. The model testing was originally to be undertaken by laboratories in the United Kingdom or Australia, but Chan suggested that the study be assigned to the local university's engineering staff, who he believed were up to such a task. His argument that this would help to upgrade local engineering capability was accepted by his partners. The university staff, according to Chan, did a good job--the operational characteristics predicted by their study were very close to those of the actual gate.

During late 1972 to 1975, many local engineers, technicians, draftsmen, and surveyors were involved in the design and construction of the dry dock, and appreciable foreign expertise was acquired by local personnel. For example, the Singapore engineers learned design methods that relied more heavily on computer utilization, and Singapore technicians and draftsmen learned more modern approaches to site supervision and design presentation for civil engineering projects.

During the commencement of the dry dock project in 1972, Chan became a full partner of the Singapore office. Upon completion of this project, the partners discussed the long-term future of the Singapore operation and decided to reconstitute the firm into its present structure (designated as CMP), with Chan as its managing partner and holding a major equity interest. All CMP engineers were graduates of United Kingdom or Australian universities, and the technicians and draftsmen were graduates of polytechnic and technical institutes.

At this juncture, it may be relevant to examine briefly the state of the engineering and construction industry in Singapore and the desirability of developing an "appropriate" technology in engineering design in Singapore. Singapore is a small island republic of 585 square kilometers with a population of 2.3 million. While it has a modestly advanced economy with a Gross National Product per capita of S $6,900[1], it has no natural resources other than its human resources. Although Singapore has virtually no unemployment, one goal is to upgrade the quality of its labor force.

Singapore depends on its strategic geographical position as a trade and transshipment center for Southeast Asia. When the country became independent in 1968, the government of Singapore was conscious that its economy must be diversified and set a goal to industrialize in as short a time as possible. Next, the concept of further developing Singapore as a service center for Southeast Asia gathered momentum; and today, besides being a trade and transshipment center, Singapore is a center for communications and medical services as well as a regional money-market center. Concurrently, the idea of making Singapore a service center in engineering design and construction was being discussed. The engineers, architects, surveyors, and contractors were being encouraged by the government to "spread their wings" outside of Singapore for commissions.

In some ways, however, Chan noted that this was a trying time for the construction industry. New projects were getting fewer, and more of the projects were being carried out by the government's own engineering teams. It was in this environment that Singapore consultants and

---

[1] One U.S. dollar is equivalent to S $2.16.

contractors began to attempt to break into the international market.

For a number of reasons this proved difficult. For some firms, this was due to limited financing capacities and experience. For CMP, however, Chan recalled that the biggest difficulty was the firm's lack of experience in forming joint ventures with contractors to offer integrated bids. Such projects were becoming more prevalent in the international market.

In Singapore, the conventional working method was for a client to first contract for a project design with a design engineering consulting firm such as CMP. After the client received the design, he would then request tenders from construction firms for that phase. Thus, in Singapore during the early 1970s, consultants and contractors tended to work independently of each other.

Because of the international trend by 1975, the Ministry of National Development requested professional organizations such as the Association of Consulting Engineers, the Institute of Engineers, and the Institute of Architects to study the implications of introducing a new system of contract tenders, know as "turnkey projects" or "package deals," with special reference to improving and upgrading the capacity of the construction industry of Singapore. The professionals recognized that the package deal system could be applied only to certain types of projects, such as the design and construction of petro-chemical plants, bridges, and wharves, where the number of component elements as well as detailed specifications were manageable and could be laid down clearly. Where applicable, such a system may forge a closer working relationship between designers and contractors in project execution and lead to stronger teams to compete for overseas projects. However, some reservations were expressed that the system would work beneficially only under certain conditions.

These reservations concerned a potential wastage of scarce manpower and financial resources. Under the conventional form of contracting, the client would select a design engineering consultant based on its credentials or reputation. The cost of the design work was then borne by the client. Under the package deal concept, design engineers and contractors would have to team up, and much of the design work would have to be conducted during the proposal stage so that the contractor could cost out the construction work.

Under the package deal form of contracting, the normal international practice is for a client to call for qualification statements. Based on these, a "short list" is prepared, and ventures on the list are asked to submit tenders. The concern of Singapore designers and contractors with this form of contracting ultimately centered on the number of ventures that the client would place on the short list. Ideally, not more than four or five ventures should be asked to submit tenders. This would avoid wastage of expensive design manpower. However, most clients tend to adopt a long list of ten to twelve ventures.

In 1975, the government of Singapore called for the construction of the 5.6 kilometer East Coast Parkway project under a package deal system. CMP was invited by a local contracting group to participate in a consortium together with a large Japanese contractor from Tokyo. The tender was a fairly large one, and its preparation required a great deal of effort from the consortium in terms of conceptual design and detailed pricing.

After one year of preparation, submission, evaluation, and negotiation, the consortium was awarded the tender at S $160 million in October 1976. The project would include a high-level bridge, four ramps, two overpasses, a viaduct, and four underpasses. The bridge would be 1.7 kilometers long and carry eight lanes of traffic. It would be about thirty meters high and have a clear span of eighty-five meters for river traffic on the Kallang River. CMP and its overseas associates began to tackle the detail designs in both Singapore and London, while the Japanese contractor had its own design team working in Tokyo. According to Chan, the interaction of these three design centers resulted in an aesthetically pleasing bridge design in prestressed concrete with the desired economy in construction.

Construction began in early 1977 with local and Japanese engineers and workers participating in different areas of work. Some of the problems encountered at the beginning included communication difficulties and the fact that different design codes were used. However, these were worked out with regular liaison meetings and discussions. As of February 1979, the bridge foundation has been completed and most of the bridge piers and trestles are ready for launching the mainbeams. The project is scheduled for completion by early 1981, when the new Singapore Changi International Airport will be ready to receive its first plane. The East Coast Parkway Bridges will form the link between the expressway from the airport and the center of the city of Singapore.

The project entails the pooling of various expertise and resources from Singapore, the United Kingdom, and Japan. Throughout the design and construction phases, an exchange of technical know-how and experience has occurred. Chan noted that the project has given careful consideration to the availability of local expertise, construction materials, and equipment. This is evident, he said, in the project cost and in the extensive use of local personnel and materials, such as sand, stone, and cement.

Chan notes that CMP is in good shape. The company has a two-year backlog of work. This may be compared with past times, when Chan did not know where the payroll was coming from three months in advance.

What of the future? Chan says that the crystal ball is rather cloudy for the construction industry in general and for the engineering consultancy profession in particular. As long as the political climate of Southeast Asia remains calm and the oil situation does not deteriorate appreciably, he anticipates that the volume of construction work both in Singapore and the other ASEAN countries will continue to grow.

For engineering consultants in particular, Chan says that the ability to secure assignments against stiff competition from local organizations and international groups will be the key to survival. Since this depends on the reputation and capability of each firm to handle the increasing size and complexity of projects, he believes that the ownership structure of engineering consultancy firms must also evolve to keep pace with changing market requirements. Public projects are funded mainly by the government, while private owners fund private projects. ADB and bilateral financing are not frequent in CMP's work but do occur at times in projects carried out by the international group.

The existing ownership of consulting firms in Singapore may be divided into three general types--local, mixed, and foreign. CMP comes under the second category. Currently, not many firms of this type are operating in Singapore, but they are on the increase.

The locally owned firms, which constitute the bulk of the engineering consultancy practice in Singapore, usually are run by a single practitioner or, at most, two partners. Large-sized local firms are very few. At the other end of the range are a number of foreign-owned firms that are branches of large international groups. These serve as local regional offices.[2]

In Chan's view, the mixed ownership arrangement of CMP has a number of advantages. First, the Group can supply CMP with the latest technology in such wide-ranging civil engineering fields as bridges, railways, harbors, dry docks, and airports. Such expertise can be at the call of CMP on short notice. Second, CMP has the local contacts and is familiar with local conditions and operations for the

---

[2]Chan estimates that the average design consulting engineering firm, taking into account all three forms of ownership, employs from ten to twelve persons. Therefore, CMP may be considered a moderately large design consulting engineering firm.

successful negotiation of future assignments. Third, the linkage with a large international practice provides a cushion against the cyclic workload that can plague many a small firm. Whenever CMP needs more design or drafting capacity, the Group can usually supply the need through its interfirm arrangement for staff secondment. On the other hand, when there is a drop in workload for CMP, outside work belonging to other offices of the Group can be siphoned to the Singapore office. For example, when the workload was low between 1975 and 1976, the Hong Kong office (which had grown very quickly since 1971 as a result of many development projects awarded by the HK government) channeled a portion of design work to be done in Singapore. Finally, the close relationship between Chan and the other overseas partners through the long years of working together creates a bond of friendship, trust, and respect that is often not found in ad hoc associations between a local and a foreign firm.

As a result of its past experience in Singapore and the other ASEAN countries, Chan believes that CMP is now in a competitive position to secure assignments both of the conventional and the package deal types. Although it cannot be said that CMP now possesses full design and drafting capability for large, complex engineering projects, Chan noted, a great deal of such capability does exist in CMP. For example, a recent turnkey tender proposal for the proposed Jurong Bridges was prepared mainly by the Singapore office with experience gained from the East Coast Parkway project. If the contract is successfully secured, detailed design will be executed by CMP's staff with only limited assistance from the Group.

Chan notes that the growth of an engineering consultancy service, like CMP, is always a slow process. This is in the nature of engineering design organizations, including their capability for transferring technology. In the use of technology, he said, it is vital that only that which is appropriate be proposed, otherwise the project execution cost will be noncompetitive. An example, provided by Chan, is the use of concrete, as opposed to steel, as the building material for the East Coast Parkway Bridge. This choice was made because of the ready availability of cement in Singapore and the corrosive environment of the site, which is exposed constantly to sea air.

Although CMP will continue to be a consultancy firm independent from contracting or manufacturing interests, Chan foresees the need to participate constantly in package deal projects. The trend is in this direction, and if Singapore is to get a bigger share of the international market, Chan notes, its construction industry must move with the times.

REFERENCES

Far Eastern Economic Review: Asia 1979 Yearbook, Far Eastern Economic Review Limited, Hong Kong, 1979.

PERSONS INTERVIEWED

Chan Chee Wah,
Managing Partner,
Chan Chee Wah Maunsell and Partners,
Singapore.

P. Arumainathan,
Deputy Secretary,
Ministry of Science and Technology,
Kay Siang Road,
Singapore 10,
The Republic of Singapore.

QUESTIONS TO CONSIDER

1. Is this example of the gradual development of a consulting engineering capability unique to the Singapore environment, or could it be introduced elsewhere in the developing world?

2. Were the circumstances surrounding the founding and success of this firm so unique as to constitute an exception rather than a model to be emulated elsewhere?

3. There are strong elements of entrepreneurship in this case history; is this a quality that occurs throughout the developing countries, or is there some type of distinction in this regard?

4. What may be considered the single most important feature of this situation, which resulted in the success of this private venture?

5. How important is the continuing relationship through the partnership to the apparent success of CMP? Is there a feasible model here for forming similar relationships between developing and industrializing countries? Or are the characteristics and motivations of the people of Singapore unique?

6. Concerning the trinationalism of the consortium described, does this point the way toward a future in

which the community efforts among the industrialized and developing countries will become more established and effective? Conversely, are there diminishing marginal benefits to such collaboration which make it a dubious means of promoting development?

7. Are there ethnic differences, or those caused by climatic conditions, which provide a basis for estimating the probability of new-venture success?

20
Sri Lanka:
The Ceylon Institute of Scientific and Industrial Research

Donald D. Evans

INTRODUCTION

The Ceylon Institute of Scientific and Industrial Research (CISIR) is a government-sponsored industrial research organization that was established twenty-four years ago to identify, acquire, develop, and apply technology for the benefit of the country. The existence of more than eighty such institutions in as many different countries forms a significant international "community of interest" dealing with the institutional approach to the application of technology to the problems of development.

SETTING

Ceylon gained its independence from Great Britian in 1948, at which time the country's name was changed to Sri Lanka. Sri Lanka has one of the highest standards of living in Southeast Asia due to the tea, rubber, and coconut plantations. The population of Sri Lanka is concentrated in the moist southwestern one-third of the 270-mile-long island. The dry zone elsewhere flourished milleniums ago under a remarkable irrigation system, but warfare led to neglect. Today, the nation works to reirrigate the dry zone and thus reduce the food that it must import to feed its growing population. The island itself is located twenty miles off the tip of India and comprises 25,332 square miles. The population, composed of 71 percent Sinhalese and 20 percent Tamils, with large Hindu, Christian, and Moslem minorities, reaches near fourteen million. Agriculture employs half of the working force, but industry is growing. The country ranks second in the world in tea production and fourth in rubber production. Gem and graphite mining is also an important industry. The major city, capital, and principal port is Colombo, with a population of 618,000.[1] It is within this environment that the CISIR is located and where it has pursued its objectives laid down a quarter of a century ago.

BACKGROUND

The International Bank for Reconstruction and Development (the World Bank) was invited by the government of Ceylon (later to become Sri Lanka) in 1952 to conduct an overall study of the economic situation of the country and to make recommendations that would lead to rapid development and relieve the residual negative effects of the colonial period. Subsequently, against a background of an unindustrialized, agricultural economy, a Bank team of eight members of varying professional backgrounds produced a comprehensive study of the Ceylonese situation. Among the more significant of the recommendations for change was the establishment of an industrial research organization.

Although organizations directed toward researching the "plantation crops" of rubber, tea, and coconut had been established under the British, in the view of the Bank team a highly pragmatic research and development, industry-oriented group was needed. The Bank report recommended an institution that would be as autonomous and free from direct governmental supervision as possible and that would be able to attain financial independence in as brief a time as feasible. It was thought at the time that such self-sufficiency could be gained in as little as five years, if the institute's development and growth were nurtured in the meantime by a government grant.

Consequently, under legislation passed in 1956, the CISIR was established with the following objectives:

(1) To undertake testing, investigation, and research in such manner as the institute may deem advisable with the object of improving the technical processes and methods used in industry and of discovering processes and methods which may promote the expansion of existing or the development of new industries or the better utilization of waste products.

(2) To advise on questions of scientific and technological matters affecting the utilization of the natural resources of Ceylon, the development of her industries, and the proper

---

[1]Adapted from the National Geographic Atlas of the World. Washington, D.C.: National Geographic Society, 1975.

coordination and employment of scientific research to those ends.

(3) To foster the training of research workers.

(4) To foster the establishment of associations of persons engaged in industry for the purposes of carrying out scientific and industrial research.

(5) To undertake to collaborate in the preparation, publication, and dissemination of useful technical information.

(6) To cooperate with departments of government, universities, technical colleges, and other bodies in order to promote scientific and industrial research and the training of investigators in pure and applied science and of technical experts, craftsmen, and artisans.

(7) To assist otherwise in the advancement of scientific and industrial research and technical training.[2]

PROBLEMS OF THE CISIR

That the CISIR may not have been totally successful in its pursuit of its objectives may be deduced from these statements: "When the CISIR was established in 1955, its principal objective was to further Ceylon's productive development through applied research and technology. However, the several directors of the late 1960s and 1970s and the successive governing boards appear to have not had the capacity to appreciate the significance of the contribution of applied research in the development of a technological capability within the country. Under their stewardship, the institute has veered away from applied research to fundamental or basic research."[3] "The impact that a single institution (the CISIR) with limited resources could have on the entire industrial sector is . . . limited. Moreover, the R&D efforts of the CISIR are mainly on a laboratory scale and do

---

[2]<u>The Economic Development of Ceylon</u> - International Bank for Reconstruction and Development, Washington, D.C., 1952

[3]Member of CISIR governing board to governing board, letter, November 1977.

not generally extend to scaling up into industrial prototypes and processes.[4]

In 1976 the chairman of the CISIR and concurrent vice chancellor of the University of Sri Lanka was quoted as saying: "The CISIR had at that time (1955) the ambitious intention of generating its own finances, by assisting local industry through research, advice, and the testing of industrial products, but for several reasons these ambitions did not materialize."

The director of the CISIR, Dr. Mervyn Wijeratne, expressed the belief that the institute had indeed failed to live up to its initial promise and that this was a great loss in view of the opportunities and needs that continued to face the nation and in which an institution carrying out the mandate of CISIR could make such important contributions. However, he expressed conviction that the organization could be brought into a much more productive role and essentially fulfill its mandate if certain internal conditions could be improved and if the environment within which CISIR existed could change.

Although the CISIR has encountered problems of organization, staffing, facilities development, information acquisition, client relationships, program planning, public attitudes, and research application--problems that are common to this community of institutions--at the present time it is especially concerned about its relationship with the government in Sri Lanka and with the attitudes of governmental officials in the Ministry of Industry and Scientific Affairs. Another difficulty of first-rank importance is the blight of "brain drain" that has afflicted the CISIR, especially in recent years. The institute is seeking ways to strengthen its most valuable resource (in the view of Wijeratne and others)--its staff-- and to reduce the rapid turnover rate. Wijeratne and his deputy, Dr. Edwin E. JeyaRaj, also are concerned with updating the facilities of the institute and with providing further education and advancement to the professional staff. Also, the institute's research output and technical capabilities need to be utilized to a greater extent by the industrial community.

Sharing in these feelings of need were the majority of the professional staff of the CISIR, as was disclosed in a series of informal meetings with small groups of the staff members.

---

[4]Sri Lanka National Paper for the United Nations Conference on Science and Technology for Development. (Colombo: Production and Printing Unit Industrial Development Board of Ceylon, October 1979): 16.

DESCRIPTION OF THE CISIR

In 1979 the institute was composed of 70 professionals and 100 supporting staff, housed in four principal buildings in a pleasant and well-established section of Colombo. Two multistory laboratory buildings had only recently been completed. Other facilities included: conventional chemical laboratory spaces and equipment; a small but diversified machine shop and fabrication facility; some pilot plant spaces; a technical information center that was reputedly one of the best in the country; and the usual ancillary spaces.

The research equipment, except for some recent additions, was rather old and obsolescent. Various members of the professional staff complained that they were unable to carry out satisfactory research because of equipment limitations and especially because of the long delays in getting replacement parts from foreign sources. Obtaining the "hard currency" needed to pay for foreign imports was vexing until recently, when the government liberalized foreign exchange negotiations. (Thanks to the foresight of the planners of CISIR, the institute had been granted relief from customs duties and taxes.) Other younger staff members complained of problems in getting colleagues to share equipment because of shortages and supply difficulties.

Other problems resulted from the deficiency of pilot plant capability. One industrialist and member of the CISIR governing board stated: "As projects mature to pilot-plant scale and to the level of development of industrial prototypes, a heavy demand will be foisted on the pilot plant and designs section. If this vital section remains understaffed all research results will be consigned to files and archives, and no take-off into industry and commerce will result."

Several staff mentioned the inadequacy of their information resources. Director Wijeratne stated that the budget for library periodicals and new acquisitions had been cut for the fiscal year 1979 by some uncomprehending official in the ministry who apparently had little concept of the great need for access to contemporary science and technology information by a professional research staff. The library itself was housed in crowded spaces but was to be relocated to more spacious quarters within a short time.

Another difficulty was maintaining adequate laboratory atmospheric control in Sri Lanka's hot and humid climate, which affected various pieces of delicate laboratory equipment and reference standards. Air-conditioning equipment was generally hard to obtain even if ministry sanction could be obtained for this "status symbol" appliance that generally is reserved only for the offices of higher-level government officials.

FIGURE 20.1
ORGANIZATION CHART OF CISIR

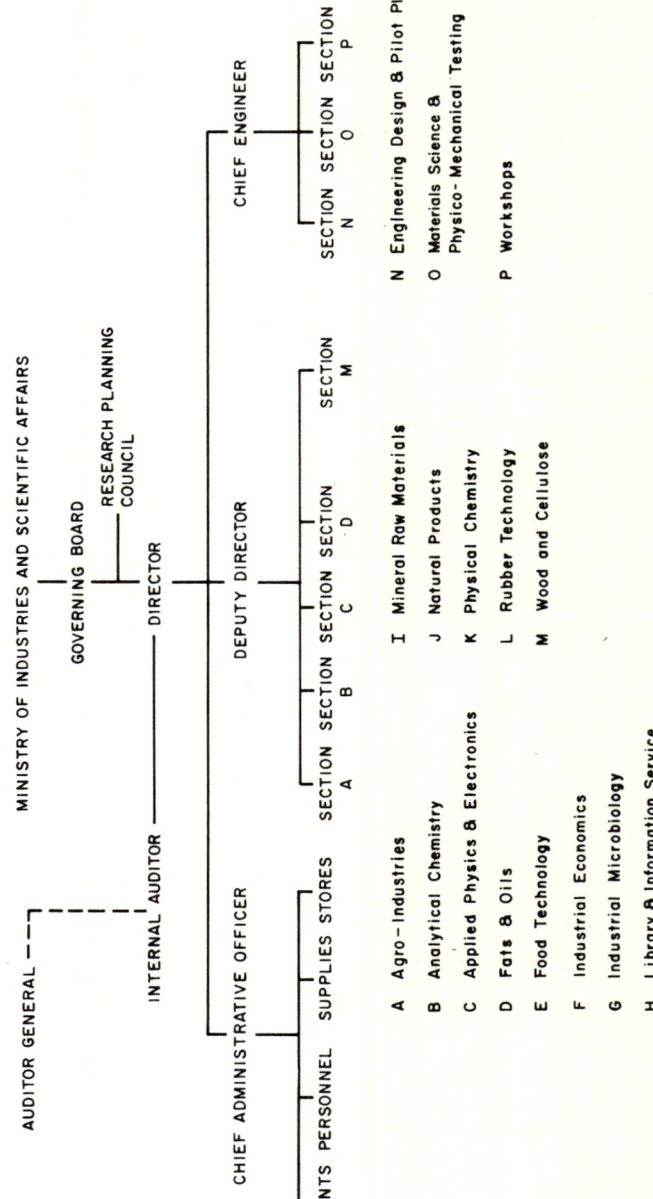

The Professional Staff

Perhaps the most significant characteristic of the CISIR professional staff was its age structure. The senior and more experienced members were both relatively few in number and separated widely in age from the far more numerous younger members. Adequate numbers of intermediate-aged individuals were lacking. This reflected the fact that virtually no research staff were hired for over twelve years after the establishment of the CISIR. Thus, in 1979 no experienced, intermediate-age cadre of qualified persons were available for management responsibilities. The result, as expressed by one of the young women professionals, was a lack of direction and participation in research activities by more senior and experienced supervisors. The director himself had been surprised by the statement of a junior staff member that it was felt he was "too lenient" and that the staff would welcome "more explicit direction." At the same time he was concerned that perhaps the staff held too much confidence in his ability to influence the external forces that affected the operations of the institute.

It was also felt by some staff members that there were far too many active projects in the institute at any one time. The ratio of staff members to projects (approximately one-to-one) meant that only insufficient and long-protracted attention could be devoted to individual projects. An informed outside observer of CISIR commented that because of such understaffing, projects were requiring six to eight years for completion instead of two to three, which should be the maximum.

Opportunities for overseas training of the staff were considered by the director and by members of the National Science Council as being too few and often inappropriate. The director explained that this was largely the result of CISIR's lack of decision-making authority with regard to how many staff members would receive overseas training and in what subject areas. Such decisions were made by the Ministry of Education and the Foreign Ministry, which administered the training assistance grants accorded to Sri Lanka from foreign sources. Wijeratne observed that the CISIR representative in the Ministry of Industry and Scientific Affairs was not of comparable rank to those who were responsible for other technical institutions in the country; therefore, CISIR did not receive a share of training opportunities commensurate with the size and needs of its staff.

On the other hand, when staff members did receive overseas training, negative consequences sometimes resulted. For instance, there was a discouragingly high rate of resignations. Notwithstanding the fact that those going on training assignments in foreign countries had to sign an agreement that they would return to Sri Lanka to serve a specific number of years at CISIR, significant numbers of Sri Lankans were managing to circumvent this provision either by paying the stipulated amount to discharge their obligation early or simply by remaining overseas. More recently, many educated and skilled Sri Lankans have been leaving the country for periods of two years and longer to take high-paying positions in the Middle East Organization of Petroleum-Exporting Countries (OPEC) nations, where they are very much in demand due to their high educational levels and their especially good command of English.

CISIR has found it particularly difficult to recruit and retain engineers, whether they are fresh out of the technical universities at home or abroad or are qualified persons with experience and knowledge of the Sri Lankan industrial environment. As noted in the Sri Lanka national paper for the United Nations Conference on Science and Technology for Development (UNCSTD): "For the LDCs, this is a significant loss of human capital and a substantial reduction in their technological capacity. For the receiving developed countries, such inflow has helped to increase and diversify their own technological manpower base. Brain drain of this type is essentially a <u>reverse transfer of technology</u>. It is ironic that through this process, the poorer LDCs pass on a substantial part of their technological assets each year voluntarily to the rich countries." "The institutions and workshops available in Sri Lanka for training this grade of manpower are quite inadequate to produce them in the numbers required to compensate the loss." (The paper proposes that some type of financial compensatory system be established to which the recipient developed countries would contribute for the benefit of the donor LDCs.)

The CISIR staff is organized in discipline-oriented groups, a fairly typical kind of organization in similar institutions of other countries. However, the director and a concerned ministry official noted that such a structure tended to discourage the interdisciplinary research that was badly needed to confront the country's increasingly complex and interactive problems. In fact, critics of the CISIR stated that the institute was too concerned with basic science, especially chemistry, and was not diversified enough in other subject areas, particularly engineering. At the recommendation of a Soviet team that asssisted the institute during 1974 to 1976, an industrial economics unit was established within the CISIR. However, in 1979 there was evidence that this activity had gained little acceptance among the "hard scientists" on the staff or among industry and other potential users on the outside.

With regard to personnel problems, probably the most difficult aspect to overcome is the markedly deficient compensation schedule for the staff. In the view of Director

Wijeratne, this underreward for services rendered has been the major cause of problems within the CISIR. He noted that part of the problem results from the circumstances of the institute's first five years. At that time the staff was paid substantially more than those employed in government agencies. When at the end of the first five years financial self-sufficiency had not been attained (by a large margin), the government political forces brought CISIR under the direct control of the Ministry of Industry. This, of course, violated the objective of the CISIR to be autonomous, and the budget for the institute was reduced to conform to that of other government agencies. A management report dated July 1978 stated: "The research staff at CISIR work under constraints of limited facilities, and yet strive to work out new processes or improve existing ones. They cannot also be expected to work for lower pay than their colleagues in other institutions whose academic achievements are often lower than their own." Notwithstanding the efforts of CISIR's management, over one year later the recommended salary increases had not been granted.

In a discussion with junior staff members it was disclosed that several of them had outside sources of income, and they felt that only because of this were they able to maintain a suitable standard of living. One stated: "I happen to own a small plantation, and even that returns me several times what I earn at the CISIR." Another said that he was unable to get married because his salary was his only income and would not allow an adequate living standard for a family. Loyalty and satisfaction in doing research even under difficult conditions were the principal reasons why the junior staff chose to remain at the institute.

Some staff members also indicated dissatisfaction over their lack of opportunity to interact with peers in other countries. They felt that, as a result, they were not able to maintain a contemporary awareness of developments in their fields.

Director Witjeratne pointed out other personnel difficulties that hindered operation of the institute. These ranged from a proscription on the payment of stipends to board of governors members (other state institutions were able to do so) to the unavailability of English-speaking/writing secretaries due to a government policy that promotes the use of the national language (Sinhala).

Nothwithstanding these critical comments, it was both stated and observed that the staff felt very favorable toward the director and his deputy, both in terms of their skill and efforts at management and on a personal basis. These two men had been at the institute since its inception, had risen "through the ranks" to their present posts, and therefore had a comprehensive knowledge of the evolution of the CISIR and of its staff members.

Project Selection and Research Operations

One knowledgeable critic of the CISIR stated: "The research planning council (an internal body of the institute comprising senior research staff members) appears to have planned the institute's research program with little concern towards the resource base of the country and the priorities in relation to national economic goals."

In the national paper prepared for the UNCSTD it is maintained that: "The scientists from Sri Lanka carrying out postgraduate research in developed countries use sophisticated and advanced techniques and naturally work on problems most often unrelated to the problems of Sri Lanka but related to advancing the science and technology of those countries. Quite often, such scientists find it difficult and disturbing to return home, where funds allocated for scientific research are low, the apparatus and equipment available are of a basic nature, and scientific literature is scarce."[5]

The program of research at the CISIR is noticeably directed toward the development of products and processes based on the country's plant resources, and the staff's professional qualifications show a predominance of organic chemists and life scientists. Following is a partial list of ongoing and completed projects, which gives an impression of the characteristic content of the research activity:

(1) Instant tea (from green leaf)

(2) Bottled coconut cream

(3) Removal of H_2S from arrack (a distilled liquor)

(4) Ebonite (latex + kaolin) as a PVC substitute for fan blades, etc.

(5) Improved stills for the distillation of essential oils

(6) Chemistry of Sri Lanka essential oils

(7) Cassava detoxification

(8) Preservation and bottling of young coconut water

(9) Elimination of Salmonella in dessicated coconut

(10) Carbonated lime tea formulation

---

[5]Ibid., p. 24.

(11) Cashew nut wine manufacture

(12) Osmotic dehydration of fruits and vegetables

(13) Technical and management assistance to the Oils and Fats Corp.

(14) Development of a method for relining electrolytic cells with rubber compound.

Despite the apparent practicality and relevance of these research subjects, a ranking official of the Ministry of Industries and Scientific Affairs noted that in the past the CISIR had been viewed as a collection of "academics" "living in white towers." He continued that it was difficult to evaluate the work of the institute because there was little planning against which to measure results. It was noted that the research tended to be too basic-oriented and that the institute had "not met expectations" in terms of its research output. He felt that the research program had not benefited from a careful examination of the national development plans and that therefore the research output sometimes tended to digress from what was most needed for the country.

In a report to the board of governors, a member of the board noted: "as at present (for year ending 1976) about 75 percent of the operational expenditure for research has been incurred on projects in the class of fundamental or basic research. This class does not in the short term assist government to mobilize national resources in a productive way." In response to this charge, the research planning council of the CISIR noted that its definition of "basic" versus "applied" research differed substantially from that of the board member. On the basis of the council's definition, "the CISIR did not exceed 10 percent of the total operational expenditure" in basic research, and "hence the council does not agree with the statement that 75 percent of the operational cost during the year 1976 has been spent on what has been referred to as fundamental or basic research."

Director Wijeratne emphasized that this difference of opinion was typical of the climate within which the Institute operated. He stated that there was a disconcerting lack of common definition of goals and objectives between the management and staff of the CISIR and those in the public and private sectors who have an interest in CISIR programs. Wijeratne continued that the management of the institution had not in the past been adequately attentive to communicating with influential persons in the public and private sectors and that he was seeking to overcome this discrepancy.

The research council's rejoinder went on to state: "If the CISIR has not made an impact on the industrial development of the country at a level envisioned, it is because of the lack of an overall plan to expand the activities of the institute to a higher scale. What is required is a fully equipped pilot plant to undertake upscaling of processes worked out in different sections such as Food Technology, Industrial Microbiology, and Natural Products. For this heavy investments are needed."[6]

Staff and management of the institute both acknowledged that insufficient attention was paid to project development and to external relations of the organization. The CISIR had no organized marketing activities, and each principal investigator was left largely to his or her own devices in terms of gaining user acceptance of research results. For example, a promising process for the manufacture of a bottled carbonated tea drink was developed after long and intensive effort by the staff. After a series of ineffectual and frustrating efforts to enlist a manufacturer for this product, including the filing of patent applications in many foreign countries, the product ultimately failed in the Sri Lanka market despite the intrinsic flavorful appeal of the beverage.

A lack of planning was viewed as another problem in research operations. In the opinion of the present management of the institute, previous directors of CISIR generally had been negligent in terms of planning for and developing basic staff capabilities; this should have been possible even with the financial constraints that were experienced intermittently by the institute. For instance, some felt that better use might have been made, of the opportunities for foreign training through various assistance programs available to Sri Lanka.

Further reference to institute planning was made during a discussion with a ministry official. He said that under the aegis of the present minister notable improvement had been made in the "team approach" to research problems, although the national development plans had not been studied sufficiently and integrated into the research plan. He said that it was necessary for the institute to "plan-up" to meet the "plan-down" efforts of the government and noted that, in general, the technological institutions of the country under his and other ministries had not been adequately involved in development plans. He did feel, however that "the role of institutions (such as CISIR) in decision making is increasing." (It was during this discussion that the ministry

---

[6]"Observations of the Research Planning Council," 24 October 1978, p. 2.

official learned from Director Wijeratne that the Finance Ministry had reduced next year's budget for CISIR from the approved (by the Ministry of Industry and Scientific Affairs) Rs. 10.6 million to Rs. 4.52 million.

Financial Matters

Under terms of an agreement between Sri Lanka and the All Union Corporation of the Union of Soviet Socialist Republics (USSR), a Soviet industrial economist, Dr. V. Lats was seconded to the institute from 1974 to 1976. In the course of his stay, Lats analyzed the income and expenditures of the CISIR for the preceding twenty years, Lats pointed out: "The CISIR was setup with an initial government donation of Rs. 5 million spread over the first five years, the bulk of which was expended on capital expenditure on building and equipment. Initially, it had been hoped that at the end of the five years, the institute would be in a position to operate on its own income by the sale of services to both the public and private sectors. However, this hope was not fulfilled and now it is obvious that it cannot be fulfilled, at least in the near future.

"Even in a highly industrialized capitalist country it is known by experience that a minimum of five to ten years is essential to make a research institute self-supporting. In a developing country an even greater period of time may be required together with an assurance of financial stability which would produce a feeling of confidence in the staff, the sponsors of research, and the whole technical community. There is no magic way of predicting the time for reaching self-supporting status of the research institute. This period would depend on a number of factors and primarily on the level of technical development in a given country."[7]

The essential correctness of his observation can be deduced from the events at the CISIR when, in 1960, the annual stipend of Rs. 5 million came to an end and the institute was required to go to the government for its annual grant. This, however, was not forthcoming for some two years. The institute consequently was placed under heavy financial strain, resulting in reduction in staff and no expansion of programs. As Lats pointed out: "From the financing standpoint, the second five-year period was the hardest time in the institute's life. Financial constraint had caused a considerable delay in its development."[8]

---

[7]Memorandum, Lats to Governing Board, 10 September 1975, p. 2.

[8]Ibid., p. 18.

Also during this period, the original autonomy of the institute was abridged significantly by legislative amendment. As a result, the government took a much more active financial and operational role in CISIR programs. According to Director Wijeratne, at this juncture the CISIR became more typical of R&D institutions in the other developing countries in terms of its relationship to and control by the government. Direct government audit of accounts was required, and new ex-officio government members were placed on the governing board. Since funding was only on a year-to-year basis, the institute was not able to initiate, with confidence of completion, any programs that could not be completed within a single budget year.

In his conclusions and recommendations, Lats stated: "For proper research planning, it is necessary that the institute should be aware of the funds (including foreign exchange) that are likely to be available to it for the next five years and should in turn communicate this information to its constituent sections." He also noted, "The budget should be 'project-oriented' and not 'expenditure-oriented', so that funding would be related to the actual project requirements."

Despite the presence of a notably efficient and comprehensive cost accounting system (given the lack of attention that this subject receives in similar institutions in most other developing countries), Mrs. da Silva, the chief accountant, related a story of failing to get the professional staff to account regularly for how their time was spent on projects. The feeling prevailed among the staff that since the institute was sustained by an annual budget, "there was no utility in determining exactly how much was spent on each project or activity."[9]

Much of the CISIR's funding comes from its charges to outside organizations (see Table 20.1). However, it was stated that the charges for these services and the development of this aspect of the institute were inadequate. It was also disclosed that payment for these services was frequently delayed; many customers felt that services should be free since the CISIR was a public-supported institution. Conversely, various of the professional staff felt that the provision of testing services by the institute was not consistent with its role as an R&D organization and that the sooner these activities were divested, the better.

In September 1975, Lats concluded: "The immediate tasks facing the institute are to continue recruitment of necessary manpower and acquisition of facilities. Emphasis should be on selection and training of qualified research

---
[9]Ibid.

personnel. The laboratory equipment and plant must be continually upgraded to enable the staff to improve its productivity." In 1979 Director Wijeratne discussed the requirements of the CISIR in these same terms, indicating that no great progress had been made toward these fundamental goals. In fact, he observed that after its auspicious initial five years, the CISIR had fallen into a mode of operation that was generally stultifying and nonprogressive, although there had been periods of greater activity. With regard to the contemporary situation, he said, "We're back to square one."

TABLE 20.1
CEYLON INSTITUTE OF SCIENTIFIC AND INDUSTRIAL RESEARCH

Year	Total operational expenditure Rs	Expenditure on salaries and wages Rs	Income from Services Rs
1955	1,004,813	21,140	3,303
1957	444,785	304,390	360,847
1958	485,935	348,755	292,088
1959	553,158	407,331	290,894
1960	719,410	537,069	429,006
1961	774,090	606,758	402,242
1961-62	1,222,750	785,806	508,175
1962-63	858,437	492,615	321,108
1963-64	900,816	519,741	322,003
1964-65	974,316	541,719	161,429
1965-66	983,757	530,304	348,000
1966-67	1,157,826	703,242	415,621
1967-68	1,426,576	874,278	524,175
1968-69	1,645,914	1,010,310	624,703
1969-70	1,670,762	1,095,817	596,529
1971	2,391,625	1,356,399	966,394
1972	2,167,528	1,264,156	281,517
1973	2,263,469	1,580,004	203,628
1974	2,565,761	1,816,025	242,318
1975	2,859,303	2,005,501	179,757
1976	4,055,002	2,199,715	219,300
1977	4,728,797	2,619,138	315,175

Foreign Assistance

"The Sri Lanka Government believes that the strategies adopted during the last decade will have to place a high priority on training and developing the human resource infrastructure necessary for full utilization of science and technology for development. This means incorporating effective training components in all plans and programmes, and making full use of experts (foreign and local) in special training programs bringing to the forefront the problems, prospects, and solutions regarding science and technology for development."

Beginning with its first director who came from the U.S., the CISIR received assistance from several foreign sources: the government of the USSR; the University of Uppsala, Sweden; the Tropical Products Institute of London; and the Indo-Sri Lanka Science and Technology Programme in collaboration with the government of India. These all benfited the institute, and the general attitude by its management and staff toward foreign assistance was favorable.

Some problems existed however. Director Wijeratne, for example, observed that the CISIR had only indirect access to the sources of such foreign assistance and that the institute had to negotiate such programs through the intermediation of governmental agencies. This inhibited the clear definition of problems and the determination of optimum assistance programs. The staff felt that intermediaries in the ministries were not sufficiently familiar with the problems of research institutions to make judicious choices of assistance options.

Concurrently, Wijeratne emphasized the importance of the foreign inputs to the CISIR, stating that the institute could not survive in a vacuum and that frequent working-level contact with peers in foreign (particularly developed) countries was vital to the professional growth of the staff. Wijeratne particularly pointed out that it was highly desirable for the institute staff to gain industrial experience. He believed that this was a serious deficiency in their experience background hampering effective relationships with industry. This was especially true of the younger professional members, many of whom had had no direct industrial experience. He felt that if the foreign assistance encompassed direct industrial experience, he would be willing to exchange formal postgraduate training for such experience. At the same time, Wijeratne pointed out that staff members with this kind of background--especially under present conditions in the country--were likely to be "pirated away" by local industry, which was experiencing a growing shortage of such qualified persons.

The CISIR and External Factors

The utilization of science and technology for the development of Sri Lanka may be summarized as follows:

"Sri Lanka is today classified as a less developed country (LDC) on the basis of its GNP per capita (Rs. 1,754 or US $200 in 1976). The rate of economic growth has been slow (3.1 percent average annual increase in the GNP in the decade 1968-1977), and there are about one million unemployed in a population of fourteen million. In contrast to this situation, Sri Lanka's Physical Quality of Life Index is much higher than that of most other less developed countries.

"Agriculture is the mainstay of the economy, accounting for one-third of the GNP, and it is mainly confined to tea, rubber, coconut, and rice. The first three crops account for the major foreign exchange earnings, while rice is the staple food. Sri Lanka is still an importer of rice; the highest annual production was in 1977, when 80 million bushels, equal to 75 percent of the national need, were produced. Research and development in agriculture will also focus on other crops besides rice. These include the spice crops, legumes, coarse grains, sweet potatoes, chillies, sugar, cocoa, and coffee.

"In the industrial sector, significant development took place only after the independence in 1948. Since then, several industries have been established (e.g., paper, tires, rolled steel, sugar, etc.), but the growth of the sector has fallen far short of planned targets. The main unsatisfactory feature of the sector is the high dependence on imported raw materials and machinery. Both research and development are inadequate in this sector, but more particularly, the development component, resulting in shortcomings in regard to the transfer, adaptation, and innovation of technology.

"The main development envisaged for the industrial sector is the establishment of the Export Processing Zone (EPZ), which is expected to attract foreign capital and would result in increased employment and foreign exchange earnings. Outside the EPZ, efforts will be made to increase production, to reduce the dependence on imported raw materials, and to effect process improvements in the industries that have already been established. The establishment of new industries which are labor intensive and suitable for setting up the rural areas will be encouraged.

"In the sphere of health, there has been a sharp decline in infant mortality (from 140 per 1,000 live births in 1946 to 43 in 1971) and an equally spectacular increase in life expectancy (from forty-four years to sixty-four in the same period). The main health problems at present are the incidence of preventable diseases such as bowel infections, diphtheria, whooping cough, tetanus, malaria, etc., and the lack of adequate potable water. Malaria, which had almost been eradicated in 1960, is of serious concern again.

"In the housing sector, the main thrust will be towards providing a greatly increased number of housing units, encouraging private house construction by providing incentives, and pursuing research on developing alternative materials for building that could be mass-produced at low cost.

"Development in the sphere of energy will consist mainly of increasing hydroelectric power generation and extending the program of rural electrification. Programs of forestation to provide the future needs of fuelwood (at present, fuelwood and agricultural residues account for 60 percent of the energy consumed) would be implemented.

"The sustained development of Sri Lanka's economy will depend heavily on the building up of a fully effective <u>indigenous scientific and technological infrastructure and capability</u>. The deficiencies that now exist, and the problems encountered in developing self reliance in S&T are as follows:

(a) Inadequate capacity in R&D resulting in deficiencies in the transfer, adaptation, and innovation of technology

(b) Inadequate managerial capacity

(c) Lack of easy access to scientific and technological information

(d) High cost of books, periodicals, and scientific equipment

(e) The heavy exodus of trained manpower-doctors, scientists, engineers, technicians, and skilled workers; the expertise available in the country is of a very high standard, but the numbers are diminishing due to the exodus ("brain drain")

(f) The inadequacy of training facilities for increasing the numbers of skilled personnel, particularly at the technician level."[11]

In this environment, the CISIR follows its mandates involving the creation and improvement of industrial processes, the development of new products and the substitution of imported ones, the training and development of manpower, the provision of testing services, the dissemination of information, cooperation with other related public and private bodies and the advancement of science and technology.

---

[11]Ibid., <i>passim</i>. This paper was prepared by L.C.A. des-Wijesinghe, acting secretary-general of the National Science Council of Sri Lanka.

In undertaking these tasks, the institute encounters a wide range of circumstances and problems that are typical of similar R&D organizations in other developing countries. These stem generally from lack of public and industry understanding of the functions and possibilities of R&D; difficulties with the political process and the governmental systems; shortages of skilled manpower, equipment, and facilities; a dearth of technical and other information; and problems related to the growth rate of the economy.

Industry, primarily the food processing industry, was the principal beneficiary of the efforts of the CISIR. Unfortunately, certain situations and attitudes that seriously hampered the efforts of the institute existed in industry. Perhaps first among these was the strong preference for foreign technology. Until recently, government policy had been to severely restrict importation of products and processes, and this had stimulated the establishment of domestic industries utilizing their own technology. However, recent loosening of restrictions on imported products and technology had resulted in their reintroduction in the domestic market. One senior CISIR staff member pointed out the effect of this had been to drive the domestic producers from the scene. He said that even with a 100 percent import duty on various items, it was still not possible for local manufactureres to match the imports either in price or quality. He attributed this to two essential facts: the domestic market was so small that it could not support an efficient scale of production for most products, and the supply and quality of components from local secondary suppliers to the final manufacturers was inadequate.

In this regard, the Sri Lanka National Paper for UNCSTD stated: "Industrialization based on import substitution may look attractive in principal, but in practice, <u>if not properly controlled</u>, can lead to a situation where industries may be developed in which the local effort consists merely of assembling components (e.g., radios) or only of packaging the finished products imported in bulk (e.g., pharmaceuticals, milk foods). This results in industries being set up where the import content is very high and the foreign exchange costs of keeping the industries going are unduly high. With regard to technology, this situation will not boost the science and technological capability of the country towards using local raw materials and/or local talent in place of imported ones. From the economic standpoint, import substitution of this kind will not bring the expected relief to the balance of payments."[12] (At another point in the national paper, however, it is stated that the official

---

[12]Ibid., p. 6.

government policy recognizes this effect and states the intention to minimize it.)

Governmental policy during recent months strongly encouraged the importation of foreign technology for certain uses, such as for export production in the recently established EPZ, where the principal domestic input would be labor. It is asserted that local industrialists and investors there show a strong preference for foreign "turnkey" technology.

In any event, as the CISIR staff member pointed out, the effect of foreign technology importation generally is to reduce the demand for indigenously produced technology. In this way, the situation of CISIR is greatly affected.

Another aspect of the institute's relationship to the industrial sector concerned a lack of confidence by industry management in the capabilities and experience of the CISIR. Director Wijeratne pointed out that the severe financial constraints of the institute make it virtually impossible to hire and retain highly qualified engineers with broad industrial experience. This was becoming increasingly important as the attitudes of younger, well-educated persons in the country leaned progressively toward professional mobility. Industry simply could provide greater compensation and material status than the research institutes. Thus, a principal problem of the CISIR was to develop career models and methods that would offer younger professionals sufficient inducements, including compensation, to attract them to and keep them within the organization.

Industrialists in Sri Lanka, as elsewhere, regarded return on investment as the principal criterion of success. (They recognized, though, that there are many other measures of organization success.) Nevertheless, the need to optimize return on investment usually corresponds to an unwillingness to invest in R&D, where the return typically is lower than other alternatives (at least in the short run), and where the results are much less predictable than when acquiring a demonstrated foreign technology (which, incidentally, frequently comes with a package of management assistance that the research institute cannot supply).

As Professor Brian Quinn of Dartmouth University has pointed out, the propensity for making long pay-back investments diminishes exponentially with the linear increase in interest rates. Thus, in the Sri Lankan relatively inflationary economy (11.5 percent per annum), investors do not have much incentive to take the risks of R&D.

One observer, a well-known scholar of Sri Lankan history said that private industrialists were prejudiced against science and technology because the traditional English manner in which many had been educated emphasized the liberal arts rather than technology. Thus, the management of domestic industry, in his view, lacked understanding of the role and utility of science, but especially of technology, in the conduct of business ventures. Evidence indicated that this attitude also prevailed in Sri Lankan higher education circles. A Sri Lankan country paper prepared for a 1976 United Nations Economic, Scientific and Cultural Organization (UNESCO) seminar in India on management of R&D institutions pointed out: "It is worth mentioning that the only University in the country has not yet produced a single graduate in chemical engineering."

Evidence in Sri Lanka business circles also indicated a high degree of circumspection, if not to say apprehension, concerning government-sponsored institutions, particularly when they function in such arcane (to the average businessman) areas as science and technology. Government traditionally has intended to influence or constrain the business community, and the CISIR as a government institution appeared to be affected by business' attitudes. As previously mentioned, the management of the institute admitted that they had not effectively directed their activities toward creating bridges of understanding and mutuality with the industrial sector.

Judging by the comments of persons both within and outside the CISIR, the institute was not and had not been sufficiently attentive to those organizations and individuals on the outside who influenced the acceptance and utilization of the institute's research products and affected operations funding. One small but significant example of the current management's interest in overcoming this deficiency was the publication of a small quarterly newsletter, with a mailing list drawn from this external community.

The same foreign observer noted that senior members of the industrial elite of Sri Lanka and other Asian countries had a traditional paternalistic, hegemonic attitude that also influenced the board of governors of the CISIR. In the past, the board consequently had gained an undue amount of operational control, which tended to diminish the influence of the director and principal management staff. This violated commonly held views of Western business theory that the board should restrict itself to policy matters. Director Wijeratne, with some reluctance, admitted that this was a notable factor in his efforts to bring the CISIR out of its past deficiencies and into a more effective role and force for development.

A foreign diplomat who had been present at the inception of the CISIR commented that the institute's first director, an American, had been very effective and exemplified the benefit to be gained through a careful matching of foreign experts with local human and material resources. He felt, however, that in the Sri Lankan society there was a certain lack of appreciation for "inventiveness." Such a lack must necessarily work somewhat against the success of an innovation-oriented institution such as the CISIR, he believed.

The foreign observer described the general atmosphere of the country as "an ambivalent, rich tapestry" of sometimes conflicting social and cultural influences, which makes the operations of the CISIR more difficult than would be the case in a more homogeneous society. Sri Lanka he characterized as a "two-tier" society with a very small economically affluent elite, in contrast to a poverty-immersed mass of the population; he expressed concern that policies relating to industrial development and the inducement of foreign investment would result in further accentuation of this stratification, as it had in many similar societies. He felt that the CISIR, ideally, could do much to spread the beneficial effects of technology broadly throughout the society, but thought that the institute theretofore had not been particularly effective in doing this.

With regard to technology, he noted that the current enthusiasm for "appropriate technology" could have a number of benefits, but he also felt that this should not be at the expense of the introduction and utilization of technology of whatever sophistication is necessary to accomplish rational development goals. "Of course job creation is of very great importance, but that does not mean that this is the sole or paramount purpose for the use of technology." In Sri Lanka, he believed, there were clearly differing views within both the public and private sectors with regard to this question. "There are those who say 'use labor at any cost', versus others whose view is that the only basis for rational choice of technology is the comparative cost/benefit ratio."

One of this expert's principal concerns was that with the liberalization of foreign investment and import controls, various important development-related decisions might be made, de facto, by foreigners whose interests were primarily in the present and who therefore were largely motivated by considerations of short-range economic return.

The charter of the CISIR provided for exchange of board of governors' memberships with the Development Finance Corporation of Ceylon (DDCF). Considering that the DDFC was continuously reviewing loan applications for projects having significant technology components, and that it was the original intent of the World Bank proposal for these two institutions to collaborate in technology-related matters, it was concluded to discuss this institutional relationship with Donald W. Kannangara, general manager/director of the DFCC.

He stated that the relationship between the two organizations had been allowed to languish and that he, for one, felt than an active association should be established, although he did not at that time mention in what specific ways he felt such a collaboration might be implemented or for what purposes. It was his feeling though, that the CISIR represented "a considerable resource of which we're not making adequate use."

The DFCC received its foreign currency from sources such as the Asian Development Bank and the World Bank. Local currency was provided through the commercial banking institutions of Sri Lanka and occasionally directly from government, but not often. The DFCC made loans only to firms that were predominately (80 percent or more) owned by private sector interests. It was noted that the government was moving to establish its own development banking institution that would serve primarily the public sector companies and help finance state-owned infrastructure elements (dams, roads, water systems, etc.).

With regard to the DFCC's involvement with questions of technology, Kannangara indicated that they were "passive," although they had a small review staff that examined projects (sometimes in conjunction with consultants) to determine if the planned technology met world standards, was in wide use, and came from a reliable source. When asked if a DFCC policy existed with regard to utilization of indigenous technology, he replied that at that time there was none but that it had been considered. He went on to say that any effort to stress the use of indigenous technology through the mechanism of the corporation's industrial loan activities would have to come as a policy decision on the part of the government.

It was stated that in receiving secondary financing from regional and international sources, such as the Asian Development Bank, they were occasionally asked to reserve a certain amount--perhaps 10 percent--to support small-scale business in the country, but that no other policy on the part of the international banking community had been established calling for specific encouragement of indigenous technology.

It was noted that by the time the loan applications reached the DFCC, the technology had usually been decided. He said that perhaps this situation called for establishing an intermediary organization, but he did not elaborate on this idea.

In reflecting on the effects that technology had had on Sri Lanka, Kannangara recalled the "chekku" units of his youth, which were buffalo-powered stone mortar-and-pestle devices used to extract coconut oil. He noted that these had disappeared entirely, to be replaced by modern, electric- or diesel-powered machines in the larger cities; he was gratified at the role played by his organization in bringing about this evolution.

With regard to the interaction of the government with the CISIR, two senior staff members of the institute voiced the opinion that far too little attention had been paid to the CISIR by the Planning Ministry and that the institute should be brought into the national development planning process on an active and continuing basis; it was believed that science and technology were not factored into national development plans to the extent necessary. It also was observed that science and technology were "magic words," and were often invoked by politicians who did not have an adequate understanding of the implications and significance.

A predisposition to create new, faddish S&T-based institutions had recently become evident with the announcement that the government was going to sponsor the creation of a basic research institute. This left questions in the minds of the CISIR staff concerning what this would mean for their own research activities. According to one senior staff person, Sri Lanka was too poor to afford much, if any, basic research; he would consider a new basic research institution to be a waste of scarce resources for a nation that could barely feed its own people.

The director and his deputy agreed that the CISIR had not functioned effectively in its past relationships with government entities that were important to the institute's growth and viability. They had not "played politics" with sufficient adroitness, although one director, in their view, had concentrated entirely too much on this aspect to the detriment of his management effectiveness. Examples were given of other government-sponsored institutions that had fared appreciably better because they had been more alert to the political dynamics and had been less self-effacing in presenting their qualifications and needs to the various entities influencing their budgets.

Concerning the lack of scientific manpower in the country, both men felt that this problem was of first-rank priority and indicated that foreign assistance for this purpose had always been welcome and would be even more so in the immediate future. Also of great importance would be the provision by the government of more modern laboratory and pilot plant equipment.

It was felt that the institute had lost considerable momentum and freedom as a consequence of its takeover by the government in 1962 and that any subsequent inadequate performance of the CISIR was largely the result of budget deficiencies and changing government policies.

Recent governmental decisions had resulted in the establishment of the Industrial Development Board and the National Engineering and Research Center, both of which had scopes of activity that significantly overlapped those of the CISIR. Some CISIR staff members felt that these were redundant and posed a threat to the CISIR's continued activity. In commenting on this situation, Director Wijeratne wrote in 1979: "The areas of activity of the three organizations therefore are fairly clearly defined even if not strictly demarcated. There is clear understanding within the organization of their objectives and functions. It is impossible, however, to prevent any of these organizations from occasionally indulging in working on ideas which may strictly speaking lie within the province of another organization, particularly when the subject is of topical interest. Such cases however are few and not of significance."

> It is not generally appreciated that applied research...is a profession in itself, to be distinguished from the more academic fundamental scientific studies. We must emphasize that neither money alone nor the mere creation of scientific councils and organizations will solve any research problems, either now or in the future. The entire success of this program, with its intangible training aspects, will depend upon the most skillful selection of personnel. Next to this in importance is the freedom of the research men to pursue their investigations to a practical conclusion without interference; it has been wisely said that 'the main job of a research director is to protect the research men from those who want to direct them.

Source: The Economic Development of Ceylon. International Bank for Reconstruction and Development, 1952.

> Scientists and technologists have always delivered the goods if there is a national recognition of the concept that the development of science and technology is essential for socioeconomic progress. The Government of Sri Lanka has accepted that our goals of economic and social development can only be achieved through the application of S&T, and will therefore do everything possible to encourage the development of a strong scientific and technological capability within the country.

Source: Sri Lanka National Paper - United Nations Conference on Science and Technology for Development - 1979.

REFERENCES

Correspondence, member of CISIR governing board to governing board, November 1977.

The Economic Development of Ceylon. Washington, D.C.: International Bank for Reconstruction and Development, 1952.

Memorandum, Lats to governing board, CISIR, 10 September 1975.

National Geographic Atlas of the World. Washington, D.C.: National Geographic Society, 1975.

"Observations of the Research Planning Council," CISIR, 24 October 1978.

Sri Lanka National Paper for the United Nations Conference on Science and Technology for Development. Colombo: Production and Printing Unit, Industrial Development Board of Ceylon, 1979.

PERSONS INTERVIEWED

E. P. Paul Perera
Deputy Director General
Greater Colombo Economic Commission

Donald W. Kannangara
General Manager/Director
Development Finance Corporation of Ceylon

E. E. JeyaRaj
Deputy Director
CISIR

Mervyn Wijeratne
Director
CISIR

Dr. E. R. Jansz
Senior Research Officer
CISIR

Dr. A. S. L. Tirimanne
Senior Research Officer
CISIR

Dr. Erma Lord
Research Officer
CISIR

Dr. Onil Perera
Research Officer
CISIR

Dr. S. Gnanalingam
Senior Research Officer
CISIR

Dr. W. Howard Wriggins
Ambassador of the United States
Sri Lanka

John Ericcson
Deputy Director
USAID
Colombo

L. J. Fernando
Additional Secretary
Ministry of Industry and Scientific Research
Colombo

Clark Billings
USAID
Colombo

Dr. P. M. Jayatissa
Research Officer
CISIR

QUESTIONS TO CONSIDER

1. Was the original concept of the CISIR too idealistic? What do you feel was the probability of gaining self-sufficiency through contract research? How does this apply in other world situations?

2. Has the institute been diversified enough in its technology? Did it have any real opportunity to extend its technological horizons given the constraints under which it operates?

3. Has the Sri Lankan government placed the CISIR and its functions in proper perspective, given the existence of other technological institutions and the limitations on resources?

4. Is the present management making the best decisions and efforts in view of the circumstances?

5. What could be recommended to the CISIR with regard to extending its contacts with the user community? What should be done in view of the recent decisions to permit foreign technology to enter the country with little restraint?

6. What approach might be taken in consideration of the staff compensation schedules? Should there be greater effort to generate outside income from sponsored projects?

7. What could be done to gain more experience for the staff in the realities of industrial operations? What formal training is indicated, if any?

8. What is the role of the industrial research institute in the appropriate technology concept, and how does the CISIR conform to this?

21
Papua New Guinea: Micro-Hydroelectric Projects for Rural Development

Ed Arata

BACKGROUND

Papua New Guinea (PNG) is a land rich in many resources, one of which is a great potential for the generation of hydroelectric power. Hydroelectric power is not new to PNG; it has been used for many years both in private and governmental sectors and is widely used at church mission outstations. This source of energy presently supplies a large percentage of the electricity produced in PNG, but most of this power is distributed in the urban areas.

In light of this urban electrification and the commitment of the PNG government to rural development, as stated in its Eight-Point Plan, PNG University of Technology (Unitech) in Lae and the Appropriate Technology Development Unit (ATDU) have considerable interest in the development of small-scale hydroelectric plants for village use and problems of rural electrification in general. As a result, these two groups coordinated the installation of a pilot hydroelectric plant near Baindoang village in the mountains of the Huon peninsula. The project was initiated at the request of the people living in the Baindoang area; Unitech and ATDU agreed to assist in the hope of gaining field experience in this type of project plus the chance to assess the impact of such a program on the community involved. A seven-kilowatt plant was installed at Baindoang over a period of two and one-half years as a village self-help project directed by technicians from Unitech and ATDU. The plant is now supplying power to the school and airstrip area, where it presently is used primarily for lighting and water heating.

THE TECHNOLOGY

The introduction of electricity into the rural areas of PNG can be accomplished in several ways: extending the existing power grid; installing small diesel or petrol engine generators; using wind-powered generators; the use of photovoltaic cells; or installing water turbines connected to generators. Engine generator sets usually require maintenance skills that are not found in the rural areas, plus fuel oils and lubricants must be purchased constantly and transported by airplane or human effort to the site. This was the case with Baindoang. The extension of existing grid systems is practical in some locations, but in the Baindoang area it would be nearly impossible to install transmission lines due to terrain. If lines were installed, the cost would never be recovered because of the low power demands of the rural sector. The use of wind power may be considered for some of the offshore islands and the Markham valley, but Baindoang, located in a sheltered mountain valley, would not be a favorable location for exploiting this energy source.

Photovoltaics are another possibility, but at the time the project was started the cost would have been prohibitive for the size of installation requested. With the projected increased availability and lowering of costs related to photo cells, they might be considered for future projects. Again, Baindoang probably would not be suited for this energy source because of the heavy cloud cover. Very few solar radiation readings are available for areas outside Port Moresby and Lae, and stations are only now being set up to gather these data. Both wind and photovoltaic systems would require power storage in the form of batteries and electrical inverters if the DC power were to be used with AC equipment. These would add costs and possible maintenance problems to proposed systems.

The choice of hydroelectric generation for the Baindoang project was fairly straightforward, since reliable water sources were available. The people at Unitech and ATDU have designed the system to be simple, yet reliable and efficient. Most components were selected for their sturdiness in an effort to minimize maintenance, but not without an eye toward cost. As many components as possible were obtained in PNG to minimize problems in replacing parts. In line with this decision, the mechanical speed governor that usually is fitted to hydroelectric turbines was omitted in favor of an electronic governor system that has no moving parts and requires little maintenance.

The university group opted to purchase a Pelton wheel turbine and generator from Gilbert Gilkes and Gordon, Ltd. in Kendall, England, in spite of the higher cost. It was believed that the added reliability of this imported set was important for the success of this pilot project. In an effort to reduce this cost the ATDU also has undertaken a development program to produce prototype Pelton wheels and propeller turbines that can be produced in PNG for use at future sites. This, in turn, will create employment not only for manufacturing the units, but also for installing and maintaining them.

IMPLEMENTATION OF THE PROGRAM

The Baindoang micro-hydroelectric project was initiated as the result of a radio program concerning a seminar on rural electrification held at the University of Technology. The broadcast was heard by the Baindoang villagers and in particular by their primary school headmaster, Johannicus Yang. Yang subsequently wrote to the university asking if it would be possible to initiate a small hydroelectric project for Baindoang. "We have a lot of water here, and the mountains are steep. I think it would be best if someone who knows a lot about these things would come and look." A visit to Baindoang was arranged in late 1975 to contact the leaders and to make a preliminary assessment of the situation. Since the villagers had just recently completed an airstrip, the trip could now be made in fifteen minutes by air from Lae rather than by the normal two-day walk over rough terrain.

Baindoang is located about forty kilometers (air distance) from Lae and lies on the southern slopes of the Saruwaged mountain range. The country is extremely rugged; the major rivers flow in steep-sided valleys, with bush-clad ridges in between rising to two hundred meters or more. The villages in this area are located in an altitude range of five hundred to two thousand meters, and generally are found on the terraces or ridges high above the valley bottom. With an annual rainfall of two hundred to three hundred centimeters, the vegetation zones range from Tropical Woodland to Mountain Moss Forest, and the upper valleys, including Baindoang are shrouded in mist and cloud for much of the time. There are thirty-two villages within the Nabak Census Division. This division had a population of 11,313 people at the time of the 1972 census. Of this total 43 percent were recorded as absentees, living either temporarily or permanently outside the area. With a population of 594 (295 of whom are absent), Baindoang is the second largest village in the area.

The first outside contacts made in this area of the Huon peninsula occurred in about 1917, when Lutheran missionaries arrived. Governmental contacts were first made in about 1929. Cargo cult activities still flourish in many parts of the region, but are not a feature of life in the villages around Baindoang. Coffee is the only cash crop in the area, but evidence indicates that some of the vegetable gardens are declining because the young people are leaving the villages. The staple diet is sweet potato, usually supplemented with yam, taro, and tinned foods. Many other European foods are grown, but air freight costs are too high to make them a viable cash crop. There are a number of cattle projects, but difficulty of access precludes the expansion of these into real business ventures.

In this context, the people of the Baindoang area first decided to build an airstrip as a focal point for their own conceived development plan and later to ask for help in installing a hydroelectric plant. The airstrip was constructed on a terrace five hundred meters below Baindoang over a three-year period by local residents. The government provided money to buy tools and some food for the workers, but the work was initiated and organized by the people themselves through their village council and headmaster. The power to be generated by the hydroelectric project was to be used by the village people directly at their school and airstrip area, and decisions concerning maintenance, distribution, and rates were to be made by the local people through their village council. Expatriate involvement was to be kept to a minimum once the initial installation and training were completed. There seemed to be little doubt among participants that the spirit and incentive existed to make a success of the airstrip and school station. The village people did much in relocating their school nearer to the airstrip and looked forward to an aid post, an agricultural depot, a trade store with cold storage, a coffee cooperative, and a poultry project, all powered with electricity from a small hydro plant. But most importantly, they wanted their school and the teachers' houses to be lighted so that their children could receive the best education possible.

The first direct contact with the people of Baindoang was made by Jack and Mary Woodward of Unitech. They spent their initial visit discussing the possibility of a hydro project with the village leaders, examining streams and their sources, and measuring stream flows and elevations. After evaluating the situation, the Unitech people felt that they had found a viable hydro site to develop and an ethusiastic group of people with which to work.

Initial planning was begun, and a group of Unitech surveying students was sent to do a more detailed survey. The most workable site in the area was then chosen. Due to the topography of the area and the widely fluctuating stream flows, it was felt that the site could be best exploited by the use of a Pelton wheel turbine with a head of approximately 180 meters. The plan provided for a small weir or dam to be built on a stream high above the airstrip, thus diverting part of the stream flow into an eighty-millimeter polyvinyl chloride (PVC) penstock that funnels the water 600 meters down the ridge before passing it through the turbine. With a flow of about eight liters per second, the electric generator would give a power output of about seven kilowatts. A low-head turbine had been considered for use on another stream closer to the station but was dropped due to intermittent flash flooding. With these decisions made, the final design process was started and materials were ordered.

Up to this point the Baindoang project had been mostly a dream for some and an engineering exercise for others, but in November 1976 a group of electrical engineering students flew into Baindoang and installed electrical wiring in the main school building. In early 1977 an order was placed for the PVC pressure pipe that would be used for the turbine penstock. This pipe was manufactured in Lae in special three-meter lengths rather than the standard six-meter size so that it could fit into a light airplane.

Construction of the pipeline was scheduled to begin in April 1977. Pipes, cement, tools, and other materials were flown into Baindoang two weeks prior to commencement of the project. This required good planning and much thought, as one tool forgotten in Lae could delay work by several days. The Unitech-ATDU working parties usually consisted of several staff members and six or eight students from the various departments of the university. Yang divided the pipeline route into four sections so that a trench for the penstock could be dug. Each section was the responsibility of a particular village or group of villages. (It should be noted that since the Baindoang or Umbang school serves not only Baindoang but about seven or eight villages, this project was a joint effort of many groups.) The excavation of each section was a frantic affair, with forty to fifty men, women, and children wielding crowbars, picks, and shovels. Large boulders were either broken up with crowbar and pick or bodily removed. Those women and children not working in the trench moved constantly up and down the mountain carrying small stones and sand in bamboo containers to be used in the concrete that would help to anchor the pipe into the hillside.

The weather conditions for this stage of the project were poor. The wet season had set in early, and a steady, soaking rain fell each afternoon. It was difficult at times to maintain a sure footing in the steep, muddy trench.

After the pipe and anchors were installed, the dirt was carefully backfilled into the trench. During this stage of installation the upper half of the pipeline was laid and tested for leaks. Several months later the bottom half of the pipeline was completed and joined to the upper half. The entire line was filled with water and tested for leaks. When the pipe was found to be leak-free, a small tap at the bottom was opened and a stream of water shot high into the air. This demonstrated some of the power that the village would soon have.

In late 1977, again during a Unitech break, a large work group flew into Baindoang. The students were ready to work on the dam at the top of the penstock and on foundations for the powerhouse at the bottom. The weather was fairly good and the work went quickly; village men and Unitech students worked together to construct the earth-fill dam and then the spillway and intake box for the pipeline. Work also progressed on excavating for the powerhouse and assembling the form work for the concrete. By the time the work party left, the pipeline was nearly finished. The dam was in place and working; the powerhouse foundations were in and awaiting the installation of the generator set and protective building. The ATDU staff then returned to Lae to begin preparing for the next visit.

Although the initial planning covered most aspects of the project, many small, unforeseen problems arose and had to be dealt with. For example, when it was decided that a small water system was to be installed for the school, many questions had to be answered: "Where do we pick up the water?" "How big a pipe should we run?" "We have extra power and a water supply--should we put in a hot water system for showers?" "Do we buy the hot water heater or design our own from parts available in PNG?" "Should we bury the power cables or run them overhead?" These and hundreds of other minor points had to be thought out and designed, and systems had to be installed.

Baindoang villagers helping to lay power cables from the power house to the school area

During 1978 the tempo of the project increased, and the ATDU staff spent much of its time preparing for trips to Baindoang or actually working at the site. A small stone and concrete dam was placed on a stream near the powerhouse; this dam would serve as the pick-up for the school water supply. Pipes were laid from this small dam down to the school site; concurrently the power cables from the powerhouse were laid in the same trench. The village carpenters worked on projects that were designed for them, including work on the powerhouse, the wash house, and the drying room. Other work was carried on without direct supervision, and Yang usually arranged for necessary work to be completed before the next work party arrived from Unitech.

The last major stage of the project was to install the turbine-generator unit, complete wiring connections, and test the system. These tasks were completed in mid-1978, and tests proved all systems to be satisfactory. A few details remained to be resolved but these could be finished and the unit started up later in the year. Since the initial demand for power was small, the unit needed to run at only about one-half of its capacity to supply the lights in use and the water heater. Since no appliances have yet beeen installed at the station, it has not been necessary to install the electronic governor. The testing program continues to evaluate various models for operation and reliability. A simple device has been installed in the present circuit that will switch the electrical load into water heaters in the turbine tailrace should the station load switch off completely for some reason. This will maintain a constant load on the generator and thus protect the turbine and generator from overspeeding.

During these later stages of construction, Allen Inversin of the ATDU also ran half-day classes with the standard five and six classes at Baindoang on basic electricity. It was believed that since most of people who would be using the power had no prior experience with it, some training and information exchange was imperative. Planning was also started concerning training that will prepare several villagers to maintain and operate the system.

After several years of hard work, the village people were more than ready for the power to be switched on. They planned the opening ceremonies months in advance: an enclosed area for traditional dancing and places for food preparation were built, and invitations were sent to all who had been involved in the project. The opening was a mixture of traditional PNG customs and new ways brought on by the development of the area. The university and ATDU staff members, as well as other invited guests, were given traditional gifts of food in bilums (string bags), but the presentations were amplified by an electrical public address system. Traditional dances continued late into the night, but some of the light was provided by colored electric lights rather than by torches or fires. The people finally were seeing results from all the time, work, and money that they had contributed for several years.

PROJECT COSTS

Costs for the project initially were estimated at K9,000, but total costs neared K10,700 (see Table 21.1). The university group was able to secure a large portion of this amount from overseas aid agencies, various private trust groups, and the churches. The people of Baindoang were asked at the outset of the project to contribute approximately ten percent of the cost. This they did by tapping resources within the village as well as by soliciting funds from village members who work in the urban areas of PNG.

TABLE 21.1
BREAKDOWN OF COSTS FOR BAINDOANG MICRO-HYDROELECTRIC PROJECT
AS OF 21 NOVEMBER 1978.
(PNG Kina = .70 Pound Sterling or U.S. $1.40)

Turbine and generator	K5436.00
Power cable, powerhouse-school	K1192.00
Power cable, school-airstrip	K 586.00
Penstock and dam	K1160.00
Washing facilities	K 508.00
Electrical components	K 455.00
Powerhouse	K 226.00
Water pipe	K 210.00
Miscellaneous fittings	K 295.00
Airfreight	K 602.00

PROJECTED IMPACT OF THE TECHNOLOGY

Initial projections called for the use of power at the school for lighting and visual aids and lighting for the teachers' and other officials' houses, at the trade store for lighting and cold storage, at the aid station for lighting and cold storage, and at the community workshop for powering small tools. Currently, the power is being used for lighting in the school, in the teachers' houses, and in the two trade stores. Also, as a result of modifications made during construction, much of the power is being used to heat water for a community shower block. As demands for power increase, needs will have to be assessed and priorities determined by the people themselves.

The underlying factors behind this electrification program are varied. The one mentioned most often is that the electric power will improve living conditions in this rural area and thus help to stem the drift of people to the urban areas. Such improvements include entertainment in the form of films or audio equipment, lights for evening community activities, and better medical attention. Schooling should improve because teachers may remain longer if they are offered a few amenities in this rural setting. The school may also be able to take advantage of audiovisual equipment not previously usable. When food storage facilities are installed, people will have access to a wider range of foods. They also will have storage for small goods or produce that they wish to ship to Lae. The availability of hot water for showers and clothes washing should bring health benefits by helping to eradicate skin rashes and other problems caused by a lack of washing facilities. Demands on firewood, used for cooking and space heating, and on kerosene also should be reduced with the introduction of electricity, although firewood will be affected only slightly. In the same context, the university staff noted the firewood shortage early in the project and initiated a reforestation program through the forestry department at Unitech.

The hydro project will only slightly touch the area concerning the creation of jobs. The program, however, was not intended to create jobs but only to provide a service to the community. Since this is a community project, it has been installed by volunteer labor and will be maintained in the same way. Several people might be employed as part-time attendants and troubleshooters for the hydro unit; also several jobs may be created in conjunction with the stores, the aid post, and the workshop that use the electricity. Since the project has been operating for just a short time, no concrete predictions are possible.

The views of the adults and children of Baindoang were expressed via a Unitech study done prior to the beginning of the project. These opinions will then be compared with the people's viewpoints after the project has been operating for some time. The study states, "Villagers are hopeful that the electricity will be used to power a variety of equipment, including grinders for sharpening tools, chicken incubators, coffee pulpers, coffee driers, and later a sawmill and welding equipment. They would also like to see lights in their church in Baindoang village. Other villages, more distant, have little chance of benefiting directly from the project, but people are grateful that their children at school will be provided with better facilities. Quite a number of Baindoang villagers are thinking of moving down into the valley and reestablishing near the school, though no definite moves have been made yet. There are a number of reasons why this move is contemplated. One is that Baindoang, up on the hill, is much colder than the school. Also, people are becoming tired of carrying goods up the hill to the village. Some said they were not interested in the idea of moving down until they saw the university students wiring the teachers' bush-material houses. Now they see a possibility of having their houses wired if they move below the powerhouse. Several obstacles face the people in such a move, the first being the more acute firewood shortage in the school area and the second being the possibility that the clan holding rights to the land near the school may not consent to the building of houses on this land."[1]

The children in grades five and six were asked to write short stories about the hydro project. They read:

> I am going to tell you a short story about hydro electricity at Baindoang. Long time ago we people in Baindoang have no hydroelectricity but in 1976 Professor Jack Woodward arrived and he was the first person to start the power project. While he was still working with us Allen Inversin join us and we all work together. Then Professor Woodward leave us and go back to his country and only Allen work hard with us. While Allen work with us there are also many visitors arrive in Baindoang to see how hard we are working on our project. Our project almost finish in 1978 and our people in Baindoang area they are very happy and they said we work hard and its going to finish this year. And myself I am also very happy about hydroelectricity in my village and also about the hot shower room in my school. And that's about what I feel. Thank you.

> When I was little boy I didn't know how electricity works. And people go to Lae, come back and tell me the story of the light and I use to wonder how I wish I could have that kind of light but now I know how electricity looks like and how it works. And I am very happy about this. I am also happy because electricity is now right in the mountains. I am happy to see University boys and Allen Inversin and others, That is all. Thank you.

---

[1] Rex Okona, "The Baindoang Project: Basic Information for a Social Impact Study of the Introduction of a Micro Hydro-Electricity Scheme in a Rural Area," September 1978.

These, then, show some of the aspirations and feelings of the village people towards their power project.

Various problems can be seen arising in connection with and because of the hydroelectric project. The first major problem probably will concern the expectations that have been created in the people of the area. For example, it was originally thought that only one or two trade stores would have power; now seven or eight groups are planning to build permanent trade stores with lights and cold storage. This problem is representative of others that might arise. The unit installed can only produce seven kilowatts, which will not be enough power for all the planned uses.

Should people move to the airstrip area hoping to have electricity in their homes, a whole host of problems may arise. As mentioned earlier, there may be problems involving the clan that holds rights to the land around the school and airstrip. If they are not agreeable to the land use or ask high compensation for the land, clan jealousies and disputes could develop. On the other hand, if a large settlement were to grow up on the land, it would probably be bigger than most present villages and would create land pressures that are not presently found in the area. Firewood, which already is in short supply, would become a bigger problem. Land use for gardens would increase, probably accompanied by a greater erosion problem. Sanitation and refuse disposal would also have to be dealt with. It also has been proposed that power be supplied to Baindoang proper. This might cause difficulties in that the original project was designed for the development of the school and airstrip area and involved seven villages; now one village only would be receiving direct power usage. Finally, if problems of power distribution occur, the project could deteriorate into a jealous feud.

CONCLUSIONS

The Baindoang micro-hydroelectric project was installed to help meet the developmental needs of the local village people and as an experimental field project for Unitech and the ATDU. The project did not require the development of a new technology or even a major modification of known technology; it required only a rational implementation of standard hydroelectric practice in a new situation. Although many unique problems had to be solved during the installation, such difficulties would have been encountered at any site using the design criteria selected by the university. Unitech and the ATDU feel that they now have gained the needed experience in installing rural hydro projects. They can pass along some of this experience to others working in the field in PNG and use it themselves on future sites when testing turbines of their own design.

The operation of the technology in its new setting under village control and the impact that the technology will have on the people and villages over time still need to be observed. Since this project has only recently been completed and put into operation, evaluation is not yet possible.

REFERENCES

Inversin, Allen R., Newspaper articles for Lae Nuis: "Progress at Baindoang," 15 July 1977; "Another step forward," 7 December 1977; "Micro hydro nears completion," 17 March 1978.

Okona, Rex, "The Baindoang Project: Basic Information for a Social Impact Study of the Introduction of a Micro Hydro-Electricity Scheme in a Rural Area." September 1978.

Woodward, Jack, and Mary Woodward, "A Possible Small Hydro-Electric Scheme and Other Developments at Umbang Station near Baindoang Village." December 1975.

Woodward, Jack and Mary Woodward, "Progress Report, Hydro-Electric Power Project at Baindoang Village, Papua New Guinea."

Woodward, Jack, "Small Low-Cost Energy Sources and Their Effect on Rural Development in Papua New Guinea." August 1975.

QUESTIONS TO CONSIDER

1. Is this case only an isolated instance of local development of a latent energy source?

2. Does the introduction of this considerable amenity in this isolated village suggest anything in terms of longer-term social study of its impact? Would such a study yield information useful to apply in other development situations, or would it be so situation-specific as to preclude this?

3. This is an example of the installation of a sophisticated and expensive technology in a situation where the resulting number of jobs created was not significant. The predominant effect was one of convenience with only peripheral economic benefit; does this, then,

constitute an example of appropriate technology? Will "productivity" be increased sufficiently to even pay the cost of the innovation?

4. What is the relationship of this project to general economic and social development plans for PNG? Is there a policy of improving the physical amenities of the small rural villages in order to stem exodus to the urban areas? Does this particular instance have larger implications and form part of an integrated plan for development?

22
The Republic of Korea: An Example of Integrated Regional Development

Donald D. Evans

A concerted effort was made between 1975 and 1977 to examine systematically the environment and economic circumstances of a small, highly differentiated rural area of the Republic of Korea. A program to apply science and technology toward the economic and social development of this area was developed, involving the integrated efforts of the population, the local government, the national government, the local university, a national association of private firms interested in export development, and a technologically diversified contract research organization--all located in the Republic of Korea. The effort resulted in both successes and failures; it involved rural citizens interacting with scientists and engineers seeking practical solutions to significant problems affecting the lives and livelihood of a contiguous community of almost one-half million persons.

BACKGROUND

Cheju Island is one of fourteen provinces in the Republic of Korea. It lies approximately 100 miles south of the mainland of the Republic at latitude 33° 27' N., at the confluence of the Yellow Sea and the Sea of Japan. It is about 600 miles due west of the Japanese island of Kyushu and lies at the head of the South China Sea. Although it is located at a maritime crossroads of East Asia, it has never been highly developed or internationalized during ancient or modern times. The maritime climate permits growth of subtropical plant species of many unique varieties. Principal crops are sweet potatoes, used for starch production; barley, used in brewing; forage crops; and, especially in more recent times, oranges. The surrounding waters yield a variety of seafood, mostly for local consumption, plus mother-of-pearl, widely used as an inlay material in traditional oriental furniture. The volcanic core of the island, Mount Holla, rises to 6,400 feet and dominates the 706 square miles of the surrounding countryside. Many small villages and clusters of houses dot the landscape. These are built for the most part in characteristic Cheju style, which utilizes the native volcanic rocks for walls and rock-weighted, woven straw ropes to hold the thickly-thatched roofs in place against ocean winds. The mild climate, excellent beaches, and spectacular scenery have made Cheju in recent years a rapidly developing regional tourism center.

Local and national governments, recognizing the unique characteristics of the island and wishing to control its development in such a manner as to not destroy cultural or scenic values, yet seeking to improve the living standards and productivity of the province, had consistently sought to foster planned, organized growth that would lead to self-sufficiency.

During a visit to the island in 1975, the Minister of Science and Technology of Korea, Dr. H. S. Choi, met with the provincial governor, and the topic of Cheju's development was discussed. Minister Choi perceived the need for a comprehensive approach to the problem, utilizing the scientific and technological capabilities of the country to study the situation and develop a program of experimentation and application of appropriate technology. Subsequently, he presented his concept to the staff of the Korea Institute of Science and Technology (KIST), which initiated a preliminary feasibility study. This ultimately resulted in a proposal to the provincial government recommending a comprehensive study and development project.

In the meantime, the minister contacted the Korea Traders Association (KTA) at the suggestion of the President of the Republic, Chun Hee Park. The KTA, a private association of manufacturers and traders in Korea, fosters regional development as a means of stimulating production of exports that its members may market abroad.

Ultimately, this coalition of interests--the provincial government, the Ministry of Science and Technology, the KTA, and KIST--resulted in the initiation of a research and development project. It was funded at KIST through contributions of (Won equivalent) $202,000 from the budget of the ministry and $165,000 from the KTA and was launched with anticipation in April 1975.

DESCRIPTION OF THE PROJECT

As a result of the preliminaries, KIST submitted a proposal for evaluating and initiating research appropriate to the Cheju situation. KIST itself represented the concept of applying technology to achieve economic and social development in that it was conceived and created as an institution for the identification, selection, adaptation, development, and application of technology in pursuit of national development goals. Based on the example of contract research organizations in the United States, KIST is a large, multidisciplinary technological institute whose staff of 1,000 members represents a wide range of scientific and engineering disciplines. It was organized to provide research and development (R&D) services to industry (both public and private). Located in Seoul, the capital city, KIST is dependent on individual contracts with private and government entities for its viability. It receives no regular stipend or grant from the government, although government entities are a primary source of its support through R&D contracts. Based on its ten years of diversified experience, KIST was a good source of concepts and projects conducive to Cheju's development.

Realizing that no development project of the type envisioned would succeed without the agreement and participation of the affected population, one of the first efforts was to involve representative members of the Cheju community in planning and analysis. The provincial government officials strongly supported the effort from the outset and were effective in identifying local residents who were willing to participate. These persons were from the agricultural cooperatives of the island and from the fishing villages along the coast. Faculty members from the national university branch located on the island provided useful intermediary functions and interpreted local conditions in terms that were understood by KIST investigators.

In the planning stage, KIST personnel met many times with island residents to identify those areas of activity that might benefit most from development efforts. It was ascertained that the following subject areas afforded the highest probability of benefiting from the project: (1) feed for livestock; (2) development of the orange-growing and processing industry; (3) supplying electric power to remote areas; (4) development of marine resources; and (5) exploitation of plant resources unique to the island.

These were arranged in three primary catagories: (1) those to receive commercial development as a result of preliminary study and evaluation; (2) those to be studied as "basic" research subjects, to be pursued possibly later as development subjects; and (3) those to be "policy studies," to be described and qualified as a guide for authorities who would frame policy for Cheju's progress.

Livestock Feed Project

The volcanic soil of the island is of such recent origin that it has not weathered into the rich support for plant life that characterizes older soils from volcanic sources. Consequently, although it can help support a livestock industry with forage crops, it is not sufficient by itself to provide all the nutrients and energy needed by cattle.

For example, during the winter months when forage growth is slowed, cattle are undernourished to the extent that brood cows develop a deficiency disease. This interferes with their fertility, resulting in calving only every other year rather than annually. The energy and nutritional deficiencies in the available cattle feed are so pronounced that over the winter months cattle lose more than 15 percent of their weight when fed solely on local feed. Some feed supplements are imported but are very expensive, resulting in such high market prices for beef that sales are restricted. Further, it is not feasible to increase the amount of rangeland or to raise additional forage because land use already is intensive. Besides, other cash crops, including sweet potatoes and barley, are more profitable.

However, surveys by KIST food scientists showed that necessary nutrients and food energy sources were present in the Cheju environment if they could be made available to cattle. They existed in the form of cellulose, bound with lignin and silica, in the plentiful barley straw and of non-protein nitrogen (NPN) contained in poultry manure and other farm animal excreta. Important elements such as calcium, phosphorus, copper, zinc, manganese, and molybdenum were also present in the poultry manure.

The KIST staff knew of experiments in Europe and the United States where bacteria had been grown successfully on a substrate of poultry manure. The protein content of the bacteria was relatively high--and cattle could thrive on a diet of ordinary forage supplemented with this high-protein bacteria. The principal problem was that the manure had a highly unpleasant odor that repelled the cattle and raised questions of possible toxicity. The digestive systems of cattle contain a bacterium that also can utilize the poultry manure, if the cattle could be induced to ingest it. To gain the optimum utilization of this food source, however, significant bacterial growth on the poultry manure had to take place externally before cattle consumed it.

The odor problem could be eliminated through forced drying of the material and pelletizing as is done commonly in large-scale "chicken ranches" in the United States and elsewhere. However, in the Cheju situation this would have added an unacceptable cost to the process, given the high price of fuel and the requirement for additional equipment and labor. Consequently, a way to use the manure in the wet condition was sought.

A threefold problem was presented to the researchers:

(1) Could barley straw be treated on Cheju farms so that the bound cellulose could be released

in sufficient degree to significantly increase the amount available to the digestive tracts of cattle, thus increasing the energy supply?

(2) Could poultry manure be used as a substrate for growing bacteria so that, when used with other contained elements, the bacteria would meet the nutritional requirements for cattle feed?

(3) Could both of these processes be adapted to the Cheju conditions, allowing farmers to produce their own feed supplements, and at a cost that permitted the product to be competitive with imported beef?

It was determined that the first problem could be solved if a strong base chemical could dissociate the silica and lignin, thus freeing the cellulose for utilization by the animal. Such a chemical was sodium hydroxide, which, when blended at a ratio of 3 percent by weight with chopped barley straw, accomplished the separation required to raise the available cellulose from 35 percent to 50 percent and more. The dissociation took from twenty-four to forty-eight hours, depending on the ambient temperature. Although highly toxic and necessitating considerable care in handling, sodium hydroxide was familiar to the Cheju farmer, having been used for many years (in a highly dilute form) for clothes washing. Also, it was cheap and readily available, even in remote rural areas.

The problem with the poultry manure was solved by a natural organic process. When air was effectively excluded from the manure, natural anaerobic bacterial growth occurred, converting the offensive uric acid to ammonia plus other, odorless, compounds. Also, the resulting bacteria were a highly available source of protein and other nutritional elements that cattle could utilize.

However, the anaerobic bacteria required yet an additional source of energy-yielding substance on which to feed. This was handily supplied in the form of rice bran cake (i.e., rice bran from which the oil has been extracted), which was essentially an agricultural by-product in Korea and was available in large quantities at low cost.

Finally, it was determined that the chemical treatment of the barley straw and the growth of anaerobic bacteria on the poultry manure could be accomplished simultaneously, thereby simplifying and accelerating the process.

The final "recipe" for the KIST-developed feed process was:

(1) Mix 50 percent chopped barley straw, 20 percent wet (70-80 percent water) poultry manure, and 30 percent rice bran (or other type of bran) cake with 3 percent sodium hydroxide.

(2) Place the mixture in fifty-kilogram plastic bags under a plastic "greenhouse" on the ground or in a plastic-lined, covered trench. Allow the mixture to stand until the manure odor has been replaced with the typical "sweet" smell of silage that is well known to the farmer. This will require from forty to sixty days depending upon the weather.

(3) Feed the product in a ratio of one-to-one with normal roughage.

The plastic bags or coverings serve not only to exclude the air, thus permitting anaerobic bacteria growth, but also to insulate the mixture, raising the temperature and promoting the chemical dissociation of the cellulose in the straw and the further growth of the bacteria in the manure. The plastic also may be used to cover and promote the growth of vegetable crops when not being used for feed production.

Feeding tests were conducted at the time of case preparation on sixty head of cattle in four controlled experiments. Instead of the weight loss normally encountered during the winter, the test animals gained an average of 500 grams per day on the 50 percent roughage substitution and 900 grams per day when the KIST mixture was substituted for 40 percent of the imported feed supplement that composed one-third of the diet of winter herds. The imported feed supplement consisted of ground corn, soybeans, domestic oil cake, and grain. When this mixture was fed alone at the one-third ratio, each animal in the test herd gained an average of 800 grams per day. Two- and three-year tests were underway at the time this case was prepared to determine the long-term effects of the KIST feed diet. Some Cheju farmers were feeding the mixture regularly to their herds and were pleased with the results.

It was found that the mixing process for the feed required the development of a special machine. The design consisted of a fixed drum with a set of rotating blades inside. It could be fabricated from locally available sheet metal and parts and was powered by easily obtained two-wheel tractors.

The cost of the KIST feed when produced on Cheju farms was less than $100 per ton, which compared with imported feed supplement at over $200 per ton--and the

KIST product gave better results. Thus, the Cheju farmers were provided with a new cattle feed supplement that they could produce economically directly on their own farms. The technology was simple, and beef production improved markedly.

At the time, KIST was planning to extend its program to the mainland by experimenting with the use of rice versus barley straw, since the latter was much more plentiful on the mainland. If the long-term feeding tests on Cheju proved the acceptability of the feed, then it was assumed that the process would be widely adopted. In discussing the project, Director D. Chun Su Kim stressed the importance of having involved the Cheju farmers at all stages of the development process, thereby overcoming their natural reluctance and suspicion of innovations.

Dr. Kim visited Australia, where the Commonwealth Scientific and Industrial Research Organization displayed considerable interest and was experimenting with the process. The West German government also showed interest in supporting the project, particularly in testing the method on the Korean mainland. Meanwhile, information arrived that similar experiments were being pursued in the United States and Europe; although, as Dr. Kim pointed out, each situation will require specific adaptation.

Orange Products and Processes

Historically, Cheju was the site of the casual cultivation of oranges, which grew well on the southern reaches of the island. In recent years successful efforts were made to commercialize their production, since a ready market existed. Oranges were an agricultural product well-suited to the Cheju environment, being fairly labor-intensive, of high value, and capable of being raised on land little adapted for other agricultural purposes due to soil characteristics and general rockiness. From 80 thousand tons in 1975, production was scheduled to increase to 300 thousand tons in the 1980s.

The KIST researchers were not plant scientists and could contribute little to the selection and propagation of orange trees; however, they believed that the need and opportunity to improve orange storage, processing methods, products, and packaging existed.

One of the main difficulties on the island was the absence of cold storage facilities for the orange crop, resulting in spoilage rates of up to 30 percent, plus the inability to hold the crop long in storage, causing market oversupply and reduced revenues. The best storage conditions were determined by experiment to be a temperature of 4° to 6°C with relative humidity of 85 percent. The KIST staff observed that, although mechanical refrigeration was possible and might even be cost-effective, the temperature gradient of the island over a twenty-four-hour period during the harvest season showed nighttime temperatures well below those required to greatly extend the storage life of oranges. They decided to try to design a storage method that could average normal day-night temperature differences. The objective was a steady humidity and temperature condition that would "smooth the curve" of daily fluctuations.

By experimenting with a variety of storage house designs that incorporated a humidifier, ventilator, insulation and locally available materials, an efficient, low-cost configuration that was capable of being constructed with local labor was finally developed. The method did involve a relatively sophisticated temperature and humidity monitoring and control system that determined when the flow of humidified air through the storage house needed to be changed; that is, cool air introduced at night with sealing off during the warm daytime hours. Without these continuous monitoring provisions, the system would not function well enough. Purely unassisted human control of the process would not be adequate.

Additionally, the KIST food scientists were able to recommend chemical treatment for the oranges which essentially eliminated the spoilage caused by bacteria and other organisims that attack the fruit. Thus it was the combination of the improved storage conditions with the chemical treatment that resulted in a dramatic increase in the time which oranges could be stored and still retain their marketability.

The three-year development of orange storage facilities was perhaps the greatest immediate economic success of the KIST project on Cheju, promoting the preparation and distribution of a design and construction manual throughout the province. It was a prominent factor in an increase over a two-year period of about $4 million annually in orange crop revenues.

Additionally, the packaging of oranges came under study. The customary method has been to utilize wooden crates, but they were expensive to produce and failed to adequately protect the product in shipment to mainland markets. Given the relatively high cost of oranges, savings in damage losses would yield a valuable reward to the processors. Ordinary cardboard containers also had been used in more recent years, but these were not durable enough.

The researchers and producers approached the problem from the standpoint of "containerization"--the complete orchard-to-market containment of the fruit. This involved the design of a packaging system using plastic wrapping and boxes, which were then placed in large metal shipping containers for transportation to the final metropolitan distribution points. The recent introduction of container ships operating to the mainland appeared to make this feasible.

This system came to be used to ship approximately 40 percent of the crop to market. Reasons for its failure to capture all of the market were that many growers still preferred the older, initially less expensive method, although it could be shown that the new system was overall more economical. One factor inducing higher initial costs was the use of the special plastic containers, which had to be cycled through the system six or seven times to make them cost-effective. This required more systematizing and handling between the orchard and the market than smaller private producers were willing or able to undertake. Additionally, container ships were not always available at the desired times. Further, distributors had a preference for the cardboard containers, perhaps because they had other uses, while the expensive plastic ones had to be returned to the shipping depots.

As of 1978, 30 to 40 percent of the annual Cheju orange crop was being packaged and shipped in the KIST-designed system. This usage was primarily by the growers' cooperatives that the government had encouraged on the island and that generally were more progressive and development-minded than the independent producers whose individual growers were larger, but who tended to adhere to conventional methods and to resist change.

The research and development teams also made a laboratory study of orange processing methods for the production of marmalade, pectin, juice, and oil from orange skins. Although they felt that they had made some improvements in these products, the commercial processors on the island, in subsequent interviews with project evaluators, stated that "there was nothing new" in what they had been told as a consequence of the KIST experimentation.

A successful laboratory process also was developed for the production of an orange-based liquor, which it was believed would find a good market both domestically and abroad. However, the product evaluation report noted that there was consumer resistance to the taste, which apparently was too different from that of traditional alcoholic beverages (derived primarily from rice). In the meantime, the project exhausted its funding, and the liquor development effort was shelved. (Also, a surplus orange problem had existed when the research was started, but this later had been solved by increased fresh orange sales.)

Wind-Powered Generators

Cheju Island experiences year-round strong and steady winds near the sea, and seasonal winds inland over its expanse. Given the high costs of imported fuel oil used to operate the island's thermal-energy-powered generating plant, it was desirable to experiment with using the free, abundant wind to drive windmills and to integrate this source of energy with the use of electric generators. Additionally, the dispersion and small size of individual dwellings around the island made the use of self-contained, local generating sources even more practical.

Consequently, the electronics and mechanical engineering sections of KIST designed and built an experimental system for testing in the Cheju environment. The basic design for the windmill was adapted from a popular Australian commercial product that had been developed and long-used in the "outback," where wind conditions were similar to those at Cheju. KIST engineers ordered a three-phase brushless generator, manufactured in Korea, to be coupled to the windmill through a gearbox, increasing shift speed five to one. Its output ranged from 100 watts at a wind velocity of 4 meters per second, to a maximum output of 2 kilowatts in a fairly strong wind of 12 meters per second. Overspeed of the windmill at rates above this velocity was anticipated, and an automatic blade-feathering system was provided.

The feathering system used initially was based on the Australian design. It was comprised of three pivoted weights around the hub of the windmill's three-bladed propeller. When wind caused the rotational speed of the propeller system to increase to the critical point, the centrifugal force of the pivot-mounted weights overcame the effect of permanent magnets that held the individual blades in the attack position; and the individual blade was free to rotate into the feathered position until the speed was reduced sufficiently to allow the magnets to again return the blade to the attack angle.

However, in Cheju's occasional high winds, this system had to operate too frequently and cyclically. Under these conditions, the system's power output was too unstable. To overcome this difficulty, KIST engineers designed a wind velocity meter and small generator that produced an output voltage proportional to the wind velocity. This voltage was used to vary the field voltage of the main AC generator in the system so that as wind velocity increased, the axial (i.e., armature) resistance to turning was increased, thus providing greater rotational resistance to the wind. This not only (in most instances) damped the propeller speed sufficiently, but increased the power output of the system.

Energy was stored in ten twelve-volt automobile batteries, arranged in series. The test unit operated over a period of six weeks, generating five to six kilowatt-hours per day. It was later determined that one unit would serve four to five households under typical daily local conditions. Weather station records and on-site observations indicated that the design should provide sufficient storage capacity for up to three low or essentially windless days. For AC operation, (some households had small television sets), the system was fitted with inverters, yielding sixty-cycle current.

The first experimental installation was provided free of charge to a group of rural homes on the flanks of Mount Holla. The wind generator was mounted on a six-meter tower; service lines were run to each participating home from the battery house. Theretofore, some of the houses had had intermittent electric service from small generators powered by their three-horsepower garden tractors, but this was noisy and expensive, relied on imported fuel, and was not widely or consistently used on the island. The new energy source was used for lighting, for operating a small hotplate on occasion, and, in some instances, for powering television sets and radios.

The users were pleased at the outset with the free service that they were receiving and seemed content with the innovation. Then lapses occurred in the continuity of service because the KIST staff could not constantly provide the intermittent but frequent maintenance that the electrical and mechanical systems required. Such maintenance included lubricating bearings (on the KIST-designed models), tying down the propeller assembly when very high winds threatened, maintaining battery water level, and so on. Mr. Lo felt that if a number of such installations were operating, then the pro rata distribution of maintenance costs would lower the per-unit-cost factor drastically.

In fact, the small group of users ultimately became very insistent that the system be constantly operative and demanded that KIST maintain it. They complained about voltage drops but refused to become involved in the system's operation, even when threatening high winds necessitated shutting down the windmill,

It was, in fact, a high-velocity typhoon wind (it was never determined just how high) that forced a close to this initial experiment when the blades of the windmill literally were blown off. These original blades had a hollow form with double thicknesses of 0.5-millimeter stainless steel--and this obviously had been insufficiently rigid. The blades subsequently were redesigned in fiber-reinforced plastic (FRP), and the unit was moved to another, less-exposed location. However, not long after the new unit was installed, high winds blew off the new-design blades.

After this experience, one of the Australian commercial units (which was warranted to withstand 26 meter per second winds) was purchased and installed at the same location. Residents who used the system were charged Won 50,000 (about $1,000) per household. It was felt that with such a significant financial involvement, the households might take more interest and even be willing to take a turn at routine maintenance. But this was not the case.

Although the third windmill operated successfully through subsequent high winds, it finally suffered a generator failure, putting the unit out of service. This happened coincidentally with the effective conclusion of the R&D project: contract funds were depleted and the system was deactivated. The residents were angry about this cessation of service just as they were getting to enjoy the convenience of electricity and were considering it a part of their daily lives. Although their Won 50,000 fees were returned to them, there was still much dissatisfaction with the entire program. One evaluator of the Cheju project observed later that resident users thought KIST had not been foresighted or properly attentive to the role of the consumer in the project and had been too occupied with the mechanical and electrical details of the experimental development.

When asked to comment on the relative economics of the KIST windmill system, Mr. Hong Jo Lo, one of the project leaders, noted that the installed capacity of the KIST system at about $1,000 per kilowatt made the capital cost of this system about twice as high as the cost of a conventional fossil-fuel-fired central generating station. However, he thought that even with relatively high maintenance costs the windmill-powered system should ultimately be economically viable when the fact that its energy source was free was factored into the analysis.

At the time of this case preparation, the wind-powered generator project was inactive although the installation still existed on Cheju Island. Meanwhile, the West German government, in collaboration with the Korea Electric Company had agreed to sponsor a similar project on a number of other Korean islands that were entirely without power. These numbered in the hundreds, and it was part of the Korean governmental policy to assist the introduction of modern amenities to them. The new program was to incorporate German-manufactured generating units combining solar- and wind-generating capabilities in ten-kilowatt units.

OTHER RESEARCH SUBJECTS

KIST efforts included several other areas of investigation. One project, the production of starch from bracken species that grew naturally on the island, was a technical success but a commercial failure. As noted previously, starch was a customary product of the island's sweet potato industry and was one of the principal exports to the mainland and to foreign users. However, the much higher yields of starch obtained from sweet potatoes made bracken an unattractive alternative, and the process was never commercialized.

The seas surrounding the island were found satisfactory for the cultivation of varieties of crustaceans, bivalves, fish, and seaweed. However, this research was undertaken only to provide useful information to others who subsequently might wish to go develop such industries.

Forty-nine species of indigenous plants were studied in the laboratory to determine their use as sources of oils and essences. The most promising was the wild rose, which grows profusely on the island and yields commercial grades of scent concentrates and pigment. The commercialization of this product was considered a good prospect if information was made available to domestic and foreign firms operating in this business.

Finally, a variety of fig that had been cultivated casually on the island and sold commercially to a limited extent was examined to determine its marketing possibilities. Improved preservation methods were devised, resulting in reduced spoilage losses--the principal detrimental aspect to the marketing of the fruit. This product also was believed to have good commercial potential if an interested producer/processor could be found.

POLICY SUPPORT STUDIES

The provincial government needed information and analyses of the island's economy to formulate developmental and regulatory policies. The KIST techno-economics section provided information on a wide range of related matters, including marketing channels for various products, price stabilization and support systems, and concepts for increasing farm income. Some of the study's recommendations were to develop better transportation systems both internally and externally, to subsidize the island's packaging industry, to increase the orange storage capacity on a schedule over the next several years, to further encourage the formation of cooperatives, and so on. These analyses and recommendations were used by the provincial government and formed an important part of the regional development plan.

PROJECT EVALUATION

The management of KIST wished to objectively evaluate the Cheju project, and assigned the task to one of the institute's social scientists who had a background in evaluation methods. Dr. Hong Ik Chung studied the documentation of the project (he had not been a member of the project team), visited Cheju Island to talk with various persons who had been involved, and conducted a similar series of interviews with participants at KIST.

Chung made several observations and recommendations. First, with regard to the utilization of the results of the study, the following were suggested:

(1) The cattle feed project results should be dispersed widely within the country, and efforts should be made to seek the utilization of this technology in other areas of the Republic through demonstration projects and further research. (As noted, the intent had been to extend this research to include rice straw as well as barley straw, recognizing the much greater availability of the rice straw on the mainland.)

(2) The information and the design for the orange storage facilities probably had application to other crops and other locations; therefore, this technology should be disseminated further.

(3) More basic research was needed to support the findings of the Cheju project, and this "directed basic research" should be supported by KIST and/or others.

(4) Notwithstanding the difficulties encountered and the doubt concerning economic viability, it was desirable to continue the wind energy research, especially considering the particular vulnerability of the Republic to interruptions in the supply of fossil fuels.

With regard to the conduct of the project itself, it was suggested that the quality and effectiveness of the integrated regional development study could have been upgraded by:

(1) Limiting the number of research subjects undertaken. It seemed that too much was attempted, given the resources and sponsorship available.

(2) Better overall management and project coordination would have been advantageous, especially considering the complexity of the project.

(3) It would have been beneficial to have given appreciably more attention to the utilization of the research results--even before those results were achieved.

(4) Dissemination to the public of more information on the project and its results would have been advantageous.

(5) Most of the project results were useful and encouraging, and the experience should be conveyed to other countries where the same approach might prove beneficial.

The KIST staff offered the opinion that this project had been a very useful learning experience for the institute and that this type of coordinated effort would not only serve the interests of sponsors, but also would help to develop a greater feeling of unity and mutuality of interest among the KIST staff. It was hoped that other opportunities would be afforded to conduct similar regional studies in other parts of the country, because KIST was particularly well-qualified to conduct them and because such projects represented one of the most effective ways for science and technology to be applied for the overall benefit of the country.

PERSONS INTERVIEWED

Dr. H. I. Chung
Techno-Economics Group
KIST

Dr. H.S. Choi
Distinguished Scientist
KIST

Dr. C.S. Kim
Food Technology Lab
KIST

Dr. T. W. Kwon
Vice President
KIST

Mr. W.J. Lee
Director
Korea Traders Association

Mr. H. J. Lo
Electronics Laboratory
KIST

QUESTIONS TO CONSIDER

1. Does this case, as described, conform to your concept of the term "integrated regional development"? If not, what elements are missing to make it so?

2. Considering the cost of the project, were the results commensurate? What other approach might have been taken to aid the Cheju economy with such funding?

3. Were the results of the wind-powered generator experiment sufficient to reach a conclusion on this subject? What next should be done if this project were to resume?

4. Does the cattle feed product appear to have applications elsewhere, or is it peculiar to the Cheju scene and not generally replicable?

5. What additional information would have been useful concerning the distribution/marketing of the orange crops? Is more research suggested here?

6. What conclusions can be reached as a consequence of the receptivity and reaction of the families that took part in the wind-powered generator project? Were there perceptions of consumer behavior here that have significance for other situations where technology is being introduced?

7. What general conclusions can be reached concerning the utility of the KIST effort? Is this type of program best managed by a technological institute? What should be the nature of an organization designed to develop and apply technology in rural areas? Is there need for special institutions for this purpose, or is such effort best assigned to generalized industrial research organizations like KIST?

Analyses of Case Histories

Various methods could be devised as the basis on which to analyze the case histories. Of course, the best method is to discuss them in an interactive atmosphere with other concerned persons, where a wide-ranging examination of their implications can occur. The value in analysis of complex socioeconomic-technological situations is largely subjective in any event.

Somewhat arbitrarily, a system has been developed for arraying these case histories according to a structure of interactions and effects.

Cases can be examined in terms of three primary elements: (1) the resources, human and material, with which they are concerned; (2) the means by which these resources are developed or utilized; and (3) the effect that this utilization has on the environment and people within it.

Resources themselves can be classified in two ways: one is in terms of natural endowments such as minerals, energy sources, organic materials, and climatic factors. The other is on the basis of human attributes, particularly analytic capability, innovation and inventiveness, and the elusive quality of entrepreneurship.

The classification of "means" includes, first of all, technology in its multiple forms. It also includes the provision of finance for development purposes. The role of government as the manifestation of social and political will plays a pivotal role. So, also, does foreign aid in the form of materiel, finance, and technical assistance.

The effects of the interaction of resources with means in the pursuit of some planned objective can be classified as: (1) economic--with a special subset concerned with foreign and domestic markets; (2) social impact; and (3) the effect on the physical environment. Within this generalized structure it is interesting to examine the various case histories.

It is not surprising that about half of the cases are concerned particularly with resource utilization. The Tunisian use of available, less costly butagaz as an alternative for gasoline is an example, as is the harnessing of hydro power in Samoa or the enlistment of human resources in the CRUTAC community development project in Brazil and in the "rural university" in Colombia. In some cases it is the use of a waste resource, as in the coffee beneficios of Central America or the generation of biogas from cow manure in Tanzania.

Perhaps most important of all the resources in the various cases is the human function of providing inventiveness, organization, and drive toward goal achievement; this element is especially prominent in fourteen of the twenty-two cases. Certainly nothing occurs in the development process without human intervention, but these qualities are particularly prominent in such examples as the ball-point pen manufacturing venture in Pakistan, the Samoan Methodist land development program, the engineering consulting firm in Singapore, and in the "coffee roads" project in Haiti. In each instance the enthusiasm and persistence of the motivated person or group played a dominant role.

In examining the means by which these projects have been accomplished (and considering that technology is the subject of this study), it is not surprising that some specific aspect of technology has been at the focus of each situation. This is especially the case in the involvement of such relatively sophisticated technologies as the explosive metalworking applications in Brazil or the use of bioengineering approaches in developing cattle feed in Korea from natural indigenous sources. At the other end of the sophistication spectrum are the design of a simple, efficient cookstove in Central America; the traditional process of preserving fish in the Philippines; or the manufacture of bricks in Malaysia using labor-intensive, energy-conserving methods. Foreign technology use as a stimulus to domestic inventiveness is a feature of the ball-point pen enterprise in Pakistan and the fungal conversion projects in Central America.

Because of the complexity, pervasiveness, and scale of the development need, it falls to governments of both the developing and the industrialized countries to take a leading role in planning and implementing appropriate programs (notwithstanding the important role played by the private sector in technology development and transfer).

Sometimes this government role can thwart the development process--it certainly does not always have a positive effect. Government action was decisive in the case of the non-use of composite flour to reduce wheat imports in Colombia, where policy action and market intervention have stopped this program. Similarly, government policy in Indonesia resulted in large dislocations in the farm labor market. In Sri Lanka, the minimal support given a research institute by government significantly inhibited its effectiveness. On the other hand, government support of a similar institution in Brazil through provision of development funds has been instrumental in revitalization and growth in that instance. Also, government interest and support were responsible for initiating the regional development program on Cheju Island in Korea.

The provision or, conversely, lack of finance will always be a fundamental factor in any endeavor. The lack of finance in the case of the entrepreneur in butagaz conversions in Tunisia was a cause of the slow development of that promising technology application. Government grants are important to the functioning of the rural university in Colombia and to the similar project in Brazil. The Korea Institute of Science and Technology (KIST), as well as the Ceylon Institute of Scientific and Industrial Research (CISIR) and most other research institutions, were funded at least initially through government grants. Private financial support makes the rural development program in Samoa possible (along with the dedication and enthusiasm of the participants, of course).

Foreign assistance from the industrialized nations is a prominent part of many development efforts. This is supplied in combinations of material products, technical assistance, and finance. In Colombia, the composite flour project received technical and financial assistance over a period of years from both the Dutch government and the Organization of American States, as well as assistance from the United States through a private research institution. The coffee roads project in Haiti received both financial and technical assistance from USAID. The Lorena cookstove project had technical advice from consultants provided by a private voluntary organization, as did the small-scale sugar processing project in Ghana and the lime kiln development effort in Honduras. Private sources assisted with equipment and technical assistance on the hydroelectric project in Papua New Guinea. One of the most intensive and sustained technical assistance programs was that provided by the Denver Research Institute in the transfer of explosive metalworking technology to the Institute for Technological Studies (IPT) in São Paulo, Brazil.

Assistance of the industrialized countries to the developing world is, of course, a central theme of the development effort, although in recent years there has been increasing emphasis on the attainment of greater degrees of technological self-sufficiency on the part of the Third World (note especially the precepts of the New International Economic Order).

In over half of the cases one of the principal outcomes of the impact of technology has been its economic effect. Thailand's main export, cassava pellets, has been heavily influenced by technology, both foreign and domestic. Coffee roads in Haiti are expected to have great economic effect on remote rural areas that the roads open up to commerce and communication. The Lorena cookstove is making possible appreciable savings in fuel costs for users. Cheju Island's economy has grown significantly as a consequence of development work based on technology application. Explosive metalworking in Brazil makes savings possible through import substitution. Sugar-processing technology in Ghana is also saving foreign exchange and creating domestic employment. Imports have been reduced by butagaz use in Tunisia and by ball-point pen manufacturing in Pakistan. Local professional engineers and architects are being used in Singapore in place of foreign experts. On the other hand, the Thai mint project, which had such optimistic beginnings, has not lived up to expectations of having a major economic effect.

Social impact of technology is sometimes difficult to assess and especially to measure; only in recent years have there been systematized efforts at technology assessment from this standpoint. Nevertheless, the cases reveal some important social results of technology utilization as in the instance of community development along Haitian roads, the changes in community life in Papua New Guinea as a result of electrification, and the social development aspects of the extension and education projects by universities in South America. Rice hulling in Java had serious impact on rural populations, especially women. Part of the disappointment in the economic performance of the Thai mint program was the result of life-style preferences of farmers in that country. The <u>absence</u> of technological innovation in the Philippines' fish-drying industry permits it to continue as a family-centered activity.

Finally, environmental effects have resulted from technology introduction in several of the examples. Cassava-pellet dust pollution has been a very important cause of technological innovation and economic impact in Thailand. The Lorena cookstove has not only reduced deforestation in Central America but has also improved the health conditions of housewives in that region. The biogas generators in Tanzania have also reduced the demand for fuel wood. And fungal fermentation holds promise of reducing the polluting effects of coffee processing wastes while creating a valuable feed supplement for livestock.

CASE HISTORIES MATRIX

This matrix is based on the analysis method presented in the text. Principal elements of case histories appear in the left-hand column; the individual cases, identified by number and country of occurrence are across the top; the numbers and country names correspond to the list of cases.

A mark was entered in a column if that specific element of the analysis format was judged to be relevant to the case. There is, obviously, a degree of subjectivity in deciding the relative importance of one of the analysis elements, but it is felt that this system is useful.

It can be used as a means of selecting cases for review, based either on country site or on identification of some particular set of characteristics. Or, it may be useful to compare to the reader's own assessment after reading an individual case.

	1 TUNISIA	2 THAILAND	3 COLOMBIA	4 HAITI	5 PAKISTAN	6 INDONESIA	7 GUATEMALA	8 HONDURAS	9 MALAYSIA	10 COLOMBIA	11 BRAZIL	12 WESTERN SAMOA	13 TANZANIA	14 BRAZIL	15 GHANA	16 THAILAND	17 CENTRAL AMERICA	18 PHILIPPINES	19 SINGAPORE	20 SRI-LANKA	21 PAPUA NEW GUINEA	22 KOREA
RESOURCES:																						
NATURAL	●		●			●		●	●	●	●			●	●						●	●
HUMAN	●			●		●		●	●	●			●	●	●	●			●	●	●	●
MEANS:																						
TECHNOLOGY	●	●	●	●	●	●		●			●	●	●	●	●	●		●	●	●	●	●
GOVERNMENT		●			●				●				●						●			●
FINANCE	●										●			●					●			●
FOREIGN ASSISTANCE		●	●			●	●				●		●	●				●			●	
EFFECTS:																						
ECONOMIC	●	●	●	●	●	●	●		●			●	●	●				●	●			●
SOCIAL		●		●				●	●	●	●					●				●	●	
ENVIRONMENTAL		●				●	●					●			●							

431

Part 3
Bibliography

LITERATURE SEARCH

This literature search and its subsequent bibliography were designed to offer an overview of the many and varying current opinions on appropriate technology (AT). Because of the multifaceted approach taken in its compilation, the bibliography represents the important, obtainable literature found on appropriate technology. Twenty-two bibliographic data bases and many published bibliographies were searched; the terms "appropriate technology," "light-capital technology," "intermediate technology," were among the components of the search strategy rather than the broader term "technology transfer," as it was felt that a narrowing of scope was necessary to create a manageable and meaningful information base.
Additional steps in compiling the bibliography included contact with some of the larger international appropriate technology centers, visits to the U.S. offices of several international organizations involved in appropriate technology, searches on Library of Congress data bases, consultation with Congressional Research Services, and considerable input from colleagues. Comprehensiveness was at times limited by lack of availability of documents. The literature focus is on definitions of appropriate technology, the choice of technologies, and opinions of policy makers from both less developed and developed countries.
The bibliography includes annotations and key terms for each entry to assist the reader's choice of pertinent publications. These were prepared by reviewers along with an extensive content analysis of each publication. The content analysis facilitated in the preparation of the "synthesis of the literature," which is designed to give the reader a stronger base for case analysis. Since all documents in the bibliography deal with appropriate technology to some degree, appropriate technology is not used as a specific key term.

ENTRIES

Agmon, Tamir, and Charles P. Kindleberger, eds. Multinationals from small countries. Cambridge, MIT Press, 1977. 224 pp.

A conference at MIT in 1976 examined the development of multinational corporations from the smaller countries. The chapter on "The Internationalization of Firms from Developing Countries" by Louis T. Wells, Jr., explores the relationships among LDCs. His chapter, plus the appended comment by Stephen J. Kobrin, suggests that investment from developing nations into developing nations is a possible avenue for the transfer of appropriate technologies.

INDUSTRIAL DEVELOPMENT; MULTI-NATIONAL CORPORATIONS

Alonso, Marcelo. Technological development: concepts and actions. Approtech: Journal of the Association for the Advancement of Appropriate Technology for Developing Countries, v. 1, no. 1, Nov. 1978: 3-6.

The director of the Department of Scientific Affairs of the Organization of American States outlines certain principles that should be considered in detail in a development strategy. He also lists five ways in which the productive sector of a nation can be persuaded to place greater confidence in the scientific and technological capacities of that country.

SCIENCE AND TECHNOLOGY POLICY: INDUSTRIAL DEVELOPMENT

Anonymous. Small is still beautiful: recent thoughts from E.F. Schumacher. Futurist, v. 11, no. 2, 1977: 93-97.

A brief philosophical exploration of Schumacher's "culture of poverty" concept--wherein simplicity and self-reliance are stressed; satisfying human needs takes precedence over feeding endless, high-energy-consumption <u>wants</u>.

SELF-SUFFICIENCY; SOCIAL EFFECTS

Asian Development Bank. Appropriate technology and its application in the activities of the Asian Development Bank. n.p., Asian Development Bank, 1977. 43 pp.

After a careful analysis of the characteristics, relevance, and disagreements over appropriate technology, this work examines activities of the ADB that promote appropriate technology. Then enumerated are policy measures of the Bank that favor the adoption of appropriate technologies.

CRITERIA FOR APPROPRIATENESS; CHOICE OF TECHNOLOGY; RURAL DEVELOPMENT; ECONOMIC EFFECTS; DEVELOPMENT BANKS; INTERNATIONAL ORGANIZATIONS

Auciello, Gay Ellen, comp. Bibliography of intermediate technology publications held at the International Development Data Center. Atlanta, GA, Economic Development Laboratory, Engineering Experiment Station, Georgia Institute of Technology, 1976. 58 pp.

This bibliography is meant to be a starting point for intermediate technology research. The publications are arranged by broad subject categories, e.g., natural resources, industry profiles, management.

BIBLIOGRAPHY

Baron, C. Appropriate technology comes of age: a review of some recent literature and aid policy statements. International Labour Review, v. 117, no. 5, 1978: 625-634.

This critique of three recent policy articles covers all the main points of appropriate technology as an overview of trends. The conclusion emphasizes the relationship between technology (and appropriate technology) and energy resources as being crucial within the next decades.

DEFINITIONS

Behari, Bepin, ed. Appropriate technology for balanced regional development. Vol. I: General consideration. New Delhi, Appropriate Technology Cell, Ministry of Industrial Development, Government of India, 1974. 211 pp.

Publishes the discussions and conclusions of the Second National Seminar on Appropriate Technology in India in 1974. Its concentration is specific to India yet also encompasses the broad areas of (1) concept and methodology of appropriate technology; (2) technology in a Five-Year Plan; and (3) institutional arrangements for the introduction and development of appropriate technology.

CRITERIA FOR APPROPRIATENESS; CHOICE OF TECHNOLOGY; INDUSTRIAL DEVELOPMENT

Bhalla, A.S., and C.G. Baron. Appropriate technology, poverty and unemployment: the ILO project. Appropriate Technology, v. 1, no. 4, Winter 1976: 20-22.

A brief recounting to the ILO World Employment Programme's investigations into appropriate technology in various areas of development. Future action by the ILO is outlined.

EMPLOYMENT GENERATION; INTERNATIONAL ORGANIZATIONS

Block, Donn. Environmental aspects of economic growth in less-developed countries: an annotated bibliography. Development Centre of the OECD, n.d. 111 pp.

In this annotated bibliography dealing with economic growth, sections relevant to appropriate technology are:

"Development and Environment," "Economic and Social Objectives," and "Environment and International Trade."

RESEARCH AND DEVELOPMENT; ECONOMIC EFFECTS

Breach, Ian. Technology and the Third World: supplying the right kind of aid. The Financial Times (London), 14 April 1977: 25.

Presents the case for utilization of resources to make financial, environmental, and social sense. Two questions are asked: (1) will emphasizing appropriate technology to the Third World set back all the technological advances made? and (2) what do the developing countries have at stake in this appropriate technology development? Breach is of the opinion that the developed countries have a lesson to learn from appropriate technology application: to contribute advanced ideas rather than equipment.

CHOICE OF TECHNOLOGY; RESEARCH AND DEVELOPMENT; POLITICAL GOALS

Brown, Martin. Appropriate technology for industry in developing countries (with special reference to scale of production and employment creation): some reflections on research implications. (Occasional Paper No. 15). Paris, OECD, 1977. 28 pp.

Deals with the issues concerning appropriate technology adaptation in industry, focusing on scale of production and labor intensity. Attempts to direct policy thinking as to the complex nature of technology choices as underpinnings of research. He suggests international cooperation in a sector-by-sector approach, utilizing industrial firms and engineering design firms in both LDCs and DCs.

CRITERIA FOR APPROPRIATENESS; CHOICE OF TECHNOLOGY; INDUSTRIAL DEVELOPMENT; EMPLOYMENT GENERATION

Brown, R.H. Appropriate technology and grass roots--toward a development strategy from the bottom up. Developing Economies, v. 15, no. 3, Sept. 1977: 253-279.

An argument for a grass-roots approach to development in poor countries seeking to avoid either dependence on foreign technology and capital or dependence on totalitarian activism. Major approaches to and criteria for national development are presented, and dominant conservative strategy is criticized. Economic nationalism is presented as a viable alternative. A technology appropriate for specific rural mobilization is outlined.

RURAL DEVELOPMENT

Bruce, John I. Report of the United States delegation to the UNIDO International Forum on Appropriate Industrial Technology at New Delhi and Anand, India, 20-30 November 1978. 1 v.

Explains the background and purposes of the forum, the work of the conference, the conclusions, and the role of U.S. participation in the forum; gives a list of participants, introductory speech, and the U.S. statement at the forum.

INTERNATIONAL ORGANIZATIONS; SCIENCE AND TECHNOLOGY POLICY

Canadian Hunger Foundation and Brace Research Institute. Appropriate technology primer. Ekistics, v. 43, no. 259, June 1977: 361-367.

An adaptation of material from A Handbook on Appropriate Technology, this article outlines criteria for the selection of appropriate technologies. It also refutes several of the more frequently heard criticisms of the use of appropriate technology. Short studies of appropriate technologies introduced in the Dominican Republic, Bangladesh, and Afghanistan are also included.

CRITERIA FOR APPROPRIATENESS; CHOICE OF TECHNOLOGY

Carr, Marilyn. Economically appropriate technologies for developing countries: an annotated bibliography. London, Intermediate Technology Publications, 1976. 101 pp.

Intended use is by individuals and groups concerned with intermediate technology and the choice of appropriate technologies. The references are listed by subject categories under such general headings as: agriculture, low-cost housing and building materials, and manufacturing and are weighted toward application rather than theory. Entries are indexed by author, country, and subject.

ECONOMIC EFFECTS; BIBLIOGRAPHY

Choice and adaptation of technology in developing countries: an overview of major policy issues. Paris, Development Centre, OECD, 1974. 240 pp.

In 1972 some fifty people from various parts of the world, with varying backgrounds and orientations, met for three days to discuss both "technical aspects of 'appropriate technology' and the broader policy environments being challenged thereby." The second part contains eight participants' viewpoints of the major issues arising from the discussions. The third section consists of condensations of twenty-one background papers submitted to the group.

DEFINITIONS; CRITERIA FOR APPROPRIATENESS; CHOICE OF TECHNOLOGY; INDUSTRIAL DEVELOPMENT; ECONOMIC EFFECTS; MULTI-NATIONAL CORPORATIONS

Congdon, R.J., ed. Lectures on socially appropriate technology. Eindhoven, Technische Hogeschool, 1975. 235 pp.

A collection of twelve lectures delivered as part of a course to provide a technically detailed introduction to socially appropriate technology. Viewpoints based on each speaker's particular area of study and interest are presented. Special emphasis is placed on the potential of technology for the developing world.

CRITERIA FOR APPROPRIATENESS; CASE STUDIES; CHOICE OF TECHNOLOGY; RESEARCH INSTITUTIONS; DEVELOPED COUNTRIES; INDUSTRIALIZED COUNTRIES

Darling, H.S. Appropriate technology and Third World agriculture. Span, v. 18, no. 3, 1975: 119-121.

What is needed in LDCs is rural development; the best way to accomplish this is by increasing productivity without displacing workers. The criteria for selection of appropriate technologies are detailed and a brief introduction to the way The Intermediate Technology Development Group's work has helped develop the appropriate technology concept is given.

CRITERIA FOR APPROPRIATENESS; GOALS; RURAL DEVELOPMENT; SELF-SUFFICIENCY

Davis, John. Technology for a changing world, compiled by Roger England from a series of papers by John Davis. London, Intermediate Technology Publications, Ltd., 1978. 58 pp.

Deals with "appropriate technology" in modern industrial countries as these countries move from valuing the quantity of consumption to valuing the quality of life. Discusses the specifics of this broad objective for Great Britain.

CHOICE OF TECHNOLOGY

Development Coordination Committee. Development issues: U.S. actions affecting the development of low-income countries: the third annual report of the President, transmitted to the Congress April 1978. 130 pp.

This 1978 report of the U.S. President gives a detailed picture of White House views on overall development issues. Presents Executive positions on the current relationships among countries of the world in response to developmental questions. Focus is on maintaining balance in the North-South dialogue.

SCIENCE AND TECHNOLOGY; SOCIAL EFFECTS; POPULATION; ECONOMIC EFFECTS; POLITICAL RELATIONSHIPS

Dickson, David. The politics of alternative technology. New York, Universe Books, 1975. 224 pp.

Discusses the social function of technology and the socially accepted legitimations placed on this function. Puts forth the thesis that technology has a political role; asserts that modern technology is now a threat to man's survival and sanity and that it must be politically changed to orient toward more intermediate technologies.

DEFINITIONS

Diwan, Romesh, and Dennis Livingston. Development strategies and technological choices in developing countries: report submitted to the National Science Foundation. Troy, N.Y., Rensselaer Polytechnic Institute, 1978. 142 pp.

Examines conventional development strategies and alternative development strategies and determines that appropriate technology, without excluding other alternatives, should be the technical choice. The final third of the paper deals with U.S. policy options.

DEFINITIONS; CRITERIA FOR APPROPRIATENESS; CHOICE OF TECHNOLOGY; BARRIERS TO AT; RESEARCH AND DEVELOPMENT

Drissell, Jean Pope. Why the world will shift to intermediate technology: an interview with George McRobie. Futurist, v. 11, no. 2, 1977: 83-89.

A discussion of appropriate technology ideas, goals, processes, aiming for technology that is small, simple, capital-saving, and nonviolent; and of the inevitability of the trend as the world's resources dwindle and population increases.

AT DEVELOPMENT ORGANIZATIONS

Dunkerley, H.B. The choice of appropriate technologies. Finance and Development, v. 14, no. 3, 1977: 36-39.

Reflects the various factors considered by the World Bank (and other multilateral financing agencies) in order to lend money for development technology projects.

CHOICE OF TECHNOLOGY; DEVELOPMENT BANKS

Eckaus, Richard S. Appropriate technologies for developing countries. Washington, National Academy of Sciences, 1977. 140 pp.

The role of technology in development and the "character and consequences" of choices of technology are examined. Eckaus feels that understanding of the interaction between the developmental processes and technological choice is limited. Appended is a statement of exception with Eckaus's basic approach by Simón Teitel.

CRITERIA FOR APPROPRIATENESS; CHOICE OF TECHNOLOGY; ECONOMIC EFFECTS; CAPITAL STRUCTURE

Edwards, Edgar O., ed. Employment in developing nations: report on a Ford Foundation study. New York, Columbia University Press, 1974. 428 pp.

A selection of twenty papers prepared for the Ford Foundation's 1973 seminars to study less developed country employment problems. Groupings are: "Basic

Discussion Papers;" "Selected Papers on Generic Issues;" and "Some Sectoral Considerations."

CHOICE OF TECHNOLOGY; RURAL DEVELOPMENT; INCOME DISTRIBUTION; EMPLOYMENT GENERATION; GOVERNMENT

Ellis, William N. AT: the quiet revolution. Bulletin of the Atomic Scientists, v. 33, no. 9, Nov. 1977: 25-29.

The author feels that the appropriate technology revolution is the most important one since the Renaissance. He sets the background, gives examples of emerging technologies, recounts varying political views of appropriate technology, and briefs the reader on the present (1977) developments shaping the future. He concludes that the quiet revolution will continue at a grass-roots level whether or not it is supported nationally and/or internationally.

DEFINITIONS; INTERNATIONAL ORGANIZATIONS

Fox, T.H. Appropriate technology for developing countries. New York American Society of Mechanical Engineers, Paper No. 76-WA/TS-11, 1976. 4 pp.

Briefly explores the development of current interest in appropriate technology. Reviews definitions. For appropriate technologies stresses the need for support from institutions and structures with influence and credibility.

CHOICE OF TECHNOLOGY

French, David. Appropriate technology in social context: an annotated bibliography. Mt. Ranier, MD, VITA, 1977. 33 pp.

The author maintains that appropriate technology can be successful if attention is paid to locally based development institutions and to specific innovation and its effect on people in a given environment. Also considered relevant is the need for local participation in decision making.

BARRIERS TO AT; RURAL DEVELOPMENT; WOMEN IN DEVELOPMENT; AT DEVELOPMENT ORGANIZATIONS

Frost, Dennis H. Appropriate industrial technology: an integrated approach. Vienna, UNIDO, 1978. 30 pp. (Document ID/WG.279/6).

Describes the conceptual beginnings of appropriate technology and some of the main developments. Suggests that the success of appropriate technology in LDC's industries will depend upon their understanding and willingness to adopt appropriate technologies.

DEFINITIONS; GOALS; CRITERIA FOR APPROPRIATENESS; INDUSTRIAL DEVELOPMENT

Fuglesang, Andreas. Doing things...together: report on an experience in communicating appropriate technology. Uppsala, Dag Hammarskjold Foundation, 1977. 108 pp.

An approach to appropriate technology focusing on the hands-on communication skills involved in transmitting appropriate technology concepts and techniques. Also considers the cultural reactions of the people as significant and demonstrates convincingly that appropriate technology must be appropriate "mythically" in order to serve the people. The entire text is a case study (of Papua New Guinea).

CASE STUDIES; CULTURAL CONSTRAINTS-LOCAL TRADITIONS

Garg, M.K. Methodology and approach to development of an appropriate technology. Polytechnic Resource Letter (Allahabad), v. 9, no. 1, Jan. 1978: 22-26.

Proposes a means of making appropriate technology a meaningful instrument in the real world. Classifies the material available and the work being carried out by various appropriate technology agencies. Objectives of the Appropriate Technology Association of India are presented. A methodology to realize those objectives is presented in four categories: survey and analytical studies; pilot projects; working of integrated package plants; and the building of research facilities.

RESEARCH AND DEVELOPMENT; ENTREPRENEURSHIP; CRITERIA FOR APPROPRIATENESS

German Foundation for Developing Countries, Seminar Center for Economic and Social Development. Development and dissemination of appropriate technologies in rural areas. Berlin, German Foundation for Developing Countries, 1972. 211 pp.

Proceedings of a 1972 seminar in Kumasi, Ghana. Contains presentations by experts from both developing and developed countries and offers insight into the question of appropriate technology, not simply for developing countries in general, but for their rural areas, where socioeconomic life can be most affected.

RESEARCH AND DEVELOPMENT; DEVELOPED COUNTRIES

Ghai, D.P., A.R. Khan, E.L.H. Lee, and T. Alfthan. The basic-needs approach to development: some issues regarding concepts and methodology. Geneva, International Labour Office, 1977. 113 pp.

Essays that outline preliminary steps toward the quantifying of basic needs. Both conceptual and methodological concepts are discussed, including a survey of several international statements on basic needs and a "simple multisector material balance model" for use in development planning.

CHOICE OF TECHNOLOGY

Goldhoff, R.M. Appropriate technology: an approach to satisfying the technical needs of developing countries. New York, American Society of Mechanical Engineers, Paper No. 76-WA/TS-12, 1976. 4 pp.

Briefly describes the work of Volunteers in Technical Assistance (VITA). Gives examples of VITA's work in gathering and packaging information to assist in the solution of specific problems. Touches on the problems encountered in this type of assistance.

DEFINITIONS; AT DEVELOPMENT ORGANIZATIONS

Goulet, Denis. The paradox of technology transfer. Bulletin of the Atomic Scientists, v. 34, no. 6, June 1975: 39-46.

Technology is the major resource for development, but it also is the tool used by the DCs to "domesticate" Third World development. Goulet feels that choice of technology is done most effectively when based upon a realistic development strategy.

CHOICE OF TECHNOLOGY

Goulet, Denis. The suppliers and purchasers of technology: a conflict of interests. International Development Review, v. 18, no. 3, 1976: 14-20.

Analyzes the interests of transnational corporations in supplying technology to developing countries; outlines the benefits that developing countries hope to gain from the technologies acquired.

CHOICE OF TECHNOLOGY; MULTI-NATIONAL CORPORATIONS

Goulet, Denis. The uncertain promise: value conflicts in technology transfer. New York, IDOC/North America, 1977. 320 pp.

Technology has solved innumerable problems and at the same time destroyed many of the cultural values that societies need to achieve a "wisdom to match their sciences." The key to creating a world of genuine development lies in the "criteria chosen to decide which values will be destroyed and which will be preserved."

CRITERIA FOR APPROPRIATENESS; CHOICE OF TECHNOLOGY; SCIENCE AND TECHNOLOGY POLICY; INFRASTRUCTURE

International Council of Scientific Unions, Committee on Science and Technology in Developing Countries. UN Conference on Science and Technology for Development: suggestions for the preparation of national papers. Madras, Macmillan India Press, 1977. 52 pp.

The Committee on Science and Technology in Developing Countries of the International Council of Scientific Unions prepared these suggestions on behalf of UNESCO. The background to and the preparations for the 1979 UNCSTD are detailed. Procedural suggestions and examples are given, as well as recommendations for the critical examination of science institutions and the impact of science and technology on national development.

SCIENCE AND TECHNOLOGY POLICY; GROSS NATIONAL PRODUCT; INCOME DISTRIBUTION; INFRASTRUCTURE

International Labour Office. Employment, growth and basic needs: a one-world problem . . . prepared by the International Labour Office the decisions of the 1976 World Employment Conference. New York, Praeger, 1977. 224 pp.

Concerns national and international aspects of adoption by each country of a "basic-needs" approach. Suggests that increasing the volume and productivity of employment and taking the national and international measures of economic and social policy will bring this about.

CHOICE OF TECHNOLOGY; RESEARCH AND DEVELOPMENT; MULTI-NATIONAL CORPORATIONS

International Labour Office. Labour costs and appropriate technology. In UNIDO industrial development survey (special issue for the Second General Conference, United Nations). New York, UNIDO, 1975. 105-118.

Broad survey dealing with questions that have direct bearing on prospects for industrialization of developing countries. Has particular relevance to the problems and issues concerning the transfer of technology and its appropriateness.

CASE STUDIES; CHOICE OF TECHNOLOGY; SCIENCE AND TECHNOLOGY POLICY; ECONOMIC EFFECTS; EMPLOYMENT GENERATION; LABOR UTILIZATION

ILO Technical Meeting of Adaptation of Technology to Suit Special Market Conditions of Developing Countries. Policies and programmes of action to encourage the use of technologies appropriate to Asian conditions and priority needs. Geneva, ILO, 1975. 51 pp.

"The appropriate technology concept is of global application." This report includes guidelines for national, regional, and international programs to encourage the use of appropriate technology. A review of macro measures adopted in several Asian countries is appended.

LDCs AND AT; SCIENCE AND TECHNOLOGY POLICY; RESEARCH INSTITUTIONS; INTERNATIONAL ORGANIZATIONS

Jackson, Sarah. Economically appropriate technologies for developing countries: a survey. Occasional Paper No. 3. Washington, D.C., Overseas Development Council, 1972. 38 pp.

Survey of issues involved in changing technologies for underdeveloped countries. Emphasis is on economic effectiveness of different kinds of technologies.

ECONOMIC EFFECTS

Jedlicka, Allen D. Organization for rural development: risk taking and appropriate technology. New York, Praeger, 1977. 170 pp.

Deals with the importance of humanistic, organizational skills in establishing institutions for rural development in LDCs. The purpose of the book is to provide some understanding of "the role management plays in effective development of rural areas." The most wasted resource, according to the author, is "women," and he suggests the role that women might play in overcoming problems in rural development.

RURAL DEVELOPMENT; WOMEN IN DEVELOPMENT

Jedlicka, Allen. The transfer of technology to women: some issues to consider. Peace Corps Programs and Training Journal. v. 4, no. 6: 14-18.

Some of the issues affecting the transfer of appropriate technology to women are the effective use of extension service, governmental policy that restricts women's use of appropriate technology, cultural constraints, and the ultimate goals of development for women. Within each of these issues, Jedlicka discusses several subissues posed in the form of questions.

RURAL DEVELOPMENT; WOMEN IN DEVELOPMENT; CULTURAL CONSTRAINTS

Jenkins, Gareth. Nonagricultural choice of technique: an annotated bibliography of empirical studies. Oxford, The Institute of Commonwealth Studies, 1975. 84 pp.

Each entry in this bibliography contains substantive notes and makes for quick reference on subjects of technology. The introduction, by F. Stewart, is exhaustive in its treatment of the subject. The last section is dedicated entirely to citations of studies of appropriate technological innovations.

CHOICE OF TECHNOLOGY

Jéquier, Nicolas, ed. Appropriate technology: problems and promises. Development Centre Studies. Paris, Organization for Economic Cooperation and Development, 1976. 344 pp.

Contains case studies of projects in specific developing countries. Presents appropriate technology issues that need resolution, including: which technology is to be defined "appropriate;" which criteria should be used to judge the value of projects; to whom programs should be directed (the very poor or the more wealthy innovators); whether new ideas should be from a less developed country or outside experts: and whether the basis of progress should be cooperation or individualism. Jéquier contends that local government agencies or private voluntary organizations should assume major responsibility for developing the "softwares" of the appropriate technology system.

CRITERIA FOR APPROPRIATENESS; CASE STUDIES; CHOICE OF TECHNOLOGY; BARRIERS TO AT; INNOVATION; ENTREPRENEURSHIP; RURAL DEVELOPMENT; ECONOMIC EFFECTS; RESEARCH INSTITUTIONS; AT DEVELOPMENT ORGANIZATIONS

Jéquier, Nicolas. Intermediate technology: a new approach to development problems. OECD Observer, v. 75, May-June 1975: 26-28.

This summary of an OECD seminar on intermediate technology notes main lines of national policies to promote intermediate technology. Contains a list of major intermediate technology centers.

BARRIERS TO AT; AT DEVELOPMENT ORGANIZATIONS

Jéquier, Nicolas. Low-cost technology: an inquiry into outstanding policy issues. World Tech Report No. 2. Minneapolis, Control Data Corporation, 1975. 48 pp.

Explores the complexities of political, sociological, economic, scientific, and technological effects of appropriate technology. Focuses on the history of technological change and emphasizes the importance of diversified/innovative solutions on a localized scale, including primary education as an appropriate technology training ground.

CHOICE OF TECHNOLOGY; SCIENCE AND TECHNOLOGY POLICY

Jéquier, Nicolas. Science policy in the developing countries: the role of the multinational firm. In The gap between rich and poor nations: proceedings of a conference held by the International Economic Association at Bled, Yugoslavia. London, MacMillan, 1972: 336-347.

Discusses problems of science policy in developing nations, e.g., the absence of a way to measure the effectiveness of a science policy. Analyzes the capabilities of multinational firms and ways in which, through national scientific and technical policy, MNCs can be induced to assist the LDCs in attaining long-term goals.

SCIENCE AND TECHNOLOGY POLICY; RESEARCH AND DEVELOPMENT; INDUSTRIAL DEVELOPMENT; GOVERNMENT; MULTINATIONAL CORPORATIONS

Johnston, Peter. Appropriate technologies for small developing countries. Brighton, England, Smoothie Publications, 1974. 38 pp.

Lists areas of specific concern from the aspect of research and development and economic analysis; cites numerous authors germane to appropriate technology development; and suggests directions in which policy decisions should be aimed. Self-determination and decentralization are stressed. Provides a long list of appropriate technology organizations.

RESEARCH AND DEVELOPMENT; SCIENCE AND TECHNOLOGY POLICY

Johnston, S.F. Intermediate technology - appropriate design for developing countries. Search, v. 7, no. 1-2, 1976: 27-33.

A brief review of the tenets of appropriate technology, with numerous examples of appropriate technology at work. The author defines development as including the satisfaction of basic human needs, and concludes that the political structure of developing countries may need to be changed before such needs can be satisfied.

DEFINITIONS; SOCIAL EFFECTS

Khan, Amir U. Appropriate technologies: do we transfer, adapt or develop? In Employment in developing nations: report on a Ford Foundation study. Edwards, Edgar O., ed. New York, Columbia University Press, 1974. 223-233.

Because of the need to fill the serious gaps on the technology shelf, appropriate technology development centers should be established by international agencies. Products that need to be developed in the areas of food, shelter, transportation, and sanitation are identified.

CASE STUDIES; AT DEVELOPMENT ORGANIZATIONS; INTERNATIONAL ORGANIZATIONS

Levin, Sander. Statement before the Subcommittee on Domestic and International Scientific Planning, Analysis, and Cooperation. Committee on Science and Technology, U.S. House of Representatives. July 27, 1978.

Defines appropriate technology and demonstrates that the concept is applicable to the AID strategy for assistance to the developing nations. Reviews the agency's involvement in appropriate technology transfer.

CRITERIA FOR APPROPRIATENESS; DEVELOPED COUNTRIES

Lewis, Jordan. Statement before the Subcommittee on Domestic and International Scientific Planning, Analysis, and Cooperation. Committee on Science and Technology, U.S. House of Representatives. July 26, 1978.

Lewis, the executive director of Appropriate Technology International, shares his perspectives on the first year of ATI's operation. Explains ATI's involvement in the extension and expansion of small-scale industry capability and interest in strengthening village-level development capabilities and "microbusiness."

INFRASTRUCTURE; DEVELOPED COUNTRIES; AT DEVELOPMENT ORGANIZATIONS

Livingston, Dennis. Little science policy: the study of appropriate technology and decentralization. Policy Studies Journal, v. 5, no. 2, Winter 1976: 185-192.

An exposition of an orientation toward science and technology policy that seeks a decentralized, growth-controlled society, this article defines 'little science policy' as being concerned with appropriate technologies. Specific reference to developing nations is minimal.

CRITERIA FOR APPROPRIATENESS; GOALS; SCIENCE AND TECHNOLOGY POLICY; INFRASTRUCTURE; DEVELOPED COUNTRIES

Lodge, Donald E., and Kay Ellen Auciollo, eds. Proceedings of the conference and seminar on techniques and methodologies for stimulating small-scale, labor-intensive industries in developing countries: prepared for the Agency for International Development. Atlanta, Georgia Institute of Technology, 1975. 314 pp. (NTIS Number PB-248-338).

The speakers in this conference, from government and private organizations, presented case histories of their experiences. The audience consisted of public- and private-sector officials and educational institution administrators from developing countries, representatives from other organizations interested in industrialization, and U.S. representatives.

CASE STUDIES; INDUSTRIAL DEVELOPMENT; EMPLOYMENT GENERATION; LABOR UTILIZATION; INFRASTRUCTURE

Long, Clarence D. Definition of light capital technology, Feb. 15, 1978. (mimeo) 4 pp.

Congressman Long's definition of light capital technology and a list of amendments concerning light capital technology that have been enacted by Congress.

DEFINITIONS

Long, Clarence D. Helping the world's poor: a new approach to foreign aid. Midwest Quarterly, v. 18, Oct. 1976: 27-35.

Ninety million dollars in foreign aid has not served the world's poor and cannot be maintained at present levels of capital expenditure. Due to looming overpopulation, the arms race, and energy shortages, appropriate technology must be given a try as soon as possible.

ECONOMIC EFFECTS; DEVELOPED COUNTRIES

Love, Sam. We must make things smaller and simpler: an interview with E. F. Schumacher. Futurist, v. 8, no. 6, Dec. 1974: 281-284.

E. F. Schumacher discusses his philosophy of intermediate technology, which serves to counter the threefold crisis that he sees in the world. He cites four trends in technology that are contributing to this crisis; giantism, complexity, capital intensity, and violence.

DEFINITIONS

McDowell, Jim, ed. Village technology in Eastern Africa: a report on a UNICEF-sponsored regional seminar on "Simple Technology for the Rural Family" held in Nairobi 14-19 June 1976. Nairobi, UNICEF.

Deals specifically with dissemination of information on appropriate technology to the rural or village populations. Contains basic guidelines on applications of appropriate technology as derived from the seminar and step-by-step discussion of these guidelines.

CRITERIA FOR APPROPRIATENESS; RURAL DEVELOPMENT

McInerney, John P. "Appropriate" technology: no miracle panacea for economic development. Report (World Bank), May-June 1977.

The focus on appropriate technology is just another "technological fix" and is not the answer for development. What is needed is a responsive technology development and adaptation capability.

CHOICE OF TECHNOLOGY; RESEARCH AND DEVELOPMENT

McRobie, George. Technologies for rural industrialization in developing countries. In The role of group action in the industrialization of rural areas. Klatzman, Joseph, Benjamin Y. Ilan, and Yair Levi, eds. New York, Praeger, 1971: 65-73.

Assuming that technology in most developing nations must be adapted to a rural environment, the author stresses the significance of forming research and development groups to assist the nations in making rational, economical choices.

CHOICE OF TECHNOLOGY; RESEARCH INSTITUTIONS

McRobie, George. Technology appropriate for development. NSDB Technology Journal (Manila), v. 2, no. 1, Jan.-Mar. 1977: 4-11.

General discussion of the gap in conventional aid and development efforts of rich countries toward poor countries. Defines criteria for appropriate technologies and gives concrete examples of how to arrive at them by upgrading traditional methods, scaling down and redesigning high-cost technologies, and designing new products.

CHOICE OF TECHNOLOGY; INFRASTRUCTURE

McRobie, George. Technology for development--"Small is Beautiful." Journal of the Royal Society of Arts, v. 122, March 1974: 214-224.

A synopsis of a lecture in which George McRobie of the International Technology Development Group (ITDG) outlines and discusses the appropriate technology philosophy, examples of practical applications and global ramifications of appropriate technology research and development, implications for rural development, and effects of foreign aid on technological choice.

RESEARCH AND DEVELOPMENT; DEVELOPMENT ORGANIZATIONS; CHOICE OF TECHNOLOGY

McRobie, George. Technology - the critical choice for developing countries: the work of the Intermediate Technology Development Group. Vienna, UNIDO, 1976. 21 pp. (Document ID/WG.233/12).

Describes the origin, structure, and activities of the ITDG and details their efforts to collect and disseminate information; involvement with the U.N. and its role in the international network of appropriate technology organizations.

AT DEVELOPMENT ORGANIZATIONS; INFRASTRUCTURE

Marsden, Keith. Progressive technologies for developing countries. International Labour Review, v. 101, May 1970: 475-502.

The appropriateness of directly transferred technology to the developing countries is examined, along with suggestions of the criteria to be used in selecting alternative technologies. Introduces the concept of "progressive technologies" and gives case study citations. The latter part deals specifically with considerations of possible types and sources of progressive technology and with examinations of policy options open to governments to facilitate selection, installation, and efficient utilization of this technology.

CRITERIA FOR APPROPRIATENESS; CASE STUDIES

Mattis, Ann. Science and technology for self-reliant development. IFDA (International Foundation for Development Alternatives) Dossier 4, Feb. 1979. 16 pp.

Citizens should be in control of science and technology for development to occur. This paper studies issues involved in creating science and technology policies for development in the light of the proposed debate at UNCSTD.

DEFINITIONS; CHOICE OF TECHNOLOGY; SCIENCE AND TECHNOLOGY POLICY

Morehouse, Cynthia T., comp. Science and technology for development: international conflict and cooperation: a bibliography of studies and documents related to the 1979 UNCSTD. Lund, Sweden, University of Lund, 1978. 3 vol.

Provides continuing bibliographical coverage of preparations for the U.N. Conference on Science and Technology. Volume 3 gives a subject index, covering approximately sixty-seven subject and geographical areas. Most literature on science, technology, and development of individual countries has been excluded.

BIBLIOGRAPHY

Morehouse, Ward, and Jon Sigurdson. Science, technology and poverty: issues underlying the 1979 U.N. Conference on Science and Technology for Development. Bulletin of the Atomic Scientists, v. 30, no. 10, Dec. 1977: 21-28.

As a means of stimulating debate on issues underlying the 1979 UNCSTD, the authors offer three proposals. They believe that developing countries should go through a period of "measured technological disengagement" as a means of strengthening the capacity for autonomous development.

RESEARCH AND DEVELOPMENT; MULTINATIONAL CORPORATIONS; INTERNATIONAL ORGANIZATIONS; DEVELOPED COUNTRIES

Morrison, Denton E. Energy, appropriate technology and international interdependence. "Prepared for the Annual Meetings of the Society for the Study of Social Problems, San Francisco, Sept. 1978." 60 pp.

A sociological approach to examining the emerging "soft energy path" in its relationship to the broader "soft technology" (or appropriate technology), and the way in which these notions attempt to address certain problems of interdependence. Describes appropriate technology as a social movement and looks at the possibilities and barriers to appropriate technology addressing the issue of international development.

SOCIAL EFFECTS

Morss, Elliott R., et al. Strategies for small farmer development: an empirical study of rural development projects in The Gambia, Ghana, Kenya, Lesotho, Nigeria, Bolivia, Colombia, Mexico, Paraguay and Peru: Volume I. Boulder, Westview Press, 1976. 347 pp.

"The primary purpose of this report is to provide information on what can be done to increase the wellbeing and productivity of the small farmer in the Third World. Particular attention is given to what, operationally, AID might contribute to the attainment of these objectives; this requires a focus on strategies for rural development, in terms of both project design and implementation." (page 2)

RURAL DEVELOPMENT; CULTURAL CONSTRAINTS; LOCAL TRADITIONS; INCOME DISTRIBUTION; CAPITAL STRUCTURE; DEVELOPED COUNTRIES

Morss, Elliott R., et al. Strategies for small farmer development: an empirical study of rural development projects in The Gambia, Ghana, Kenya, Lesotho, Nigeria, Bolivia, Colombia, Mexico, Paraguay and Peru. Vol. II: Case studies. Boulder, Colo., Westview Press, 1976. 444 pp.

Prepared for AID, this volume presents numerous case studies of strategies for small-scale rural development.

CASE STUDIES; RURAL DEVELOPMENT

NBS/AID UNCSTD seminar: The technological knowledge base for industrializing countries. Washington D.C., National Bureau of Standards, 1978. Draft, 13 pp.

Excerpts from NBS/AID UNCSTD seminar reports covering metrology, standardization, quality control, technological transfer, and technical management. Participants from a variety of national and international organizations and agencies.

CHOICE OF TECHNOLOGY; INFRASTRUCTURE

National Research Council. U.S. science and technology for development: a contribution to the 1979 U.S. conference: background study on suggested U.S. initiatives for the UNCSTD, Vienna, 1979. Washington, D.C., Department of State, 1978. 212 pp.

By subject area, this work sets forth numerous policy suggestions made by national experts. A timely representation of U.S. attitudes that will be presented at the conference, dealing mainly with specific development needs rather than the choice of technology.

SCIENCE AND TECHNOLOGY POLICY; INDUSTRIAL DEVELOPMENT

Norman, Colin. Soft technologies, hard choices. Worldwatch paper no. 21. Washington, Worldwatch Institute, 1978. 48 pp.

Emphasizes the author's four main concerns for technological choice--employment, equity, energy, and ecology.

CHOICE OF TECHNOLOGY

Nwosu, Emmanuel J. Some problems of "appropriate" technology and technological transfer. Developing Economies, v. 13, no. 1, 1975: 82-93.

Appropriate technology in LDCs should mean <u>either</u> intermediate or advanced technology, depending upon the natural resources of the country, the economic needs of its work force, and the quality of research and development organizations. Economic progress can only follow on the heels of scientific and technological progress. Therefore, technological education and research agencies should be given priority support by MNCs and DCs rather than the "protectionist and extortionist trade policies" that have dominated technological cooperation.

CRITERIA FOR APPROPRIATENESS; RESEARCH AND DEVELOPMENT; INDUSTRIAL DEVELOPMENT; ECONOMIC EFFECTS

O'Kelly, Elizabeth. Appropriate technology for women of the developing countries. Peace Corps Program and Training Journal, v. 4, no. 6: 10-13.

Cultural constraints and local tradition that determine the division of labor in developing countries must always be taken into account in a development program. Rural women's organizations are an effective means of getting women to participate in rural development programs. In order to raise the standard of living for the community, the women's potential must be used.

WOMEN IN DEVELOPMENT; CULTURAL CONSTRAINTS

O'Kelly, Elizabeth. Intermediate technology as an agent of change: simple technologies, high employment. In Proceedings: The World Food Conference; 1976, 505-513.

Presents appropriate technology as an effective change agent for rural areas' quality of life. Cites several examples of inappropriate technology and suggests that the role of women as social change agents has been traditionally underrated in planning appropriate technology programs.

WOMEN IN DEVELOPMENT

Owens, Edgar. The debate about development and appropriate technology, statement before the House Subcommittee on Domestic and International Scientific Planning, Analysis and Cooperation, Committee on Science and Technology. 25 July 1978.

The development officer of Appropriate Technology International states that technologies appropriate to the circumstances of developing nations must be used if they are ever to get off the "international welfare system" and become self-reliant. He discusses three development strategies, "trickle-down GNP," "trickle-down basic needs," and appropriate technology.

TECHNOLOGICAL CHOICE; AT DEVELOPMENT ORGANIZATIONS; SELF-SUFFICIENCY

PASITAM. Private voluntary organizations and appropriate technology. In Proposal for a program in appropriate technology, rev. ed. U.S. Agency for International Development. Washington, D.C., U.S. Govt. Print. Off. 1977: 78-113.

Briefly examines the role of private voluntary organizations (PVO) and their part in developing appropriate technology, pointing out how institutions and aid-giving agencies can assist the PVOs to best improve the use of appropriate technology.

CRITERIA FOR APPROPRIATENESS; RESEARCH AND DEVELOPMENT; AT DEVELOPMENT ORGANIZATIONS

Pack, Howard. Policies to encourage the use of appropriate technology. In Proposal for a program in appropriate technology, rev. ed. U.S. Agency for International Development. Washington, D.C. U.S. Govt. Print. Off., 1977: 192-248.

Indicates policies that an "LDC government may choose to alter the relative cost of using the various factors of production." How much is empirically known about the efficiency of such changes is revealed. Considers how aid to less developed country governments could best be pursued.

LDCs AND AT; BARRIERS TO AT

Pickett, James, D.J.C. Forsyth, and N.S. McBain. The choice of technology, economic efficiency and employment in developing countries, World Development, v. 2, no. 3, March 1974: 47-54.

A fairly technical analysis of the choice of technology for sugar and footwear industries in Ghana and Ethiopia. Examines the efficiency and employment potential of Third World technologies in economic terms.

TECHNOLOGICAL CHOICE; EMPLOYMENT GENERATION

Program on Policies for Science and Technology in Developing Nations. Science and technology for international development: a selected list of information sources in the United States and bibliography of selected materials. 2nd. ed. Ithaca, Cornell University, 1975. 117 pp.

The first section lists information sources. Full contact information, the collection's strength, publications, and information services are detailed. The second section is a forty-eight page bibliography, arranged by subject areas.

SCIENCE AND TECHNOLOGY POLICY; INDUSTRIAL DEVELOPMENT; INFRASTRUCTURE

Rahman, Anisur. All India convention of people's science movements. IFDA Dossier 4, Feb. 1979. 10 pp.

Intimate interaction between the intelligensia and the common people in India as a means of raising mass consciousness to make a social revolution was the subject of this convention in November 1978.

RURAL DEVELOPMENT; CULTURAL CONSTRAINTS--LOCAL TRADITIONS; INFRASTRUCTURE

Ranis, G. Appropriate Technology in the context of the redirection of lesser developed country industrial development strategy: concepts and policies. Vienna, UNIDO, 1978. 45 pp. (Document ID/WG.279/2).

The ability of less developed countries to develop industries on a decentralized pattern seems to lead to overall growth with equity. Appropriate technology of itself is not the solution--rather, selecting and adapting appropriate processes and specifications is the key. Governments need to assist this operation, not hinder it.

CHOICE OF TECHNOLOGY; BARRIERS TO AT; RURAL DEVELOPMENT; INDUSTRIAL DEVELOPMENT

Rosenblatt, Samuel M., ed. Technology and economic development: realistic perspective. Boulder, Colo., Westview Press, 1979. 208 pp.

The premise of this book is that a gap exists between the LDC's desire for modern technologies and the DC's newly acquired preference for appropriate technology. It suggests a compromise utilizing modern technology as it already exists to convey the more appropriate technologies.

DEFINITIONS; CHOICE OF TECHNOLOGY; BARRIERS TO AT; IMPORT SUBSTITUTION; EMPLOYMENT GENERATION; MULTINATIONAL CORPORATIONS

Schacht, Wendy H. Appropriate technology: technology for growth in the developing nations. Washington, Congressional Research Service, 1979. 14 pp. (Issue Brief number IB77092).

Prepared for the U.S. legislature, this briefly states the issue, and defines problems involved in the adoption of appropriate technology by developing countries. Traces the recent U.S. involvement in facilitating appropriate technology utilization in developing countries.

BARRIERS TO AT; DEVELOPED COUNTRIES

Schacht, Wendy H. International technology transfer to the developing nations and the role of appropriate technology: a background paper. Washington, Congressional Research Service, 1977. 23 pp.

General information on concepts of and mechanisms for technology transfer and the difficulties in application to LDCs. An alternative approach of appropriate technology is discussed, developing both the limitations and advantages of appropriate technology.

BARRIERS TO AT

Schlie, Theodore W. Appropriate technology: some concepts, some ideas, and some recent experiences in Africa. Eastern Africa Journal of Rural Development, v. 7, nos. 1-2, 1974: 77-108.

Addresses some of the complexities involved in economic/environmental technology decisions. Discusses several appropriate technology case studies and makes observations concerning future implications of technology exchange.

CASE STUDIES; ECONOMIC EFFECTS

Schroeder, Dennis. In search of an "appropriate" technology. International Development Review (Focus), no. 3, 1976: 3-7.

Reviews the development of the concept of appropriate technology and the arguments against it. Advocates a new developmental approach that would unite modern research and development with traditional experience.

DEFINITIONS; CHOICE OF TECHNOLOGY; RESEARCH AND DEVELOPMENT

Schumacher, Ernst F. Intermediate technology. Center Magazine, v. 8, Jan.-Feb. 1975: 43-49.

Discusses how organizations dealing with intermediate technology can be set up and managed. Gives specific examples demonstrating the success of cooperative efforts among the industries and universities of the LDCs.

CASE STUDIES; CHOICE OF TECHNOLOGY; INNOVATION; COOPERATIVES

Schumacher, E.F. Small is beautiful: a study of economics as if people mattered. London, Blond and Briggs, 1973. 305 pp.

Contains a chapter explaining the rationale of intermediate technology. Schumacher is impatient with the emphasis placed on the big, the new, and the modern. He submits that economics may be as appreciatively viewed from Gandhi and Buddha as from the point of view of western economists. In its entirety, the highlighting of the human element in economics and technology gives a new perspective.

ECONOMIC EFFECTS; CHOICE OF TECHNOLOGY; SOCIAL EFFECTS

Schumacher, E.F. The work of the Intermediate Technology Development Group in Africa. International Labour Review, v. 106, no. 1, July 1972: 75-92.

An account of ITDG's work program in Africa, which is intended to serve as a model for larger-scale implementation of intermediate technology.

RESEARCH AND DEVELOPMENT; EMPLOYMENT GENERATION

Singer, H.W. Appropriate technology for a basic human needs strategy. International Development Review, v. 19, no. 2, 1977: 8-11.

An assessment of present world situation--economic, technological, resources; discussing the relationship of appropriate technology and human rights to world finance and nuclear proliferation, suggesting a wide range of policy directions.

GOALS; CHOICE OF TECHNOLOGY; INTERNATIONAL ORGANIZATIONS

Singer, H.W. Technologies for basic needs. Geneva, International Labour Office, 1977. 158 pp.

Suggests new criteria for "establishing socially oriented technology policies in the developing economy." Aims for a reasonable balance between capital-intensive and labor-intensive techniques. Expands upon ILO future plans for making technology more responsive to the basic needs of all population sectors.

CHOICE OF TECHNOLOGY; SCIENCE AND TECHNOLOGY POLICY

Snyder, A.E. The transfer of intermediate technology: a challenge to the multinational company. New York, American Society of Mechanical Engineering, Paper No. 76-DET-81, 1976. 8 pp.

A policy that would allow MNCs to operate in a less-developed country in exchange for aid in intermediate technology programs is advocated. An international planning agency for studying such programs also is proposed.

CRITERIA FOR APPROPRIATENESS; RURAL DEVELOPMENT; INDUSTRIAL DEVELOPMENT; MULTINATIONAL CORPORATIONS

Standke, Klaus-Heinrich. Utilization of new technologies in developing countries. In Materials and society, v. 1. London, Pergamon Press, 1977: 45-49.

Presents the problems of choice of technology for developing countries by systematically defining developing countries and technology. Examines the two main schools of thought regarding choice of technology.

CRITERIA FOR APPROPRIATENESS

Steffens, J. Development and technology transfer. New York, American Society of Mechanical Engineers, Paper No. 76-WA/TS-15, 1976. 7 pp.

Three examples of technical aid that was not successful due to inappropriateness are given. Discusses agricultural development, health and housing, natural resources, and industrial growth and seeks solutions to these development problems.

CASE STUDIES; CHOICE OF TECHNOLOGY

Steinberg, Robert M. Devising a system for appropriate technology. Report (World Bank), July-August 1977.

Designed as a refutation of John McInerney's remarks in a previous issue of the World Bank's *Report*, this article argues that appropriate technology should be a conscious part of LDCs' development strategies. Advocates an international system to identify the options on the technology shelf.

CHOICE OF TECHNOLOGY

Stewart, Frances. Technology and underdevelopment. Boulder, Colo., Westview Press, 1977. 303 pp.

Explores the impact of technology on development, considering connections between poverty and maldistribution of income in poor countries and the condition of technological dependence.

CHOICE OF TECHNOLOGY; EMPLOYMENT GENERATION

Timmer, C. Peter, et al. The choice of technology in developing countries: some cautionary tales. Harvard studies in international affairs no. 32. Cambridge, Harvard University Press, 1975. 114 pp.

Four specialists focus on unemployment in developing countries. Each paper shows why "intermediate technology using less capital and more labor" may be a partial answer to unemployment.

CHOICE OF TECHNOLOGY; BARRIERS TO AT; INDUSTRIAL DEVELOPMENT; IMPORT SUBSTITUTION; EXPORT GENERATION; EMPLOYMENT GENERATION

United Nations. Advisory Committee on the Application of Science and Technology to Development. Appropriate technology and research for industrial development. Document No. UN-ST/ECA/152. New York, 1972. 51 pp.

Analyzes the problems of technological choice, of taking action toward applying appropriate technology, and of improving product and plant design. Presents problems in providing industrial research services and makes recommendations for allaying such problems.

CHOICE OF TECHNOLOGY; RESEARCH AND DEVELOPMENT; INDUSTRIAL DEVELOPMENT; INTERNATIONAL ORGANIZATIONS

United Nations. Advisory Committee on the Application of Science and Technology to Development for the Second United Nations Development Decade. World plan of action for the application of science and technology to development. Document No. E/4962/Rev 1. New York, 1971. 286 pp.

Identifies priorities in development problems; gives detailed proposals concerning the development of indigenous science and technology research institutions.

SCIENCE AND TECHNOLOGY POLICY; RURAL DEVELOPMENT; INDUSTRIAL DEVELOPMENT; RESEARCH INSTITUTIONS; INTERNATIONAL ORGANIZATIONS; DEVELOPED COUNTRIES

United Nations. Conference on Technical Cooperation Among Developing Countries for Joint Economic Growth, Effective North-South Dialogue, and More Self-reliant Development. Case studies on TCDC. Folder, 1978.

Contains twenty-five case studies of "actual and potential collaboration among developing countries" that were identified by the United Nations Development Programme during 1976-1977. This compilation was prepared for the U.N. Conference on Technical Cooperation among Developing Countries in Buenos Aires, 1978.

CASE STUDIES; RURAL DEVELOPMENT; INTERNATIONAL ORGANIZATIONS

United Nations. Conference on Trade and Development. Major issues arising from the transfer of technology to developing countries. Document No. TD/B/AC.11/10/Rev. 1. Geneva, 1974. 81 pp.

Data gathered from technology-receiving and technology-supplying nations defined the major issues in the transfer of technology to developing countries. Action at the national and international levels is proposed.

INDUSTRIAL DEVELOPMENT; GOVERNMENT; INDUSTRIALIZED COUNTRIES

United Nations. Conference on Trade and Development. Transfer of technology: action to strengthen the technological capacity of developing countries: policies and institutions. Document No. TD/190/Supp 1. 1976. 44 pp.

A synthesis of two UNCTAD progress reports on the establishment of technology transfer and development centres, analyzes the functions of such centres, especially in relation to a nation's stage of development. The problem of the need for information systems to assist in technology transfer and responses to this problem is discussed.

SCIENCE AND TECHNOLOGY POLICY; RESEARCH AND DEVELOPMENT; INFRASTRUCTURE; INTERNATIONAL ORGANIZATIONS

United Nations. Department of Economic and Social Affairs. The impact of multinational corporations on development and on international relations. New York, United Nations, 1974. 162 pp.

Discussion of the role that multinational corporations play in development. Supports recommendations that host countries evaluate suitability of products, require multinationals to contribute toward innovations and research, and explore alternative ways of importing technology.

CHOICE OF TECHNOLOGY; MULTINATIONAL CORPORATIONS

United Nations. Economic and Social Commission for Asia and the Pacific. Guidelines for development of industrial technology in Asia and the Pacific. (Document E/CN.11/1273). Bangkok, United Nations Center, 1976. 225 pp.

Proposes measures that governments of developing countries could consider in the development of technology. Expresses the need to adopt national science and technology policies. Based on experiences in India and the Republic of Korea, techniques for science and technology planning are suggested and assessed.

CHOICE OF TECHNOLOGY; SCIENCE AND TECHNOLOGY POLICY; RESEARCH AND DEVELOPMENT

United Nations. General Assembly. Preparatory Committee for the United Nations Conference on Science and Technology for Development. Preliminary draft programme of action: note by the secretary general of the conference. Document No. A/CONF. 81/PC.21. New York, 1979. 52 pp.

Contains a "theoretical and conceptual framework and recommendations for concrete measures for action at the national, subregional, regional, interregional and international levels" covering the six "target areas." These target areas are "(1) the sharing of knowledge and experience by all members of the international community; (2) increasing the capability of policy-making in science and technology in the framework of general development planning; (3) transfer of technology for the benefit of development; (4) enhancing endogenous capabilities in a context of national self-reliance; (5) promoting collective self-reliance through cooperation among developing countries; and (6) strengthening the role of the UN in the field of science and technology cooperation.

SCIENCE AND TECHNOLOGY POLICY; RESEARCH AND DEVELOPMENT; SELF-SUFFICIENCY; INFRASTRUCTURE; DEVELOPED COUNTRIES

United Nations Industrial Development Organization. Cooperative programme of action of appropriate industrial technology: report by the executive director. Document No. ID/B/188. Vienna, UNIDO, 1977. 28 pp.

After an introductory discussion of concepts and approaches, this paper develops specific programs for consideration, including: evaluation and comparison of alternative industrial technologies; promotion of research; data collection; application for rural development; alternate energy sources; national and international policies for appropriate technology; institutional infrastructure; and training.

RURAL DEVELOPMENT; INDUSTRIAL DEVELOPMENT; INTERNATIONAL ORGANIZATIONS

United Nations Industrial Development Organization. International Forum on Appropriate Industrial Technology, New Delhi/Anand, India 20-30 November 1978. Draft report of the Technical/Official Level Meeting to the Ministerial Level Meeting. 22 pp.

A summary of the principal considerations and conclusions reached at the technical/official level meeting.

SCIENCE AND TECHNOLOGY POLICY; INDUSTRIAL DEVELOPMENT

United Nations Industrial Development Organization. International Forum on Appropriate Technology. Anand, India, 28-30 November 1978. Report of the Ministerial-level Meeting. 29 pp.

Deals with strategies for appropriate technology and industrial development, governmental policies and technologies in developing countries, and measures for international cooperation. A program of action for implementation of the plan is recommended. Special emphasis is given to the role to be played by UNIDO.

CHOICE OF TECHNOLOGY; RESEARCH AND DEVELOPMENT

United Nations Industrial Development Organization. International Forum on Appropriate Industrial Technology. Report of the Technical/Official Level Meeting to the Ministerial Level Meeting: Part Two: Summary of the twelve industrial sectoral working group reports on appropriate industrial technology held at New Delhi, India, 20-24 November 1978. UNIDO. 87 pp.

The UNIDO report of the twelve working groups that examined specific sectors of industry (1978). The sectors investigated reflect the needs of the poorer sectors, those sectors that contribute to better use of natural resources, and those sectors that upgrade skills and manufacture of basic inputs. Each group presents its findings in three parts: Summary, Policy Aspects, and Programme of Action. A large number of DCs and LDCs were able to examine the information together, and make important conclusions.

INTERNATIONAL ORGANIZATIONS; SCIENCE AND TECHNOLOGY POLICY; CHOICE OF TECHNOLOGY

United Nations Industrial Development Organization. International Forum on Appropriate Industrial Technology, New Delhi/Anand, India, 20-30 November 1978. Working Group on Conceptual and Policy Framework for Appropriate Industrial Technology. An approach to the development of appropriate technology: background paper, by the Appropriate Technology Development Association, Lucknow. Document ID/WG.282/10. 8 pp.

The technologist's answer to how to reduce the scale of present-day technology while maintaining quality and not raising costs is a pilot project approach. Stages and steps in the process, as practiced by ATDA, are examined.

CRITERIA FOR APPROPRIATENESS; CHOICE OF TECHNOLOGY; AT DEVELOPMENT ORGANIZATIONS

United Nations Industrial Development Organization. International Forum on Appropriate Industrial Technology, New Delhi/Anand, India, 20-30 November 1978. Working Group on Conceptual and Policy Framework for Appropriate Industrial Technology. Conceptual and policy framework for appropriate industrial technology in developing countries: discussion paper, prepared by the secretariat of UNIDO. 31 pp.

Partially based on the recommendations of the first and second consultative meetings on appropriate technology convened by UNIDO. Stresses the need for mechanisms to gather, analyze, and disseminate information concerning technological alternatives. The possibility of an International Center on Appropriate Technology (ICAT) is proposed.

DEFINITIONS; CHOICE OF TECHNOLOGY; SCIENCE AND TECHNOLOGY POLICY; INDUSTRIAL DEVELOPMENT; INTERNATIONAL ORGANIZATIONS

United Nations Industrial Development Organization. International Forum on Appropriate Industrial Technology, New Delhi/Anand, India, 20-30 November 1978. Working Group on Conceptual and Policy Framework for Appropriate Industrial Technology. Industrial development strategies and choice of appropriate technology in developing countries: background paper, prepared by the secretariat of UNIDO. Document ID/WG.282/113. 21 pp.

"To achieve the major objectives of development...a framework of appropriate economic and technology policies (and planning) has to be established. Within such a context the selection of appropriate technologies becomes an essential and all-embracing element of the development strategy." (Paragraph 31)

DEFINITIONS; CRITERIA FOR APPROPRIATENESS; SCIENCE AND TECHNOLOGY POLICY; INFRASTRUCTURE

United Nations Industrial Development Organization. International Forum on Appropriate Industrial Technology, New Delhi/Anand, India, 20-30 November 1978. Working Group on Conceptual and Policy Framework for Appropriate Industrial Technology. Institutional development of appropriate industrial technology in developing countries: research and development policies and programmes: background paper, by W. A. Fischer. Document ID/ WG.282/90. 37 pp.

General suggestions of infrastructure development in research and development capabilities and educational programs are given. The role of R&D by multinational corporations in developing nations is explored.

RESEARCH AND DEVELOPMENT; MULTINATIONAL CORPORATION

United Nations Industrial Development Organization. International Forum on Appropriate Industrial Technology, New Delhi/Anand, India, 20-30 November 1978. Working Group on Conceptual and Policy Framework for Appropriate Industrial Technology. Management of appropriate technology: background paper, by V. K. Chebbi. Document ID/WG.282/3. 36 pp.

Steps in the development of appropriate technologies within developing countries are given with a "systems approach" to the assessment of technology for developing countries. The author favors changes in labor laws to help more labor-intensive industries and the establishment of appropriate technology development cells.

GOALS; CHOICE OF TECHNOLOGY; AT DEVELOPMENT ORGANIZATIONS

United Nations Industrial Development Organization. International Forum on Appropriate Industrial Technology, New Delhi/Anand, India, 20-30 November 1978. Working Group on Conceptual and Policy Framework for Appropriate Industrial Technology. Operational and policy choices for technology and industrialization in developing countries: background paper, by I.H. Abdel-Rahman. Document BP/2. 14 pp.

Technological development in free market and centrally planned economics are examined and related to change in the less developed countries. An "integrated technology development system" is necessary.

CRITERIA FOR APPROPRIATENESS; CHOICE OF TECHNOLOGY; DEFINITIONS; CASE STUDIES

United States. Agency for International Development. Proposal for a program in appropriate technology. Rev. ed. Washington, D.C., U.S. Govt. Print. Off., 1977. 382 pp.

AID's proposal for carrying out Section 107 of the Foreign Assistance Act of 1961 to "support an expanded and coordinated private effort to promote the development and dissemination of technologies appropriate for developing countries." The program is divided into five categories: communication and coordination, national policies and appropriate technology, appropriate technology projects in less-developed countries, education, and U.S. business involvement. Recommended is the organization of a new, independent, nonprofit organization, now ATI. Several attachments follow the proposal.

CRITERIA FOR APPROPRIATENESS; CASE STUDIES; CHOICE OF TECHNOLOGY; RESEARCH AND DEVELOPMENT; SOCIAL EFFECTS; ECONOMIC EFFECTS; INTERNATIONAL ORGANIZATIONS; DEVELOPED COUNTRIES

United States Congress. House. Committee on Science and Technology. Subcommittee on Domestic and International Scientific Planning, Analysis and Cooperation. Appropriate Technology. Hearings, 95th Congress, 2nd Session, 25-27 July 1978. Washington, D.C., U.S. Govt. Print. Off., 1978. 1289 pp.

A comprehensive compilation of various opinions and actions regarding appropriate technology; contains testimony from many experts, an annotated bibliography, and case histories.

CASE STUDIES; DEFINITIONS, SOCIAL EFFECTS; ECONOMIC EFFECTS; INFRASTRUCTURE; DEVELOPED COUNTRIES

United States. Department of State. Science and Technology for Development: United Nations Conference 1979: U.S. National Paper. U.S. Dept. of State, 1979. 35 pp. (International Organization and Conference Series 139)

A brief reflection on the U.S. experience in applying science and technology to development. Analyzes the vital role that science and technology have played in the evolution of the U.S. economy and society and in the U.S. contribution to economic and social growth in the developing countries. Suggests steps that could improve the application of science and technology to problems of development.

CHOICE OF TECHNOLOGY; SCIENCE AND TECHNOLOGY POLICY; RESEARCH AND DEVELOPMENT; INFRASTRUCTURE; POLITICAL GOALS/RELATIONSHIPS; INTERNATIONAL ORGANIZATIONS; DEVELOPED COUNTRIES

United States. Library of Congress. Congressional Research Service. United Nations Conference on Science and Technology for Development--a background paper, prepared for the Committee on Foreign Affairs. U.S. House of Representatives. Washington, D.C., U.S. Govt. Print. Office, 1979. 127 pp.

Gives a historical setting for the conference; an overview of the present UN institutional arrangements for science and technology; briefly discusses issues that may appear during preparations for the conference.

DEFINITIONS; SCIENCE AND TECHNOLOGY POLICY; INTERNATIONAL ORGANIZATIONS; DEVELOPED COUNTRIES

Villegas, Bernardo M. For an appropriate technology that uses the resources of the Third World by substituting labor for capital. Ceres, v. 7, no. 3, May/June 1974: 44-47.

Explores briefly the relationship between labor and capital and suggests that the responsibility for progress lies in innovative thinking on the part of LDCs.

CRITERIA FOR APPROPRIATENESS; RESEARCH AND DEVELOPMENT; IMPORT SUBSTITUTION

Wakefield, Rowan A., and Patricia Stafford. Appropriate technology: what it is and where it is going. Futurist, v. 11, no. 2, April 1977: 72-76.

The appropriate technology movement is part of a larger movement that seeks to improve the quality of life while solving the problems of environment and employment. The philosophy of and the worldwide interest in appropriate technology are explored. Criticisms, both practical and theoretical, are examined.

DEVELOPED COUNTRIES

Weiss, Charles. Mobilizing technology for developing countries. Science, v. 203, 16 March 1979: 1083-1089.

New technological policy is needed to create productive jobs and provide minimum public services at a cost and level of sophistication that can actually reach the poverty areas in developing countries. This policy needs to include both the hardware such as tractors or waterless toilets and software such as the training of numerous supervisors to implement improved technologies.

DEFINITIONS; SCIENCE AND TECHNOLOGY POLICY; CHOICE OF TECHNOLOGY

Wells, Louis T., Jr. Encouraging appropriate technologies in developing countries. Statement before the Committee on Science and Technology. U.S. House of Representatives. 26 July 1978. 9 pp.

A presentation of appropriate technology for developing countries with a strong emphasis on labor-intensive technology. Presents the view that additional technological inventions are not really necessary because existing ones have not been exhausted. Offers suggestions for promoting adaptations of labor-intensive techniques and the role of U.S. and international agencies.

CHOICE OF TECHNOLOGY

White, Lawrence J. Appropriate factor proportions for manufacturing in less developed countries: a survey of the evidence. In Proposal for a program in appropriate technology, rev. ed. U.S. Agency for International Development. Washington, D.C., U.S. Govt. Print. Off., 1977: 114-191.

Strongly suggests "that greater labor intensity in lesser developed country manufacturing is feasible and would be efficient" especially through well-directed research and development. More studies are needed, but the conclusions point to policy implications--such as proper factor prices, ceasing of discrimination against smaller firms, and tougher bargaining by LDCs to limit possible abuses by MNCs. Further cautions that appropriate technology is not a quick and easy solution for LDCs.

LDCs AND AT; RESEARCH AND DEVELOPMENT

Winrock International Conference Center. Preliminary report from a Workshop on Appropriate Technology, 1-5 December 1978. Morrilton, Ark. Winrock, 1978. 27 pp.

A preliminary report from a workshop in preparation for the UN Conference on Science and Technology aimed at guiding U.S. policy-makers on appropriate technology issues for the conference.

SCIENCE AND TECHNOLOGY POLICY; DEVELOPED COUNTRIES; GOALS

Workshop on Appropriate Agricultural Technology, February 6-8, 1975. Proceedings of the Workshop on Appropriate Agricultural Technology. Dacca, Bangladesh Agricultural Research Council, 1975. 312 pp.

In attacking the issues and problems in appropriate agricultural technology, the workshop studies background papers by experts in several areas, such as grain drying and storage, mechanization, and irrigation. Then the results of an appropriate technology survey of Bangladesh were reported. Finally, the Workshop created a list of recommendations including a guide for evaluation of technology and specific recommendations for research and practices in Bangladesh. An Appropriate Technology Division within the Bangladesh Agricultural Research Council was strongly urged.

CRITERIA FOR APPROPRIATENESS; CHOICE OF TECHNOLOGY; BARRIERS TO AT; RURAL DEVELOPMENT; RESEARCH INSTITUTES

World Bank. Appropriate technology in World Bank Activities. 1976. 90 pp.

Selected examples of World Bank-financed projects in appropriate technology, and description of Bank efforts to assess, evaluate, and, when necessary, promote technological innovations by specific sectors (i.e., small farm, water projects, etc.).

DEFINITIONS; CRITERIA FOR APPROPRIATENESS; BARRIERS TO AT

World Bank. Central Projects Staff. Appropriate technology and World Bank assistance to the poor. 1978. 112 pp.

Describes the World Bank's use of appropriate technology in developing member countries, organized into specific sectors of bank operations and citing specific sectors of research.

RURAL DEVELOPMENT; INDUSTRIAL DEVELOPMENT